A
MANUAL OF BEE-KEEPING
for English-speaking Bee-keepers

BY

E. B. WEDMORE
C.B.E., M.I.E.E. F.Inst.P.

MEMBER OF THE APIS, PRESIDENT OF THE BRITISH BEE-KEEPERS' ASSOCIATION,

MEMBER OF THE SCOTTISH BEE-KEEPERS' ASSOCIATION AND

VICE-PRESIDENT OF THE KENT BEE-KEEPERS' ASSOCIATION

THIRD EDITION

ILLUSTRATED

Northern Bee Books

© Mrs E. B. Wedmore, 1975

First published 1932 Reprinted 1942, 1943
Second Edition (revised), 1945
Reprinted with corrections 1946
Reprinted 1947

Reprinted 1975 Reprinted 1979 Reprinted 1982 Reprinted 1984
Reprinted with Research paper and illustrations 1988
Reprinted 1997

ISBN 978-1-908904-76-8
This edition published
in 2015 by
Northern Bee Books
Scout Bottom Farm
Mytholmroyd
Hebden Bridge HX7 5JS (UK)

Artwork, D&P Design and Print
Printed by Lightning Source (UK)

A

MANUAL OF BEE-KEEPING
for English-speaking Bee-keepers

BY

E. B. WEDMORE
C.B.E., M.I.E.E. F.Inst.P.

This Edition edited by John Powell with assistance from
Daniel Basterfield N.D.B., John Chandler, Margaret Davis, Dr. Ivor Davis N.D.B., Kim Flottum,
Margaret Ginman, Dr. Roger Pool , Graham Royle N.D.B., Professor Tom Seeley,
The late His Honour Judge David Smith QC., Paul Smith,

Northern Bee Books

The Reader should be aware that Wedmore was originally published in 1932 and was revised in 1946. Since then there have been changes in legislation, understanding of bee diseases and most importantly in the use and associated dangers of some of the chemicals originally recommended. Wherever possible the Editor and Contributors have drawn attention to these changes. Most of the beekeeping techniques however remain good practice and it is for this reason that Northern Bee Books, with the agreement of David Wedmore have issued this revised edition. The original review of paragraphs that were in need of updating was undertaken by John Powell, (JP) who should be recorded as the lead Editor of this edition. He was assisted by the following, whose contributions are identified at the paragraph end by their initials

Daniel Basterfield N.D.B. (DB)
John Chandler (JC)
Margaret Davis (MD)
Dr Ivor Davis N.D.B. (ID)
Kim Flottum (KF)
Margaret Ginman (MG)
Dr. Roger Pool (RP)
Graham Royle N.D.B. (GR)
Professor Tom Seeley (TS)
The late His Honour Judge David Smith QC. (DS)
Paul Smith (PS)

Because of the reference construction of *A Manual of Beekeeping* by paragraphs rather than chapters the publisher has maintained this method of reference and the existing original numbered paragraphs. The new updated paragraphs appear before the original, which are printed in italics, and use the suffix (a)

The publisher would welcome advice on any other appropriate updates for paragraphs which could be included in a future volume. They should be sent to jerry@northernbeebooks.co.uk with the appropriate paragraph number

PREFACE TO THE SECOND EDITION

The inclusion in this edition of several hundred items of information, culled from the developments and experience of the last twelve years, without material expansion of the volume, has necessitated extensive revision and some further condensation.

Opportunity has been taken to re-cast completely the sections dealing with management, and, in conformity with the advice to seek simplicity, some of the less advantageous variations in technique have been cut out, as well as some theory. On this basis, the author's theory of swarming and evolution of the honey bee and the honey comb, detailed in the former edition, has been greatly curtailed. Those interested should consult the earlier edition.

The control of swarming and management for honey production have been treated together (Section xiii) as a single subject. This has made possible, for the first time, an orderly, critical, presentation of the numerous systems of management and control. Artificial swarming (**1129-38**), the colony (**1150-56**) and control of mating (**176-7**) have been dealt with in a novel manner.

The original sections on disease did not find favour in certain quarters, but while twelve years' further experience has brought to light some new items, the author has not found it necessary to correct what was given in the first edition; neither has any criticism of his theory of swarming led to any constructive modifications. There are still many who seek an impossible simplicity in what is, in nature, a complex instinct, variable in its manifestations; but all will agree as to the merits of seeking simplicity in methods of control.

Among the many fallacies still current in the literature of beekeeping, one of the most persistent is that small bees will be and are in fact produced by omitting to change the combs sufficiently frequently, with consequent thickening of cell walls, although who is supposed to attend to this in wild life does not appear to have been considered. The author has made many enquiries and examined many frames, some of which had been in use without attention for periods up to 18 years, and has not yet found cell side walls thicker than six thousandths of an inch. The bees won't allow it. They strip all thicker walls and rebuild. Small bees do appear, but not due to thickening of cell walls by deposit of larval skins and excreta.

The author had hoped to have added a chapter on genetics, but this must await the production of a book of a different character, for, unlike all other breeders of live-stock, practical bee-breeders have had to proceed largely without that guidance that the science of genetics can alone be expected to give.

How long is the average individual drone allowed to live? When and why is he sometimes fed by the workers, and under what circumstances does he feed himself? Why do workers sometimes ball their queen? Under what circumstances do they fail to raise a needed queen, although they possess eggs or larva of suitable age and drones are available? Can a female bee be raised in fact by artificial fertilization of a drone egg? Do bees in the colder climates exercise any selective action in favour of the production of dark workers when selecting for queen raising?

Under what circumstances does supersedure lead to swarming? In what way is a non-swarming honey-gatherer more "contrary to nature" or less desirable than a pipless orange? On all these, and many other questions, most pertinent to practical bee-keeping, we have at present only conflicting evidence, and no final answers to any of them will be found in the present edition of this Manual.

    E. B. WEDMORE

Note.-For particulars of British Standards for hives, appliances, etc., agreed since the matter for this issue was prepared in 1946, application should be made to the British Standards Institution, 28 Victoria Street, London, S.W.I.

PREFACE TO THE FIRST EDITION

This work is not intended to replace the beginner's Guide Book. It is intended for the bee-keeper who already has an elementary knowledge of bee-keeping. He cannot get all the additional help he needs, either from the current journals or through bee-keepers' societies, for, frequently, he needs guidance forthwith and generally with more explanation than can be given in the correspondence columns of a journal, or obtained at a general meeting of a society. The expert replying in the journal may say "raise a new queen" or "use the Demaree method." Now highly condensed information on Queen-raising occupies no less than 31 pages in this Manual and the Demaree system is treated in detail under no less than ten heads, with many cross-references to other parts of the work. The weekly or monthly journal is invaluable and cannot be replaced by a manual, but needs supplementing. Even the experienced bee-keeper will find in this manual many things he has forgotten, or on which he cannot readily lay hands, and for which he will seek in vain in the current textbooks.

The majority of advanced books on bee-keeping are written for the conditions prevailing in a particular country. Now, the advice given and methods described, not only need modification for use in other countries, but even for application in the writer's own country, in parts where the conditions he deals with do not happen to prevail. This work is written for the English-speaking bee-keeper, wherever found, and is intended to give him, not only the most reliable and latest information in every branch of practical bee-keeping, but such guidance as will enable him to select the best methods for his particular circumstances, be he north or south of the Equator; in warm, cold, wild or cultivated regions; whether he be keeping bees for a living or as a hobby; at home or in out apiaries; attending to them every day in the week, or only during week-ends.

In planning such a work a radical departure has been necessary from the familiar form. Written in the familiar form, this work would occupy five or six volumes, and its cost would be prohibitive. All historical matter and references to the great names of history have been omitted from the body of the work. If Blank were alive now and as lively as he used to be, surely he would bring his "method" or "system" up to date in some particular. Instead, therefore, of describing Blank's method or system under his name, with many others of equal merit, and equally limited in application - a course which would involve much repetition and confusion-the author has taken from each worker the contributions of novelty and permanent value in his work and presented these features in an orderly arrangement, each in its appropriate place

To make this mass of information intelligible and useful, guidance on the selection of method is given in each section.

The author, therefore, begs to make acknowledgment once for all to the many workers of the past upon whose knowledge he has so fully drawn, and upon whose experience and efforts the whole of modern bee-keeping rests.

Much unnecessary repetition has been avoided also by giving detailed instructions for all the common manipulations in one section, thus securing at the same time that those bearing on management and other large matters shall convey in brief compass the concentrated wisdom of the masters, undiluted and unconfused by reference to minor, though essential, details of procedure.

In a book which answers probably 10,000 questions every word should count. The beginner is warned, therefore, of the importance, where precise instruction is given, of observing every word. In the many cases where alternative methods prevail it has been the author's endeavour, supported by a long experience of critical work in other fields of applied science, to give only the best method, and to rely mainly upon evidence and very little upon opinion.

The work has grown out of an attempt, commenced some years ago, to jot down the established facts of bee-keeping, and consists mainly of facts or well-established procedure. A departure is made, however, in the introduction to the Section on Swarming. The justification for the introduction of the brief essay there appearing is given in its opening sentences. Here and there also new suggestions are made, but in such a way that they cannot be confused with what is well established.

A bee-keeper desiring illustrations of the numerous appliances he would like to see and handle can obtain them from the appliance makers, and in purchasing this book he does not have to pay for what he may obtain free of charge elsewhere. Illustrations, mainly diagrammatic, are included, however, where clarity requires.

No attempt is made to deal with certain specialist subjects, such as anatomy, bacteriology and microscopy, but essential information, within the capacity of any practical bee-keeper to master and apply, is given on the diagnosis and treatment of disease.

E. B. WEDMORE

x

CONTENTS

SECTION   PAGES

DETAILED SECTIONAL CONTENTS xii

LIST OF TABLES xii

LIST OF ILLUSTRATIONS xxiv

RESEARCH ON METHODS OF MANAGEMENT OF BEES xxvi

I   WORKERS, DRONES AND QUEENS 1

II   QUEEN AND DRONE RAISING 32

III   COMB, WAX AND PROPOLIS 57

IV   HONEY-PART I 77

V   HONEY-PART II 92

VI   THE APIARY - MOVING BEES 125

VII   HIVES AND THEIR ACCESSORIES 135

VIII   FRAMES, SECTIONS AND FOUNDATION 167

IX   APPLIANCES AND THEIR USE 182

X   SEASONAL MANAGEMENT 195

XI   SPECIAL MANAGEMENT 211

XII   SWARMS AND SWARMING 219

XIII   HONEY GETTING AND SWARM CONTROL 231

XIV   FEEDING – ROBBING - PACKAGE BEES 251

XV   MANIPULATIONS. 262

XVI   DISEASES AND PESTS 282

XVII INVENTIONS AND DISCOVERIES INDEX 297

       THE NEW CLP REGULATIONS 301

       INDEX 306

SECTIONAL CONTENTS

(The numbers refer to paragraphs)

SECTION I

WORKERS, DRONES AND QUEENS

Stages of Development

1, General; **4**, Period of Leaving the Hive; **5**, Importance of Development Periods; **6**, Bees of Mixed Sex

Worker Bees

7, General; **8**, Development of Worker Bees; **12**, Weight of Worker Bees; **15**, Length of Tongue; **17**, Honey Sac Content; **18**, Length of Life; **19**, Effective Life; **20**, Importance of Longevity; **21**, Assessment of Longevity; **22**, Conservation of Energy; **23**, Duties of Worker Bees; **26**, Hive Temperatures

Drones

31, Utility of Drones; **32**, Potency and Flight of Drones; **33**, Location of Drones in the Hive; **34**, Starvation of Drones; **35**, Reduction of Drones

Queens

36, General; **41**, Queen Calling; **42**, Longevity of Queens; **44**, Development of Queen Pupæ; **43**, Points of a Good Queen; **45**, Egg-laying and Development of Larvæ; **47**, Royal Jelly; **48**, Size of Eggs; **49**, Rate of Laying; **53**, Growth of Brood Nest; **54**, Queen Failure and Supersedure; **56**, Balling of Queen; **59**, Finding the Queen; **65**, Catching and Holding a Queen; **66**, Clipping the Queen's Wing; **69**, Marking the Queen; **73**, Supersedure of Marked Queens; **74**, Loss of Queens and its Detection; **77**, Tests for Queenlessness; **78**, Re-queening; **84**, Introduction of a Local Queen; **85**, Introduction of a Travelled Queen; **89**, Introduction of a Virgin Queen; **92**, Use of Nucleus in Queen Introduction; **93**, Methods of Queen Introduction; **94**, Use of Travelling Queen Cage; **97**, "One-Hour Method"; **99**, Direct Introduction; **100**, Use of Water for Introduction; **101**, Use of Artificial Swarm; **102**, Introduction through Screen; **104**, Queen Run In at Entrance; **105**, Introduction by Use of Anæsthetics; **106**, Two Queens in One Colony

Laying Workers

107, General; **111**, Destruction of Laying Workers

Races and Strains of Bees

116, General; **118**, Points of a Good Strain; **124**, Racial Characteristics; **129**, Italian Bees; **135**, Cyprian Bees; **138**, Carniolan Bees; **141**, Caucasian Bees; **143**, British Black Bees; **144**, French Bees; **145**, German Swiss Bees; **146**, Dutch Bees; **147**, Spanish Bees; **148**, German Bees; **149**, Egyptian Bees; **150**, African Bees; **151**, Syrian Bees; **153**, Eastern Races

SECTIONAL CONTENTS

SECTION II

QUEEN AND DRONE RAISING

Introduction

154, General; **155**, Attention to Detail; **156**, Small Scale Working

Queen Raising without Interference

157, General; **158**, Under Swarming Impulse; **161**, Under Supersedure Impulse; **162**, To Replace Lost Queens; **165**, Queenright Supersedure; **166**, How to Disinguish the Impulse

Queen Raising to Plan on a Small Scale

171, General; **172**, Reducing the Swarming Instinct; **173**, Time for Re-queening; **175**, Importance of Parentage; **177**, Drone Raising; **178**, Raising a Few Queens; **183**, Special Modifications; **188**, Queen Raising in or for Out Apiaries; **190**, Raising Queens in Nuclei

Queen Raising to Plan on a Large Scale

191, Introduction and Key Diagram; **192**, Parentage; **202**, Laying and Hatching of the Egg; **205**, Formation of Queen Cells where the Eggs are Laid; **209**, Raising Queen Cells on Prepared Edge of Comb; **210**, Raising Queen Cells on Prepared Strip of Comb; **213**, Raising Queen Cells on Prepared Surface of Comb; **215**, Transfer of Individual Eggs or Larvæ for Queen Raising; **217**, Transfer of Individual Larvæ, "Grafting"; **223**, Transfer of Individual Cells with their Larvæ; **226**, Starting and Raising Queen Cells; **230**, Developing Queen Cells in Queenless Colony; **238**, Swarm Box Method of Starting Cells; **242**, Partition Method of Raising Cells in Queen-right Colony; **246**, Emergence in General; **252**, Preparation for Emergence of Cells Raised on Comb; **256**, Protection of Queen Cells; **257**, Preparation of Colonies for Emergence; **258**, Emergence in Strong Colony; **262**, Emergence in Nuclei; **267**, Use of Small Frames in Nuclei; **270**, Emergence in Nucleoli, "Baby Nuclei"; **274**, Use of Incubators; **275**, Nursery Cages

Fertilization

276, Conditions Favourable; **283**, Procedure, and Period, of Fertilization; **287**, Queens Going Astray; **288**, Egg-laying

Drone Breeding

290, General; **292**, Selection of Drones; **295**, Control of Mating; **299**, Mating at Abnormal Hours; **303**, Mating out of Season; **306**, Mating in Enclosures.

Tested and Untested Queens

307, Classification and Definition

SECTION III

COMB, WAX AND PROPOLIS

Comb Building

308, Method; **312**, Building Drone and Worker Comb; **316**, Brace and Burr Comb; **318**, Inclination and Disinclination to Build Comb; **324**, Cell Walls; **326**, Cell

Cappings; **329**, Dimensions of Cells; **333**, Importance of Dimensions; **335**, Size of Combs and Method of Using; **342**, Use of Old Combs; **345**, Cleaning old Pollen-clogged Combs; **348**, Weight of Combs and Foundation; **350**, Repairing and Cutting Comb, **353**, Use of Foreign Substances in Comb Building; **357**, Preservation of Combs (Wax Moth); **361**, Remedies for Wax Moth

Beeswax

370, Production; **375**, Physical Properties; **382**, Chemical Properties; **387**, Tests for Adulteration; **389**, Bleaching Wax

Wax Rendering

390, General; **394**, Wax Extractors; **400**, Wax Cappings; **401**, Wax Moulding; **402**, Wax for Exhibition; **403**, Judging Wax

Propolis

404, Use of Propolis; **405**, Source and Colour; **406**, Propolis and Wax; **408**, Removing Propolis

SECTION IV

HONEY—PART I

Crops and Honey Gathering

409, Times, Distances and Temperatures of Flights; **413**, Daily, Climatic and Seasonal Changes in Nectar Flow; **417**, Weights of Honey Taken; **422**, Weight of Honey in Comb; **423**, Yield per Acre; **425**, Yield from Use of Drawn Combs and Foundation; **428**, Statistics of Honey Production

Composition and Properties of Honey

431, Water Content and Density of Nectar; **435**, Water Content, Refractive Index and Density of Honey; **439**, Changes with Composition and Temperature; **441**, Use of Hydrometers; **442**, Chemical Analysis; **445**, Food Value and Medicinal Value; **450**, Acidity

Honeydew

451, Sources; **452**, Characteristics; **453**, Collection and Use

Sources of Honey and Pollen

454, Introduction; **455**, Principal Sources of Honey and Pollen

Sources of Honey in Different Countries

460, General; **461**, Great Britain and Ireland; **463**, U.S.A. General List; **464**, Eastern States; **465**, Southern States; **466**, California; **467**, Northern States; **468**, Canada; **469**, Australia; **470**, New Zealand; **471**, South Africa; **472**, Tropics and Sub-Tropics

SECTIONAL CONTENTS

SECTION V

HONEY—PART II

Taking and Grading Honey

473, Taking Comb Honey; **474**, Extracted Honey; **477**, Use of Super Clearers and the Like; **483**, Supers Stuck Together; **484**, Supers Containing Brood; **485**, Supers Containing Pollen; **486**, Grading Honey; **489**, American Standards; **490**, Canadian, Australian, Great Britain, Irish Free State and New Zealand Regulations

Extraction of Honey, Chunk Honey

491, Extracting Honey, Temperature; **492**, Uncapping Knives; **493**, Uncapping; **495**, Centrifugal Force; **497**, Cappings; **498**, Use of Ripener; **499**, Bottling; **500**, Care of Extractor; **501**, Ling and Heather Honey; **504**, Chunk Honey

Granulation and Fermentation, Heating and Sterilizing

507, Granulation of Honey; **508**, Production of Large Crystals; **509**, Production of Fine Crystals; **510**, Rapidity of Granulation; **511**, Tendency to Granulate; **512**, Prevention of Granulation; **514**, Treating Honey already Granulated; **515**, Whipped Honey; **516**, Fermentation; **517**, Prevention of Fermentation; **523**, Use of Fermented Honey; **524**, Heating and Sterilizing Honey; **525**, Time and Temperature; **526**, Destruction of Bacteria; **527**, Putting up for Sale; **532**, Destruction of Diastase

Storing, Packing, Selling and Showing Honey

533, Storing Honey; **537**, Packing Section Comb and Chunk Honey; **539**, Packing Extracted Honey; **543**, Labels for Tins and Bottles; **547**, Selling Honey; **551**, Blending for Sale; **558**, Pure Food Laws; **559**, War-time and Post-War Regulations; **560**, Honey for Show—Extracted Honey; **564**, Section Honey; **569**, Points for Judging

Uses of Honey

571, Use in Cookery and Confectionery; **572**, Recipes for Cooking, Using Honey; **573**, Sugar Equivalent; **574**, Neutralizing Acid; **575**, Relative Sweetness; **576**, Icing; **577**, Preserves, Sauces, Drinks; **578**, Ice Cream; **579**, Toilet Preparations; **580**, Antifreezing Mixtures

SECTION VI

THE APIARY—MOVING BEES

The Apiary

581, Location near Honey Sources; **583**, Pollen and Water Supplies; **584**, Prevailing Wind; **585**, Floods, Fires and Vibration; **586**, Fencing and Shelter; **588**, Liability to Neighbours and Others; **591**, Out Apiaries; **592**, Apiaries in Orchards; **594**, Payments by Fruit Growers; **595**, Package Bees for Orchard Work; **596**, Spraying in Orchards; **597**, Town and Garden Sites; **598**, Preparation of Stands; **599**, Arrangement of Hives; **605**, Honey House or Shed; **606**, Windows; **607**, Walls and Floor; **608**, General Arrangement; **611**, Winter Cellars

SECTIONAL CONTENTS

Starting an Apiary

612, Starting in a Small Way; **613**, One-man Apiary; **614**, Essentials of an Apiary; **615**, Influence of Good and Bad Years; **616**, Obtaining Bees; **617**, Colonies and Nuclei; **619**, Partnership and Renting

Moving Bees

620, General; **621**, Aids to Noting Location; **622**, Influence of Weather; **625**, Long Distance Moves; **627**, Critical Distance; **628**, Preparation of One-Piece Bodies; **629**, Preparation of Hives with Outer Covers; **630**, Securing Frames; **632**, Heavy Hives; **633**, Cart, Motor and Railway Transit; **634**, Moving to the Heather; **635**, A Swarm

SECTION VII

HIVES AND THEIR ACCESSORIES

Hives and Their Parts

636, Use of Skeps and Early Types; **637**, Size of Hive in General; **638**, Hives Too Small and Too Large; **639**, Size in Relation to Management; **642**, Types of Movable Frame Hive; **647**, Perpendicular *v.* Parallel Way (Cold *v.* Warm Way); **650**, Horizontal *v.* Vertical Extension; **652**, The "Long Idea" Hive; **655**, Single *v.* Double Walls; **658**, Outer Cases; **661**, Fillets or Plinths

Parts and Fittings

664, Floor Boards; **666**, Legs and Stands; **668**, Alighting Boards; **669**, Special Floor Boards; **673**, Entrances; **677**, Brood Chambers; **679**, Shrinkage of Wood; **680**, Supers; **681**, Frame Runners; **683**, Propolizing of Lugs; **684**, Metal Runners; **685**, Section Racks; **686**, Dummies; **687**, Division Boards; **689**, Quilts and Inner Covers; **690**, Material for Quilts; **691**, Glazed Inner Cover; **692**, Hive Roofs; **694**, Keeping out Water; **696**, Keeping out Bees; **697**, Escape Boards and Super Clearers; **698**, Return Hole; **699**, Faulty Bee Escapes; **700**, Ventilated Escape Boards; **701**, Canadian and Improved Pattern; **702**, Queen Excluding; **703**, Queen Excluders; **706**, Use of Excluders

Packing, Ventilation and Temperature Control

708, Packing in General; **709**, Too Much and Too Little Packing; **710**, Degrees of Protection; **714**, Packing at Top; **717**, Packing at Sides; **719**, Wooden Winter Cases; **723**, Tarred Paper Covering; **725**, Packing at Bottom; **728**, Ventilation in Cold Weather; **732**, Ventilation in Warm Weather; **733**, Floor Ventilators; **734**, Temperature Control

Hive Patterns and Materials

736, Particular Makes of Hive; **737**, Standardization of Dimensions; **738**, Present-day Practice; **741**, The Langstroth Hive; **743**, Dadant Hives; **745**, British Standard Hives; **746**, British National Hive; **750**, Hives to Old B.B.K.A. Standards; **751**, The W.B.C. Hive; **753**, The "Single-walled" Hive; **754**, Other British Hives; **756**, Observatory Hives; **757**, Use of Glass Walls in Hives; **758**, Top and Middle Entrance Hives; **762**, Materials Used in Hive Construction; **764**, Objects of Hive Painting; **765**, Keeping out Damp; **769**, Painting Hives; **773**, Re-painting a Hive; **776**, Colour of Hives

SECTIONAL CONTENTS

SECTION VIII

FRAMES, SECTIONS AND FOUNDATION

777, Frames in General; **778**, Preference for One-piece Combs; **782**, Combs in a State of Nature; **783**, The Form of the Brood Nest; **786**, Unimportance of Cubical Form; **787**, Further Advantages of Large Frames; **788**, A Working Compromise; **789**, Form in Relation to Management; **790**, Frames for Supering; **791**, Dimensions of Frames; **795**, Comb Area in Frames; **796**, Spacing of Combs; **803**, Improved Metal Ends; **804**, Divided Frames; **805**, Assembling Frames; **807**, Fixing Foundation; **808**, Split Top Bar; **809**, Top Bar with Wedge; **810**, Special Top Bars; **811**, Fixing with Molten Wax; **813**, Useful Tips; **815**, Cutting Foundation; **816**, Wiring Frames; **818**, Methods of Wiring; **820**, Embedding the Wire; **821**, Wired Foundation; **822**, Dimensions of Sections; **825**, Bee-way and No Bee-way; **826**, Folding Sections; **827**, Inserting Foundation in Sections; **832**, Choice and Care of Foundation; **836**, Foundation Making in General; **838**, Preparing Wax Sheets for Foundation; **843**, Rolling Foundation

SECTION IX

APPLIANCES AND THEIR USE

846, Protection of Person; **849**, Veils; **852**, Gloves; **853**, Subduing Bees; **857**, The Smoker; **861**, Use of Carbolic Cloth; **862**, Use of Chloroform; **863**, Use of Ethyl Chloride; **864**, Use of Temporary Covering; **865**, Remedies for Stings; **867**, Treatment of Serious Cases; **868**, Internal Remedies; **869**, Destruction of Toxins by Heat; **870**, Anti-Sting Ointment; **871**, Sting Preventives; **872**, Bee Stings and Rheumatism; **873**, Sundry Minor Tools and Devices; **874**, Hive Tool; **875**, Bee Brush; **876**, Vaseline Pot; **877**, Indicators and Cards for Notes; **878**, Apiary Barrow; **879**, Comb Holders; **880**, Scale Hive; **881**, Cheap Alternative; **884**, Feeders; **888**, Construction of Travelling Boxes; **890**, Use of Travelling Boxes; **892**, Shipping Bees in Skeps; **893**, Shipping Swarms; **894**, Drone and Queen Traps; **896**, Robber Screen; **897**, Care of Appliances and Cleaning Metal Parts; **900**, Coating Wooden and Metal Vessels with Wax

SECTION X

SEASONAL MANAGEMENT

Introduction

902, The Problem of Presentation; **903**, The Problem Treated; **904**, Systems of Management; **906**, Single *v.* Double Brood Chambers; **911**, Seasonal Management

Spring

914, Springtime Management in General; **915**, Early Flights; **916**, External Examination and Re-arrangement; **920**, Use of Candy and Spring Feeding; **921**, First Internal Examination; **926**, Further Examination; **930**, Attention to Stock of Combs; **931**, Stimulating Brood Production; **934**, Importance of Maintaining Stores; **935**, Preparations for Swarming; **936**, Onset of Early Harvest

Summer

937, Summer Management—Principal Object; **939**, Removal of Surplus; **943**, Making Increase; **944**, Merits of Extracted, Comb and Chunk Honey Production; **948**, Management for Extracted, Comb and Chunk Honey; **955**, Close of Nectar Flow

SECTIONAL CONTENTS

Autumn

956, Early Autumn Examination and Manipulation; **959**, Autumn Feeding; **961**, Stores suitable for Winter Use; **964**, Weak and Queenless Stocks; **966**, Old Combs for Wintering; **967**, Amount of Stores for Wintering; **972**, Arrangements for Wintering

Winter

978, Winter Conditions; **979**, Generation of Heat and Size of Cluster; **981**, Causes and Prevention of Winter Losses; **983**, Keeping out Damp; **985**, Production of Damp; **987**, Freedom from Disturbance; **988**, Moving Hives in Winter

Cellar Wintering

989, Conditions; **990**, Construction of Cellar; **994**, Cellar Temperatures; **995**, Moving Bees into the Cellar; **999**, Conditions in the Cellar; **1002**, Removal from Cellar

SECTION XI

SPECIAL MANAGEMENT

1003, Introduction; **1006**, Relation of Management to Flow Periods; **1011**, Influence of Longevity of Bees; **1014**, Working to Produce Bees *v.* Honey; **1015**, Importance of Keeping Stocks Strong and Timing Development; **1016**, Special Cases of Large Apiaries; **1018**, Influence of Disease; **1023**, Classification of Districts; **1025**, Bees suited to the Conditions; **1026**, Principal Flow Early; **1033**, Principal Flow Late; **1038**, Early and Late Flow with Gap Between; **1043**, Prolonged Heavy Flow; **1048**, Irregular and Uncertain Flows from Spring to Autumn; **1052**, Flow continued in Winter; **1054**, Week-end Bee-keeping

SECTION XII

SWARMS AND SWARMING

Theory of Swarming

1058, Introduction; **1059**, Evolution of the Colony; **1061**, Development of Modified Females; **1064**, Advancement of the Imperfect Female; **1066**, Co-operative Flight of Workers and Queen; **1067**, The Trigger; **1070**, Drone Raising the First Act; **1074**, Conclusions on Swarming

Formation, Flight and Settlement of Swarms

1075, Preparation for Swarming; **1082**, Queen Cells for Swarming and Super-sedure; **1083**, Issue of the Swarm; **1086**, Issue of Casts; **1087**, External Signs of Issue; **1088**, Delayed Swarms; **1091**, Settlement of the Swarm; **1094**, Bee-bobs; **1096**, Attracting a Swarm; **1099**, Final Flight of Swarm and Legal Ownership.

Taking and Hiving Swarms

1101, Settlement of the Cluster; **1103**, Taking a Swarm in a Skep; **1106**, Swarms in Inconvenient Places; **1109**, Use of Bag Net; **1110**, Box for Taking Swarms; **1111**, Temporary Resting-place; **1113**, Sending Swarms Long Distances; **1115**, Hiving a Swarm; **1120**, Manipulation of Supers

SECTIONAL CONTENTS

Miscellaneous Notes

1121, Clipped Queens; 1122, Swarm Returning to Parent Hive; 1124, Signs of Cessation of Swarming; 1125, Factors Tending to Encourage Swarming; 1127, Factors Tending to Hinder Swarming

Natural and Artificial Swarms

1129. Characteristics of Natural Swarms; Making an Artificial Swarm

SECTION XIII

HONEY GETTING AND SWARM CONTROL

Introduction

1139, General; 1144, How Surplus Honey is Got; 1149, How Swarming is Controlled; 1150, What is a Colony?

Honey Production and Swarm Control

1156, General; 1159, Management of Swarms; 1161, Checking Swarming; 1162, Artificial Swarming; 1163, Doubling; 1166, Utilizing an Upper Brood Chamber; 1170, Transfer of Chambers; 1171, Snelgrove System; 1173, Swarm Control with Single Brood Chamber; 1176, Management for Honey Production; 1177, Management of Two Queens in one Hive

Manipulations with Swarms

1181, Discovering the Parent Hive; 1183, Preventing Casts; 1185, Preventing Early Swarms Re-swarming Later in the Season; 1186, Delaying Flight of Swarms; 1187, Stopping Swarming by Destroying Queen Cells; 1188, Returning Swarm to Parent Hive; 1192, Returning Swarm to Parent Hive, and Requeening; 1193, Hiving Swarm on the Old Stand, without Increase; 1197, Hiving Swarm on the Old Stand, with Increase

Manipulations for Swarm Control

1200, Doubling without Increase, with Natural Swarms; 1204, Use of Upper Chamber in Hiving Natural Swarms; 1207, Stopping Swarming by Destroying Queen Cells and Requeening; 1211, Checking Swarming by Removal of Queen; 1213, Checking Swarming by Removal of Brood; 1216, Checking Swarming by Removal of Brood and Queen; 1219, Doubling without Increase, with Artificial Swarming, Plan I; 1224, Doubling without Increase, with Artificial Swarming, Plan II; 1226, Use of Upper Chamber with Artificial Swarming (Demaree); 1227, Notes on the Demaree System; 1232, Finding the Queen; 1234, Use of Upper Chamber and Transfer of Brood (Modified Demaree); 1235, Repeating the Operation (Re-Demareeing); 1240, Pseudo-Demaree; 1241, Preparing a Strong Stock for Demareeing; 1242, Demareeing when Working for Section Honey; 1248, Queen Supersedure and Demareeing; 1249, Use of Upper Chamber by Transfer of Brood Chambers; 1258, Use of Upper Chamber by Transfer of Shallow Brood Chambers; 1263, Rauchfuss Method of Swarm Control; 1264, Use of Two Queens in one Hive

SECTIONAL CONTENTS

SECTION XIV

FEEDING—ROBBING—PACKAGE BEES

Feeding

1265, General; 1266, Water Supply; 1270, Giving Water; 1275, Pollen Supply; 1277, Pollen Substitutes; 1282, Honey Supply; 1285, Feeding Honey for Storage; 1286, Sugar Feeding; 1290, Note on Sugar and Use of Acid; 1291, Feeding Sugar Syrup for Storage; 1294, Stimulative Feeding; 1299, Stimulative Food Recipes; 1301, Summer Feeding; 1303, Winter Stores; 1307, Outdoor Feeding; 1308, Making Candy; 1316, Making Queen-Cage Candy

Robbing

1319, General; 1322, Detection; 1324, Hindering Robbing; 1327, Stopping Robbing

Package Bees

1332, Use of Package Bees; 1334, Bargain with Producer; 1339, Shipment of Package Bees; 1340, Treatment on Receipt; 1349, Driven Bees

SECTION XV

MANIPULATIONS

Common Manipulations

1351, Manipulations described elsewhere; 1354, Examining a Frame of Bees; 1356, Examining the Combs in a Brood Chamber; 1361, Shaking Bees off a Comb; 1363, Brushing Bees off a Comb; 1364, Destroying All Queen Cells; 1366, Destroying all Queen Cells but One; 1367, Changing the Strain Throughout the Apiary; 1372, Treatment of Combs containing Dead Bees; 1376, Re-arrangement of Brood Combs before the Main Honey Flow; 1381, Crowding Bees on Eve of Honey Flow; 1384, Getting Comb Honey from Shallow Frames; 1387, Preparation of Section Racks; 1388, Supering with a Single Brood Chamber for Extracted Honey; 1391, Supering with a Single Brood Chamber for Section Honey; 1394, Adding an Additional Brood Chamber; 1400, Supering with a Double Brood Chamber for Extracted Honey; 1402, Supering with a Double Brood Chamber for Section Honey

Uniting

1404, General; 1407, Uniting Scented Bees; 1408, Uniting after Exposure to Light; 1409, Newspaper Method; 1413, Use of Super Clearer for Uniting; 1414, Uniting Swarms; 1415, Uniting Driven Bees to an Established Stock; 1416, Adding Young Bees to a Stock or Nucleus; 1418, Adding Flying Bees to a Stock

Manipulations giving Increase

1424, When Required; 1425, Natural v. Artificial Increase; 1426, Giving Queens v. Queen Cells; 1431, Feeding for Increase; 1435, Avoiding Loss of Harvest; 1437, Seasonal Differences; 1444, Increase by Robbing Stocks; 1446, Spring Division; 1448, Spring Division, Rauchfuss Method; 1453, Increase with Artificial Swarming;

SECTIONAL CONTENTS

1454, Making Three Stocks from Two; **1455**, Multiple Increase; **1456**, Continued Increase from One Stock; **1459**, Nuclei from Stock Superseding; **1460**, Continuous Production of Nuclei; **1464**, Combining Increase with Queen Raising; **1466**, Rapid Autumn Increase; **1468**, Use of Pollen Combs and Driven Bees when Making Increase; **1469**, Using Old Bees for Autumn Increase; **1471**, Nuclei in Cool Weather; **1473**, Increase when Using Upper Brood Chamber for Swarm Control

Transferring

1475, General; **1476**, Bees Transferring Themselves from Skep or Box Hive; **1478**, Transferring from Skep or Box by Driving; **1488**, Combination Method; **1489**, Transferring Bees from Roofs, Old Trees, etc.

SECTION XVI

DISEASES AND PESTS

Diseases

1493, General; **1497**, Examination for Disease; **1502**, Examination of Dead Bodies; **1503**, Brood Diseases in General; **1504**, Distinguishing Symptoms of Brood Diseases; **1508**, Incidence and Effects of Brood Diseases; **1512**, Destruction of Germs and Virus; **1513**, Treatment of Brood Diseases in General; **1514**, Treatment of Sac Brood; **1515**, Treatment of European Foul Brood; **1516**, Treatment of American Foul Brood; **1518**, Saving the Bees; **1520**, Formalin Treatment; **1524**, Formalin and Chlorine Gas; **1526**, Minor Brood Diseases; **1527**, Diseases of Adult Bees in General; **1528**, Dysentery; **1529**, Acarine Infection; **1535**, Diagnosis of Acarine Infection; **1537**, Treatment of Acarine Infection; **1539**, Frow Treatment and the Like; **1543**, Methyl Salicylate Treatment; **1545**, Nosema; **1547**, Amœba Disease; **1548**, Paralysis; **1549**, Poisoning by Fruit Spraying and Otherwise

Pests

1552, General; **1553**, Braula coeca; **1559**, Moths; **1560**, Birds; **1563**, Toads, Dragonflies, Wasps, Spiders and Lizards; **1564**, Ants; **1568**, Rats and Mice; **1569**, Drosophila ampelophilos

SECTION XVII

INVENTIONS AND DISCOVERIES

1570, Foreword; **1571**, Introduction; **1573**, Natural History of the Honey Bee; **1575**, Queen Raising; **1579**, Wax, Comb and Honey; **1582**, Hives, Frames and Appliances; **1589**, Bee Diseases

LIST OF TABLES

TABLE   PARAGRAPH

I   Stages of Development of Bees 1

II   Period of Development for Outside Duties 4

III   Development of Worker Bees 9

IV   Weight of Worker Bees 13

V   Tongue-Length of Honey Bees 16

VI   Normal Duties of Worker Bees in Relation to Age 25

VII   Number of Frames in Brood Chamber in Relation to Egg-laying

        Capacity of Queen..49

VIII   Key Diagram to Queen Raising.. 194

IX   Points in Judging Beeswax 403

X   Weight of Combs of Honey 422

XI   Yields of Honey per Acre 424

XII   Water Content and Average Densities of Honeys. 440

XIII   Densities of Honey Solutions at 20° C 441

XIV   Safe Speeds for Extractors 496

XV   Points used in Judging Honey 570

XVI   Dimension of Langstroth Hives 742

XVII   Dimensions of Modified Dadant Hives 744

XVIII   Dimensions of Frames... 791

XIX   Comb Area of Frames, in Square Inches 795

XX   Number of Sheets of Foundation per Pound 833

XXI   Destruction of Disease Germs and Virus 1512

LIST OF ILLUSTRATIONS

Figures in the Text

1   Finding the Queen 61

2   Queen-raising Appliances 219

3   Cappings Basket 400

4 & 5 Honey Sheds 607

6   Hive Fillet and Alternative Construction 661

7   Combined Floor Board and Stand 670

8   Special Floor Board for Nuclei 672

9   Mouse-proof Entrance Block 676

10   Frame Supports 681

11   Canadian Bee-escape Board 701

12   Improved Cleaner Board 701

13   Dimensions of American Hives 742

14   Design of British Brood Chambers 745

15   British National Hive 746

16   Design of W.B.C. Hive, with Outer Case 751

17   Irish and Scottish Hives .755

18   Top-entrance Hive 757

19   Middle-entrance Hive 757

20   Forms of Brood Nest in Springtime 784

21   Dimensions of Frames .792

22   Improved Metal End 803

23   Wiring Frames 819

24   Weighing a Hive with a Spring Balance 881

25   Ventilating Covers for Travelling-box 889

26   Doubling without Increase, with Natural Swarms 1200

27   Doubling without Increase, with Artificial Swarming, Plan I 1220

28   Doubling without Increase, with Artificial Swarming, Plan II 1224

29 Re-Dressing 1239

30 Use of Upper Chamber by Transfer of Brood Chamber 1250

31 Use of Upper Chamber by Transfer of Shallow Brood Chamber 1263

32 Using Two Queens in One Hive 1264

33 Adding Flying Bees to a Stock-Multiple Hives 1422

34 Adding Flying Bees to a Stock-Pair of Hives. 1423

35 Spring Division 1447

36 Rauchfuss Method, Swarm Control with Increase 1449

37 Increase by Artificial Swarming 1453

38 Making Three Stocks from Two 1454

RESEARCH ON METHODS OF MANAGEMENT OF BEES*

By E. B. Wedmore

I. INTRODUCTION

In beekeeping, and especially among amateur beekeepers, there are scores of methods of management in wide use; some are simple, others relatively complicated, but each consists of a series of steps with or without precautionary measures, each step based on some expectation of the way bees will behave in given circumstances. These expectations are put forward as being the result of experience and even of established laws of behaviour. One finds, however, not only that the evidence on which they are based frequently falls far short of proof but also, what is far more serious, that different experts give different and conflicting advice as to details of procedure. This not only confounds the beginner but leads to widespread promulgation of error.

When two opinions differ, it is possible that one is right and the other' wrong; or it may happen that both are right under different circumstances. The actual evidence needed for their clarification is often right in front of beekeepers year by year, but the problems await solution until'such time as beekeepers will make the necessary detailed observations and pool their results.

Methods of management therefore afford an important and promising field for research. With the support of the B.B.K.A. Research Committee, a small beginning was made in the 1949 season; two simple and two more complicated methods of management were used for this first attempt.

From crop reports obtained in the Birmingham district it was found that a number of beekeepers declared themselves users of two methods of swarm control, viz. "Cutting out queen cells" and "Nucleus. method". Upon request eleven users detailed their procedure; they were found to use eleven different methods under these two headings. It immediately became evident that before any method could be investigated, the intended procedure must be formulated in sufficient detail to distinguish it from other more or less similar methods. Instructions were therefore drafted, based on data received, but including certain alternatives for investigation. Owing to a late start. it was not practicable to secure the help of a sufficient number of investigators to assess the value of the methods, or to ascertain working

---

*The new introduction was passed to us by the I.B.R.A. library when sorting out old papers, so we have decided to include it here, although it refers to some research by Wedmore and was about the last paper he wrote on beekeeping.

difficulties, but a study of the records received has made possible a much more effective start for 1950.

A similar course was followed with a method devised by Mr. A. Worth and known as the "Worthwhile" method and with a method introduced in the north of England by Mr. R. W. Wilson and here called "Screened nucleus" method. In each case it was found that individual users made variations for reasons of their own; instructions allowing for these were issued to a number of users and to some others interested. A number of reports have been received.

It should be understood that in selecting certain methods for these first investigations and in detailing procedure, it must not be assumed that they are being recommended for use in preference to other methods at this stage; Method III is, however, thought to be superior to Method II.

2. METHOD I. CUTTING OUT QUEEN CELLS

On its face this appears to be a simple procedure, but it raises most complicated questions.

Most commercial beekeepers avoid the use of strains predisposed to swarm, use procedures minimizing preparations for swarming and, on finding preparations started notwithstanding, successfully postpone and divert the intention of the bees by cutting out queen cells prepared for swarming, with other steps on occasion. Needless to say in their hands the procedure is commercially satisfactory, but its success depends only in small measure on cutting out cells.

Some amateurs hope to avoid swarming by cutting out all queen cells at sight and continuing to do so even where the real intention of the bees is to supersede their queen, or until the bees are reduced to a state of demoralization. Such procedure, considered as a method of management, can only lead to losses.

Thus "cutting out queen cells" cannot in itself be regarded ·as a method of swarm control or of management, and the worst offenders do not prove useful observers. Further investigation is deferred.

Reports received cover only fifteen colonies. In nine cases reported, no queen cells were raised. In three others in which no swarming occurred, the evidence strongly suggests that the very small number of cells started were in fact supersedure cells. A fourth observer reporting cutting out cells, had no swarms and very little honey but reported in insufficient detail for further comment. There was one alleged loss by swarming following the building of a number of cells under conditions of unavoidable neglect.

3. METHOD II. SEPARATE NUCLEUS

Method II consists in brief in forming a separate nucleus about the time there is danger of preparations for swarming and allowing the nucleus colony to raise a queen, which is generally used to requeen, the original colony, but sometimes for increase. The swarm control features are "removal of brood" and sometimes "requeening with a young queen" before a principle flow.

So few reports have been received that it is impossible to reach any conclusions as yet, but as the method is known to be in wide use, the research is being continued with the intention of clearing up some difficulties and working details. This method also raises some questions of great importance dealt with below under the heading "Queen raising in small colonies".

4. METHOD III. SCREENED NUCLEUS

This method consists in brief in removing the queen and most of the brood into an upper chamber, thus forcing the bees below to raise queen cells. The colony is then rearranged; the original chamber with the queen is brought down and the lower chamber made into two nuclei and put on top, so that two new queens are possibly raised. The whole procedure is carried out in one hive and in such a way that it is not necessary to find the queen unless or until she is to be superseded. Drones fly without hindrance and robbing of nuclei is eliminated. The method is used for increase, queen raising and swarm control.

Reports were received on fifteen colonies, five in the Birmingham district and the remainder in Northumberland and Cumberland. The results and observations are set out below in a form to match the Instruction Sheet. They will be readily understood by those who participated and by others who read them after grasping the instructions. Tests are required on a much larger scale before a true picture can be obtained, but observations to date show that the method has considerable promise.

Insertion of supers. Reports show that in the Midlands some first supers were put on at dates from the end of April to the beginning of June, as in Method IV below and in accordance with the instructions; the results do not disclose any utility in this step. Others, and all northern users, omitted this step and put on first supers when the bottom chamber was arranged for queen raising.

First major operation. It was hoped that this could be associated with the first gerieral blooming of white horse chestnut, but reports from a given district show so much variation that no useful correlation has been possible.

In the north it is the practice to await the appearance of drones to determine the date for the first rearrangement of brood, and this should prove a more satisfactory criterion for other parts of the country.

Brood chamber X. In this chamber, which contains the queen and loses its flying bees, no attempt was made to raise queen cells in any of the fifteen colonies under observation.

T he queens were reported as having had adequate room. It should be noted that this chamber was on top only for 9 to 15 days and was in no danger of congestion from incoming nectar. All went according to plan save in one case in which the queen became a drone layer. She was however successfully replaced by another of the new queens raised.

Some observers have not given full information as to dates, but the records indicate that this chamber was brought down again between the middle of May and late June (usually well before the main flow), and that no attempt was made in any of the fifteen colonies to start preparations for swarming either while the chamber was above or after it had been brought down. The suggestion has been made that the instinct to raise queen cells has been satisfied and that the bees never reach the stage of desiring to swarm. Observations on a much larger scale are necessary to establish fully the effectiveness of the method as a swarm preventer. The explanation of the findings is likely to prove elusive.

Several observers reported that the bees occupied the brood chamber within an hour of shaking, but Ii hours was usually allowed. Further records will be useful.

Brood chamber Y. Swarming occurred in one colony only (out of 15); this colony was in an out-apiary and did not receive attention in accordance with the instructions.

One observer reported some difficulty in feeding bees in this chamber.

In the Instruction Sheet stress is laid upon the importance of providing enough stores when this chamber is first installed.

Generally the bees were allowed to raise new queens from their own larvae, but/an observer having several colonies maturing at different dates gave the later ones the benefit of spare queen cells raised in the more advanced colonies.

Nuclei. In each case two nuclei were formed, but in three cases one of them failed to produce a queen; one queen cell was damaged, one queen died in the cell and one queen was lost. Mr. Wilson avoids these difficulties by leaving all queen cells and letting the bees choose for themselves. He reports that under this method of working very few of these small nuclei in the north of England swarm.

All the observers used the method for making increase as well as for queen raising and supersedure. In the north of England, where there is a good late harvest to be got from the heather, it was found possible to build up the original colonies and the new ones raised from the nuclei, so that both gave a good return at the moors, all of course with young queens.

Apparently the method of letting the. bees in the upper chamber find their way to the main entrance (when the upper side entrances are closed) worked in every case, but two observers reported some delay, although not enough to justify modification of the instructions.

So far the observers have used only 14 x 8 1/2 in. combs. Ignoring the outermost two, the queens had the use of only 12 or 13 of these combs for eggs and brood. The method is clearly applicable, without any special plan to more prolific bees by the use of Langstroth or the new Modified Commercial frames. It is hoped that some users of these larger frames will tryout this method in 1950.

Screen-boards and uniting. Hive-boards of thin wood have been used by some in place of boards with openings screened with perforated metal or gauze, the thin wood allowing the passage of some heat. It is believed that the use of perforations facilitates uniting, but if newspaper is used no difficulty is expected.

Mr. Wilson has had much experience in uniting these nuclei, and his method differs from usual practice. It is detailed in the Instruction Sheet as a supplementary item and it is hoped that it will be tried extensively.

Modifications. Only one observer reports using any modification not provided for in the Instructions issued. Instead of using a single hive and stand, chamber X was set out to one side after the workers had sorted themselves out. Later when this chamber was returned, chamber X was put to one side in its place. This has the slight advantage of making both lots more accessible at all times but the disadvantages that it requires more equipment, and that it hinders the flying bees from the nuclei finding the main colony when the nuclei are broken. When however the main object is increase, the modification may have some utility.

5. QUEEN RAISING IN SMALL COLONIES

Reports show that both Methods II and III have been used largely for queen raising as a primary object, rather than for swarm control with increase as a secondary object. This raises an important issue as to the desirability of raising queens at all in small colonies.

Some experienced beekeepers claim that Method II in particular leads to the production of scrub queens and frown on small cells such as are found under Method III. On the other hand many beekeepers claim to use the method successfully and without drawbacks. When looking for fine large queens it is quite natural to welcome the appearance of fine large cells. Indeed the feeling against small cells is such that in Method III, under original plan A, the first cells raised were destroyed in favour of the second batch because some extensive trials had shown the latter to be larger cells. On the other hand plan B, under which the first cells raised are retained, has been used for several years by a beekeeper in Cumberland with a well established reputation as a successful and consistent honey producer. In connection with comparison of plans A and B, it is interesting to note that one observer reported smaller cells raised in the second batch.

There is no doubt that scrub queens have been raised in small colonies and even in large colonies, but there is extensive evidence that excellent queens can be raised from queen cells that look small, and in small colonies. The writer's own experience, supported on enquiry by others, confirms this, but an excellent example is detailed below which has come to light through the present enquiry. It seems to be generally agreed that most cells raised in small colonies are small in appearance, having but little wax in their walls.

An observer reported in great detail the unintentional raising of a queen in a two-frame (14 x 8 ½ inch) nucleus which was intended to be used only for mating purposes. This queen was raised in 1947 from a "miserable little queen cell" overlooked in a pop-hole. The nucleus was built up in stages, going into winter with nine frames of brood and scanty stores. Next year the colony built up so well that it soon filled two chambers and a third had to be added. Finally it was found that there were patches of brood on all six outer combs, making 33 combs occupied by the queen's activity. The total brood and eggs appears to have corresponded to a laying capacity of 5,000 eggs per day continued as an average for 21 days at least. The actual number is not important, as it must be beyond question that this queen could not be described as in any sense a scrub queen.

It is known that supersedure cells raised in small numbers in large colonies are very generally characterized by large dimensions and an exceptionally heavy external covering of wax. Excellent queens are frequently raised from such cells, but some beekeepers of large experience claim that such queens are in no way superior in development to those raised by the beekeeper by artificial manipulation or by the bees under the swarming impulse.

The genetical constitution of the queens is in no way affected by the size of cell or the amount of wax deposited on it so any lack of development must be due to nurture rather than nature. If the resultant queen is well developed, nurture cannot have been at fault, so one concludes that an apparently "miserable little queen cell" may yet have sufficient internal capacity to accommodate the queen and an adequate supply of royal jelly.

In the light of all the above it seems clear that it is not safe to judge the prospective value of a queen entirely from the size of the outside of her cell, whether large or small. One will have to dig deeper to ascertain the conditions determining whether the resultant queen is likely to be a good one or a scrub queen. Continuation of the study of both Methods II and III is likely to throw further light on this important question.

The above mentioned two-frame nucleus was formed of combs having brood in all stages as well as eggs, and an ample supply of honey and of pol/en. Additional bees were shaken in to allow for flying bees returning home. The cell was raised with others forthwith and therefore while the colony was in a state of balance save for a temporary shortage of flying bees and having no work for comb builders. One would expect the inclusion of wax-workers but there would be no extensive production of wax. Why the cells were not better covered in this case or generally under Methods II and III is as yet by no means clear.

6. QUEENS RAISED UNDER PANIC CONDITIONS

This is an offshoot of the problem discussed in the preceding part.

Whether or not bees can experience panic is a question for the psychologists, but they certainly appear to do so. Therefore when a colony, large or small, is suddenly deprived of its queen, beekeepers expect panic attempts to raise a new one, in which to save time the bees may use a larva too old to give a satisfactory queen. For this reason instructions are often given to destroy all cells found sealed on the fourth day after queenlessness.

Where sealed cells are so found, it seems possible that cells not then sealed may also have been started later from larvae also too far advanced. The writer has not found positive information bearing on this. It might be got by ascertaining on the first or second day the position of all cells started ' and then checking up on the fourth day; any additional cells found then would be questionable. Alternatively the position of larvae of suitable age at the time the colony was made queen less could be noted, and the bees even assisted in utilizing them by cutting away surrounding wax. This would be a useful subject for investigation.

Particular observation is required to ascertain under what conditions bees raise such panic queens, or what strains do it. They do not always do so; the queen raised in the two-comb nucleus was hardly a panic queen. None such were raised under Method III.

7. METHOD IV. WORTHWHILE METHOD

Instructions for this method were first issued without a Method reference number. It is now numbered IV for convenience.

In brief the method consists of giving the queen a fresh start with a little brood as in the Demaree method, but with the difference that the chamber containing the queen is put immediately above the original chamber containing most of the brood instead of below. The new brood is thus well protected from cold. An excluder is placed between the two chambers and removed two to three weeks later. The bees cannot swarm unless and until a new queen has been raised in the bottom chamber, but it is found that they do not raise queen cells there (apparently discovering the exclusion of the queen too late).

Reports were received from nine observers who used the method on a total of 28 colonies in Warwickshire, Monmouthshire, Hertfordshire, Kent, Somerset and Cornwall. The scale of records is too small to admit of any but tentative conclusions, but useful information has been obtained.

First operation: supering. Supering was done early to assist in keeping the brood nest clear of nectar. In Kent and Herts. this was done after the middle of April, in Cornwall at the end of

April and elsewhere at the turn of the month or just after, in May.

In only one case had a food chamber been left below. In only 3 cases were food chambers left above, and as none of the hives used was larger than National, it would appear that the bees used throughout were not of the prolific type, although giving good honey returns (up to 116lb). There are some indications that some of the users could have used larger frames with advantage. The method is known to be in use where 16 x 10 in. frames are employed.

The average number of combs containing brood at the time of this operation was 7.

Second operation: adding second brood chamber and putting up queen. In carrying out this operation, the queen was found by inspection and apparently without difficulty in all cases except one, for which plan B was used, the bees being shaken or brushed in front of the hive. Since the colony is small at this time, the use of plan B should rarely be necessary except for persons not having very good eyesight.

This operation was carried out on the average 16 days after the first operation, being necessarily earlier when the colony built up quickly in relation to the size of the hive used, and later when development was delayed, or when it suited the convenience of the operator.

Third operation: removal of excluder and disposal of the lower chamber.

None of the observers used the original plan A. In fact the operation was carried out on the average no less than 22 days after operation II, and some left it dangerously long. Plan A, calling for an interval of less than a fortnight, seems unnecessarily conservative and has now been dropped.

Unfortunately the details reported about the amount of brood (even when measured only by combs occupied) is incomplete and somewhat ambiguous, but there are indications suggesting a rather early check on queen laying. This will be investigate in later tests provided observers complete the forms, as they should do.

Honey returns seem to have been good but do not admit of statistical treatment yet.

Swarming. It is important to note that the 1949 season was one generally characterized by relatively little swarming.

Reports on the 28 stocks managed by Method IV show that there were two swarms, swarm cells found in two other colonies and supersedure cells in another. The reports show however that in three of these cases operations were delayed much beyond the periods recommended. In one of the other two cases, supersedure cells appeared 18 days after operation II. The queen was replaced.

A MANUAL OF BEE- KEEPING

SECTION 1

WORKERS, DRONES AND QUEENS

Stages of Development

General

1. The following table gives representative figures for the number of days required for each stage in the development of queen, worker and drone bees respectively, from the time the egg is laid to the emerging of the perfect insect:

TABLE I

Stages Of Development Of Bees

Stage.	Queen.	Worker.	Drone.
Egg	3 = 3	3 = 3	3 = 3
Larva-before sealing	5	5	6
sealed and spinning	1 }= 8	2 }= 10	3 }= 8
quiescent	2	3	4
pupa-transformation to sylph	1 }= 4	1 }= 8	1 }= 8
quiescent	3	7	7
Total from laying of egg to emergence of image or perfect insect	15 days	21 days	14 days

2. While the figures given above are the common experience, the, exact periods vary somewhat with the race, but more considerably with the conditions under which development takes place. Good feeding causes speeding up, while low temperature and shortage of food cause delay.

Eggs have been observed to hatch in 66 hours (2 ¾ days), but if subjected to cold, may be delayed several days. Eggs have been hatched after a delay of several weeks. It has been suggested that the eggs are not hatched until they have been surrounded with food.

In temperate climates selected eggs may be sent by post [1] and good queens raised from them by the recipient.

---

[1] They should be newly laid and packed in the piece of comb in which they were laid, wrapped in several thicknesses of soft crepe paper and inserted in a light wooden box for transit.

The cells containing the larva of worker bees may be sealed under favourable circumstances in 4 ½ days, sealing taking about half a day. Thus cells may be sealed 7½ days from the laying of the egg, but again sealing may be delayed until say the twelfth day.

Some shortening of the pupal stage of worker bees may occur so that the total period of development may be reduced from 21 to 19 days, especially if the temperature is high and the larvae have had a full share of royal jelly. On the other hand, the total period may be extended to at least 26 days.

3. The development of queens may be reduced to 14 days or extended to 17 days. No doubt slower development would be possible, but, on emergency occurring, the queen cells are better looked after by the bees than are the worker cells.

Period of Leaving the Hive

4. Bees do not, as a rule, leave the hive for several days after emergence. Workers do not become field bees normally until the fourteenth day **(20)**. Queens are ready for mating about the fifth day and drones are potent in about 12 to 14 days. We thus obtain the following representative figures:

TABLE II
PERIOD OF DEVELOPMENT (IN DAYS) FOR OUTSIDE DUTIES

Period.	Queen.	Worker.	Drone
Period prior to emergence Interval after emergence	15	21	14
Interval after emergence	5	14	13
Total (days)	20	35	37

Importance of Development Periods

5. The above figures are of great importance in the economy of the hive and in most manipulations.

From the above it will be seen that of the 21 days required to raise worker bees, the egg stage lasts 3 days, the unsealed larvae 5 days and the cells are sealed for 13 days. Now if the queen has been laying uniformly for at least 21 days, the number of eggs, visible larvae, and sealed cells must be in the ratio of 3 : 5 : 13. There would be nearly twice as much open brood as eggs and nearly three times as many sealed cells as open cells containing larvae. The eggs would occupy one-seventh of the cells in use for breeding, open cells one-quarter and sealed cells more than half. If this is not the case we can deduce as follows:

If the rate of laying is increasing there will be more eggs in proportion, perhaps one-third instead of one-seventh of the total. Fewer eggs than one-seventh indicate a diminished rate of laying, at any rate in the last 3 days as compared with the average of the previous 18. Fewer larvae with a proper ratio between eggs and sealed cells would indicate a reduction of laying somewhere between 4 and 8 days previously. Fewer ,sealed cells would only indicate recently increased activity of about a week's duration. A queen that can lay to cover both sides of two combs with eggs can build up a fourteen-comb brood nest, and if well covered, with British standard combs, this may represent 60,000 brood and eggs or, with 5 weeks' average life, a colony of 100,000 bees **(19 and 50)**.

The same kind of observation can be made relative to drone breeding, noting that the ratios of eggs to exposed larvae and sealed cells are 3 : 6 : 15, when breeding has proceeded steadily for at least 24 days.

Bees of Mixed Sex

6. Certain defective queens produce bees intermediate between workers and drones. Such bees, called gynandromorphs, are a nuisance, but fortunately they are uncommon. The workers drive them aside, and where they collect they consume the larval food, causing the production of dwarfed workers. They will even rob queen cells located at the bottoms of the frames.

Frequently, the departure from normal form is not considerable, but may generally be observed by examination of the hind legs, which are, of course, markedly different in the normal worker and drone, and comparing them with the normal.

A queen producing gynandromorphs should be replaced forthwith. No queens should be raised from her eggs. Her drones and drone brood should be destroyed outright. Her sisters should be replaced as ,soon as practicable, and the source of the strain regarded with suspicion.

A research station would probably offer something for a queen producing abnormal offspring.

Worker Bees

General

7. The worker bee is a female with sexual organs imperfectly developed (**107**), and does all the work of the hive except the laying of eggs, fertilizing of queens and destruction of rival queens. Workers control the queens and the drones, possess stings, and defend the hive against robbers and other enemies. Worker bees normally keep to their own colony. If they attempt to enter other hives they will be driven off unless they arrive loaded with honey, or have mistaken their way. Worker bees drift in this way by accident and in strong winds. Thus black bees may be found in a colony of Italians (**620**).

Note, however, that a yellow queen of mixed parentage or cross mated, producing mainly yellow bees, may also produce a certain proportion of black bees and a larger proportion of black drones, or a proportion of black drones only.

Development of Worker Bees

8. Plate III, facing p.5, gives a clear idea of the size reached by larvae during the first few days of their existence and are especially useful for reference in selecting larvae for queen raising (Section II).

The egg stands on end when laid. On the second day it slopes considerably, and on the third, lies on the cell bottom. The age may thus be estimated by careful observation.

9. The figures in the first column below have been given by Nelson & Sturtevant and are here presumed to refer to a strain of American-bred Italian bees:

TABLE III
DEVELOPMENT OF WORKER BEES

Stage.	Weight in Milligrams.	Number per Ounce.
Weight of egg, new laid	0.132	215,000
just before hatching	0.08-0.10	315,000
larvae, end of first day	0.67	42,300
second day	4.5	6,280
third day	25.0	1,130
fourth day	95.0	299
fifth day	150.0	189
Weight of larva, sealing commenced	136.0	204
completed	157.0	197
on ninth day reduced to	133.0	213
of Bee on emerging	112.0	253

10. From the above table one may calculate that a British Standard Frame capable of holding, say, 5,800 cells from corner to corner and carrying, in fact, say 5,000 eggs laid in 3 days, should increase in weight by 1¾ lb. in 5 days from the laying of the last egg. The nurse bees have to provide food weighing considerably more than this for this one comb. When the queen is laying steadily for each 1000 eggs per day, the nurses must find food for the larvae to cover an increase of weight of over 5½ oz. per day (160 grams). One nurse can attend to the feeding of several larvae.

11. On occasion the ages of pupae may be approximately determined by their coloration from the following table:

13th day Pupa is white all over
14th day Eyes show pale red
15th day Eyes show red
17th day Eyes purple and thorax shows yellow
18th day Abdomen shows yellow
19th day Antennae darken
21st day Wings extended and emergence due.

Weight and Size of Worker Bees

12. Worker bees vary in weight according to race, age and condition, nutrition and contents of honey sac. They vary somewhat in size according to race and nutrition, small bees being produced at low brood nest temperatures. The weight of individual bees of a given race or strain varies more than does the average figure for a given race or strain. By the use of foundation having cells somewhat larger than usual, it is possible to produce bees of a size larger than normal, and *vice versa*, but the larger bees are of lighter build than normal and show no advantage (see also **342**).

13. The following are representative figures:

TABLE IV
WEIGHT OF WORKER BEES

Stage.	Weight in Milligrams.	Number per Ounce.
Weight of bee on emerging	110	4,100
eighth day	158	2,900
at 3 weeks old	100	4,500
	80	5,700
Average weight of field bee empty	95	4,800
loaded	133*	3,400
Weight of old starved bee, say	60	7,600

* This figure was probably taken during a good harvest. The general average would be much below this (see 17).

14. The round figure of 5,000 bees per pound is clearly over-estimated, especially as applied to bees in a swarm. Package bees vary from about 2,600 to 4,500 per pound, averaging say 3,500. Probably a figure of 3,500 to 4000 is about right for a swarm, varying with the strain and the number of hours the swarm has been out.

Length of Tongue

15. Short-tongued bees cannot reach the nectaries in certain flowers, notably some of the clovers, and are at a disadvantage. The tongue-length may be measured by anatomical dissection, but there appears to be no common agreement as to precisely what should be measured. For practical comparative purposes, a device is used on the Continent of Europe, consisting of a honey trough with sloping bottom covered with wire gauze and having a scale showing the distance from gauze to bottom at points along the incline. Honey is put in and the bees to be tested are given access to it. They will clear all they can reach, thus removing the honey to a certain level which may be read on the scale down the incline. The device is known as a glossometer.

A simple alternative consists in the use of a small glass tube with an internal diameter of about 2 mm., say one twelfth of an inch, filled with candy or granulated honey. This is placed in the hive and the bees lick out the contents as far as their tongues can reach, then the depth is measured from the outside.

Mistakes have been made in the past in breeding for length of tongue at the expense of other qualities, but in places where certain crops abound requiring more than the average tongue-length, the long -tongued bee has an advantage.

16. Skorikov gives the following average figures for tongue-lengths, showing a maximum figure for bees from the southern end of the Caucasus:

TABLE V
TONGUE-LENGTH OF HONEY BEES

Central European	60-65 mm.
Italian	6.43
Cyprian	6.73
Grusinian (Caucasian)	6.89
Abchasian (Caucasian)	7.01
Mingrelian (Caucasian)	7.10

Other Caucasian bees lie in the middle range. Probably the British black had a shorter tongue.

The Cyprian stands high, having regard to its small size, and possibly controlled hybridization with a larger long-tongued bee might produce a new maximum.

Honey Sac Content

17. A bee can load up possibly 100 milligrams of honey, but the average load of nectar deposited per journey is more like 20 to 40 mg., corresponding to 12,000 to 24,000 journeys per 1 lb. nectar. Probably1 lb.of honey represents 10,000 to 50,000 bee journeys, depending upon the nectar and the conditions, there being a large loss of weight when nectar is converted to honey (**433**). Representative water loads vary from 20 to 60 mg., requiring 8,000 to 24,000 journeys (average say12000) per pound of water. Hence the importance of water supply (**1266-74**).

A loaded bee carries its hind legs well forward under the body. A bee with empty honey sac stretches its hind legs out behind.

The honey sac will hold several days' supply for individual use within the hive and will hold pollen as well, the use of which is under control of thebee. Pollen once entering the stomach is digested and reaches the colon in about 21/2 hours (see also **1305**). A flying bee probably has to find its fuel as it goes, from hour to hour.

Length of Life

18. The life of a bee is not determined so much by the passage of days as by the activities of the bee. It lives many months in cold weather, but in the height of the season the strain of gathering water, and nectar, but especially the strain of brood raising, greatly shortens the life. It takes 21 days to produce a bee from the time the egg is laid ; thus at a time when the colony is not increasing or decreasing, the ratio of the number of bees to the quantity of brood and eggs depends upon the average life of the bees under the conditions then prevailing. For example, if the average life is six weeks, or twice 21 days, then there %vdI ire present in a developed colony twice as many bees as there are brood and eggs.

Some bees die young through disease; others are caught by birds or die through stress of weather; so, to maintain the average, some must live considerably longer than the average.

With some modern strains, individual bees emerging in May have been found alive in September, and September-hatched bees in a normal good colony have been found alive in the following June.

Effective Life

19. The 42 days in the above example may be described as the effective life. Some Italian strains imported to Great Britain have shown an effective life as low as five weeks, whereas the British black had an effective life nearer eight weeks, and some modern strains show a still better figure.

Importance of Longevity

20. Longevity is obviously a characteristic of great value to the beekeeper, especially in an uncertain climate. Bees work within the hive generally for the first two weeks, and take up field duties after that period. Thus a bee with five weeks' effective life will only put in three weeks on outside duties, whereas bees with eight weeks' effective life cost no more to produce, but can put in six weeks' harvesting. Moreover, during an unexpected short flow one is much more likely to have harvesters available with the long-lived than with the short-lived strain.

Hence the great importance of stamina and freedom from disease, and of habits conducive to longevity, such as the habit of sitting pretty when there is no profitable flow and of working like demons when there is anything doing.

Assessment of Longevity

21. A colony of long-lived bees may be readily distinguished from short-lived neighbours, because it will maintain a larger number of bees in relation to the size of its brood nest. One colony will maintain a hive full of bees for some weeks on a brood nest of 30,000 where another may require 60,000. With the latter, on a prolonged break in the flow, the stores harvested just disappear, for it takes roughly a pound of food to raise 1,000 bees, or 20 lbs. per week to maintain a brood nest of 60,000.

Generally the long-lived strains have queens that are long lived but somewhat less prolific. Nevertheless the ideal bee is both prolific and long lived, maintaining a large supply of harvesters. The two virtues are by no means mutually exclusive.

Conservation of Energy

22. Bees instinctively conserve themselves, adjusting their activities to the conditions with which they are faced. Thus a check on the food supply is at once answered by a reduction or cessation of breeding and by general inactivity. Again, with only a small supply of food in reserve, the colony will not build up in the spring until a honey flow is manifest. A stock abnormally without brood has been known to last in this condition, maintaining a precarious existence which came suddenly to an end in about 12 months.

Duties of Worker Bees

23. The schedule on the following page gives the approximate periods of life which bees in an active colony in mid-season will devote to various purposes.

24. Nevertheless, in the complete absence of older bees, bees 5 and 6 days old will gather water and nectar and build comb. Bees emerging too late to be required as nurses in the autumn can do good nursing in the spring months, but in the height of the season are of little use as nurses after 3 weeks, and bees 5 or 6 weeks old, hard-worked, cannot raise a queen.

25. Bees engaged in one duty, say as water-gatherers, or as guards, will keep to that one duty for many days together, as will bees gathering honey from a particular source, but the number so engaged is changed continually according to requirements. It is believed that bees once engaged for several days as harvesters do not normally return to nursing. Every bee normally follows roughly the above time-table, proceeding from stage to stage but sometimes sooner or later than indicated, according to needs.

TABLE VI

NORMAL DUTIES OF WORKER BEES IN RELATION TO AGE

Period.	Duties.
First three days	Employed in cleaning cells and incubation.
Third to sixth day	Commence to feed old larvae.
Fourth day	Fully developed stinging power.
Sixth to tenth or fifteenth	Feeding young larvae.
About seventh day	First flight to use their wings if weather suit able, but may occur as early as the third, but normally about the tenth.
Eigth to sixteenth	First receive honey and propolis and pollen brought in by field workers.
Ninth to eighteenth	Cease to act as receivers of goods brought in.
Twelfth to eighteenth	Engaged in wax-making.
Fourteenth to twentieth	First engaged as entrance guards and in for aging for honey and pollen, which activities are continued until death occurs.
About twenty-fifth	First engaged in gathering propolis.

Hive Temperatures

26. The temperature in the hive may be above or below that without. The bees raise the temperature by the consumption of food and lower it by the evaporation of moisture. The figures given below should be taken as indicative and subject to modification with more precise, observation and the use of better apparatus. For hive temperatures in relation to design see Section VII.

27. If a bee's temperature is reduced below 45° F. (7° C.) its life is endangered. Bees will fly with a shade temperature below 45° F., but only in the sunshine and at considerable risk. Within the hive bees form a cluster in cold weather to maintain their temperature within safe limits, with a minimum consumption of food. The cluster is packed closer and closer as the temperature falls. The cluster is formed if the bee's temperature falls to about 57° F. (14° C.), the temperature outside being perhaps 48° F. When the outside temperature rises to about 45° to 50° F. in the shade, activity commences and flights may occur for clearing the bowel and later for fetching water, then pollen and nectar, as conditions improve.

28. The working temperature in the hive for general duties is from, say, 60° to 80° F.,
averaging 70° F. (21° C.), but higher temperatures are necessary for deposition of eggs, care of brood and wax-building.
Egg-laying requires a temperature of 85° F. to 93° F. (29 ½ -34° C.) The normal temperature in cells containing worker larvae is in the neighbourhood of 92° F. (33 ½° C.) and possibly 1 F. lower in drone cells. After sealing a lower temperature is permissible, and it may fall to 80 F. (26° C.) without injury to emerging brood.

29. A temperature of 95° to 98° F. (35-37° C.) is required for wax secretion, but sustained temperatures of 98° F. upwards (37° C.) are fatal to brood, especially if much moisture is present.

30. During swarming a general temperature of about 95° F. (35° C.) is reached. Serious disturbance in the hive may cause a local temperature as high as 103° F. (40° C.). If through exposure to sunshine or other cause the temperature approaches 140° F. (60° C.) collapse of combs and other serious trouble may be confidently expected.

Drones

Utility of Drones

31. It is believed that the sole useful purpose of the drone is that of fertilizing the queen, which, as a rule, is done once only, during the marriage flight, after which the drone dies. Drone-raising commences before queen cells are built, drone eggs being laid generally 2 or 3 weeks earlier than eggs are laid in queen cells. The bees like to maintain a continuous small supply of drones from about April until late summer. If there is much drone comb many thousands may be produced. The reasons for this are not fully understood. It is likely also that the numerous drones flying around while the queen is out greatly reduce the chance of her being snapped up by birds.

Potency and Flight of Drones

32. Drones fly about a week after emerging but cannot fertilize the queen until the twelfth or fourteenth day. For black bees 10 to 12 days has been given. Drones fly only in bright weather, mostly between 12 noon and 4 p.m. (G.M.T.) and for periods up to about 45 minutes, average say 25 minutes. They may fly up to 2 miles or more for exercise, but mating generally occurs quite near the hive. There is no foundation for the idea that one of the strongest drones out of hundreds or thousands flying can alone catch the queen.

The drone dies after mating, but not always before or on being parted from the queen.

Location of Drones in the Hive

33. In the hive, drones are generally found on the outside of the brood cluster where the honey is, and where they are out of the way of those engaged in the business of brood-raising.

Drones generally return to the hive in which they were bred. Very little drifting takes place. Flying drones may join a swarm, and stray drones are not refused entrance to any stock that is tolerating the presence of drones. A queen producing none but banded workers may at the same time produce a certain proportion of black drones; these may be wrongly supposed to have arrived by drifting.

Starvation of Drones

34. Drones are driven out to die on signs of a serious shortage of food. Towards the close of the summer season, when there is no chance of starting new colonies, they are driven out even from hives well stocked with food, but exception is made in case the hive is queenless. Very strong stocks have, however, been observed to suffer the presence of a few drones in the winter.

Reduction of Drones

35. If left to themselves, bees will raise many thousands of drones. This is most wasteful and should be prevented by the use of worker foundation and certain details of management. It is, however, good to let each colony raise some drones, as bees are apparently uneasy if they have none and are liable to tear down good worker comb to start drone-breeding.

Reduction of drone comb is important, for nurse bees will require about 1 ¾ lb. of food to produce 1,000 drones, and these in turn will consume about ¼ lb. of honey per day.

Drones can feed themselves, but nurse bees have been seen feeding them. The average length of life of a drone is not known, but the author suggests that the length of life of the self-fed honey-fed drone is short, and maybe that nurse feeding is necessary to secure virility.

Queens

General

36. The main business of the queen is egg-laying, her capacity to do so being controlled by the extent of the food given her by the worker bees. The queen alone destroys rival queens, but whether she is permitted to do so depends upon the worker bees. It is doubtful if the queen exercises any control at all in the hive. Queens take no part in the other activities of the hive, but hav been observed modifying a queen cell to their liking before laying therein. The queen usually flies for mating when about 5 to 8 days old, if the weather be favourable, but may fly on the third day or may be much delayed. Queens have been mated successfully up to four weeks after emergence, in hot weather, and up to six weeks after emergence, in the cooler weather of spring or autumn.

A quiet day with shade temperature above 60° F. (16 C.) is desirable for successful mating.

37. The queen commonly commences to lay on the third day after mating. She may commence as early as 36 hours after mating. She may be delayed to the fourth day or later by cold. It is said that if mated late in the year, just before a cold spell, she may be delayed even until the early spring. Generally in favourable weather, the queen will be found laying 10 days after emerging. She may lay on the seventh day, but will be much later if mating is delayed.

38. The queen leaves the hive for mating or to go with a swarm. She will make a few short excursions to familiarize herself with the location of her home before leaving for the mating flight. If there are but few drones about she may make several flights before being mated. Queens have been observed to make a second mating flight after the first mating, and presumably on account of the first mating being imperfect, perhaps insufficient. A well-mated queen should be capable of laying, say, 1 million eggs.

39. A queen returns to the home colony after the mating flight. If she misses her way and reaches a strange hive, she will almost certainly be destroyed and cast out. When a hive is opened the queen endeavours more or less actively to hide herself. Queens of some strains are most successful in hiding, a most inconvenient characteristic. Others stay quiet on the combs.

40. The queen bee has a long curved sting which, with very rare exceptions, she uses only to destroy rival queens.

Queen Calling

41. On a queen emerging, if the bees are preparing for a second swarm, they will retain certain virgins in their cells by thickening the wax around the tip. The imprisoned virgins will make a call, known as piping, being a sound like "peep-peep." The emerged queen will reply on a lower note, louder and harsher. These sounds may be clearly heard in the evening when all else is quiet.

Longevity of Queens

42. A good queen of prolific strain should be serviceable for two full seasons. Less prolific and queens of long-lived strains give good service for three to five years, and have indeed been known to continue for eight or nine. A queen that has given good service may be retained another year for queen breeding by removing her with some combs of workers and food to a nucleus hive, the colony being too small to reach swarming strength or over-tax the queen.

Points of a Good Queen

43. A good queen should be large and active with long, deep abdomen, not wide and flattish, or short and blunt. The abdomen is, of course, smaller in a virgin or in a fertile queen not in full lay. In the Italian variety the colour should not be too yellow and there should be but little darkening at the tall. In a hybrid more colour will show on the tail and back. The whole queen may be darker, more copper-coloured or even black with copper-coloured rings on the segments.

In the black races colour will be found only in the legs and perhaps one or more rings on the abdomen. Purity of strain shown by uniformity of character of offspring is also a valuable feature.

The quality of a queen should be judged finally, however, by observing her performance and that of her offspring and near relatives.

Development of Queen Pupa

44. On occasion, the ages of queen pupae may be approximately determined by their coloration as scheduled below, the observations having been made on certain bees of Italian origin:

11th day Pupae white all over.
12th day Compound eyes red. Simple eyes white.
13th day Compound eyes purple. Simple eyes red.
14th day Thorax and antenna brown, abdomen yellow.
15th day Abdomen yellow brown, remainder brown to black. Wings fully extended and emergence about due.

Egg-laying, and Development of Larva

45. After successful mating, a vessel in the queen's body, called the spermatheca, is found charged with the male element, spermatozoa. The eggs laid in drone cells do not receive the male element; thus the drones hatching from them are derived from the queen, and her ancestry alone, and are unaffected by the race or qualities of the drone with which the queen has mated. All eggs laid in worker cells and in queen cells except those laid by a failing queen, are, however, normally fertilized at the time of layingy by the male element, a spermatozoon entering the egg as it is laid, through a minute orifice at the upper end, which orifice closes shortly afterwards by hardening of the mucus. The eggs laid in queen and worker cells are identical, their development into queens or workers depending entirely on the after treatment of the larvae. An unmated queen lays after a while, but drone eggs only.

46. Queen larvae are fed entirely on a concentrated food, known as royal jelly. Workers and drone larvae are fed on a similar substance only until the third day, after which, worker larvae are fed on honey and predigested pollen or honey and pollen, and drone larvae receive honey and pollen.

Royal Jelly

47. The royal jelly is believed to be a glandular secretion of the nurse bees. That fed to worker larvae in the first 3 days is richer in proteins and contains less fat than the average fed to queen larvae. That fed to drone larvae contains more proteins and fat and less sugar. In queen raising there is some important change made in the royal jelly on the third to fourth day, essential to queen raising but which has hitherto resisted analysis.

Size of Eggs

48. A queen lays eggs uniform in size and shape, save that an old queen lays a much thinner egg. The egg is normally about one-sixteenth of an inch long (1.4 mm.) and about one-sixty-fourth of an inch (0.4 mm.) maximum diameter. The egg tapers slightly and is slightly curved. The largest diameter occurs near the end last to appear. The smaller end adheres to the cell base.

Rate of Laying

49. Queens have been observed laying 6 eggs per minute which is at the rate of 7,200 per day, but probably 5,000 is a maximum for sustained laying. Queens are frequently met with, laying 3,000 per day for 21 days or more, but 2,000 is a good performance for the period of heaviest laying and represents a maximum colony of 84,000 bees on an average effective life (see **19**) of 6 weeks.

TABLE VII

NUMBER OF FRAMES IN BROOD CHAMBER IN RELATION TO EGG-LAYING
CAPACITY OF QUEEN

Sustained Daily Average.	British Standard.*	Langstroth.	British Deep.	Modified Dadant.
1,200	7-8	5-6	5-6	4-5
1,500	8-10	6-8	6-7	5-6
2,000	10-13	8-10	7-9	6-8
2,500	12-16	9-12	9-11	8-10
3,000	14-18	11-14	10-13	9-11
3,500	16-20	13-16	12-15	10-13
4,000	18-23	14-18	13-16	12-15
5,000	23-28	17-21	16-20	14-17

* The frame here assumed is one with a 3/4 inch top bar.

50. The preceding table shows the approximate number of combs required in the brood chamber, according to the average continuous daily rate of laying of the queen. Combs in excess of the upper figure provide unnecessary storage room, while the minimum number represents a certain amount of crowding.

For example, it will be seen that a brood chamber with 11 M.D. frames provides for a queen laying 3,000 to 3,500 per day on an average for 3 weeks, and allowing a very good queen helping out one not so good with a frame or two of brood. Again, a single 12-frame chamber of British standard frames will accommodate a queen laying 2,000 eggs per day, while two 10-frame British bodies tiered will accommodate rather more than 11 M.D, frames.

51. This table thus shows the necessity of providing for double brood chambers except with the largest hives, such as the M. D., and even with such, good queens will fill their own hives and provide brood to help out the weaker stocks. The British standard and Langstroth brood chambers are only suitable for the double brood chamber system of working (**1158**). Even the M.D. may require helping out with a shallow chamber.

52. Noting that the daily average rate of laying must be multiplied by 21 to give the total of eggs and brood and by, say, 40 to give the strength of the colony in worker bees, it will be readily understood that the bees in full strength will occupy a far larger number of combs than are represented by the figures for brood chamber in the table (**18**).

Growth of Brood Nest

53. When the bees are closely packed in cold weather, the first egg-laying will be confined to a small patch in the centre of the cluster. This patch is extended with the coming of spring, the extension continuing for about 3 weeks, at the end of which time the centre cells will be vacated and the queen will work from the centre outward again, making a further extension when the outer margin is reached. This process can be clearly followed by examination of the combs in the case of bees kept quiescent in a cellar and brought out in the early spring, as the centre patch is clearly defined. The third expansion is generally the largest, and its termination represents the period when queen cells are most likely to be started. Where bees are wintered outdoors in milder climates the development is somewhat different. The close winter cluster formation is only required until early spring, up to which time the brood nest retains the shape of a Rugby football with its longer diameter running through the combs. The cluster is broken when the nights become warmer, and the combs themselves are used more for protecting the boundaries of the brood nest, the bees packing the edges to maintain the temperature (see also 785). Plate IVa, facing p. 150, shows a good comb of brood.

Queen Failing and Supersedure

54. The laying of eggs in irregular patches and especially the laying of more than one egg in a cell is definite evidence of failing, provided the queen is not short of empty cells and provided there are sufficient bees to cover the brood. A failing queen has been known to lay as many as twenty eggs in one cell. If the queen is insufficiently fertilized (i.e. carries insufficient of the male element) she may fail suddenly at any time to produce worker bees, all her eggs producing drones only, as first shown by the cell cappings (326). This may occur at any time of year, but failure through ageing is more likely to show at the close of a period of heavy egg-laying. The bees are generally aware of impending failure before the bee-keeper, who should therefore take warning from signs of intending supersedure. A marked reduction of egg-laying due to a failure of food supply must not be mistaken for a sign of failure in the queen. Races accustomed to accommodate their requirements promptly to a fickle climate may bring egg-laying almost to a standstill in a few days, on the sudden cessation of a honey flow in the spring (5).

55. The bees will supersede a damaged queen, though one suffering the loss of only a hind leg is frequently tolerated. Damage causing disturbance of laying power or damage of a foreleg or antenna, leads to supersedure.

Marking the queen may lead to supersedure (73).

Balling of Queen

56. In certain circumstances worker bees crowd round and enclose the queen. This is believed to be a panicky attempt to protect her. It is liable to occur if the queen is frightened. It is most likely to occur during or following manipulation in the spring in poor weather when stores arc short, in small lots, and during robbing; in other words, in time of stress and at times when the queen could not be replaced by the bees. Old bees, long queenless, are liable to ball a new queen on introduction, or a virgin queen on her return to the hive.

57. When balling, the bees form a ball with the queen at the centre.

Balling is accompanied by a distinctive hissing sound in the note of the colony. The ball will become tighter if the operator endeavours to break it by hand or by the application of smoke, and the queen will probably be damaged or even suffocated. If balling is seen, clos~ the hive at once, if practicable, and await more favourable circumstances. The ball can be broken up by dipping in water, but the queen should then be caged for a time and the condition of the stock seen to.

58. To avoid balling make the minimum of disturbance in the necessary to take into consideration the size of the colonies. A weak colony never develops as fast as a strong one (**922**). spring, making any necessary full examination when honey is coming in or after feeding, and stop if signs of balling occur. The more excitable dark races are the most likely to ball their queen.

Finding the Queen

59. Queens of the yellow races and hybrid queens are always lighter in colour than the worker bees and are thus more readily distinguished than those of the dark races. In the latter there is very little colour difference and the principal feature to look for is the long abdomen extending well beyond the wing-tips, and for colour in the legs. With dark bees especially it is of assistance to mark the queen (69-73).

60. The queen is most readily found in the middle of the morning when many of the bees are foraging. Open the hive quietly after using only a little smoke and pausing a minute or two. Remove an outside comb or two so that the others may be well separated, then examine one by one, looking first on the face which was covered just before removal, and while lifting out one comb glance at the face just exposed of the next. The queen will probably be found on a comb in which eggs are being laid, so give special attention to combs with vacated cells and young brood. The queen endeavours to keep out of sight away from the light, and if disturbed or frightened, may hide amongst the bees, or around the edge of the comb between the comb and frame.

61. While removing a comb for examination glance down over the surfaces of the comb, looking first at the surface just uncovered. The queen stands higher on her legs than the workers do, and unless the comb circular route examine the bottom edge again and then across the middle. The path the eye should follow is indicated in Fig. I.

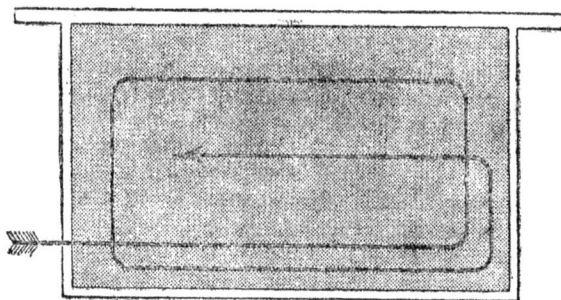

FIG. 1.—FINDING THE QUEEN.

62. The presence and activity of the queen may generally be judged by examining the brood and eggs, and it is not necessary to see the queen herself. If, however, she must be found and this proves troublesome, remove and examine the combs one at a time, putting them into a spare body. Examine also the empty hive in case the queen is hiding amongst bees on the walls. If the queen is not found, shake or brush the bees off 3 or 4 combs into the spare body and return these combs to the hive and cover with excluder with an empty body or super on top. Shake or brush all the bees on to the excluder and watch for the queen, who cannot go down. Finally, return all the combs in their original order.

63. In a large and difficult stock with dark queen, insert a frame of foundation and 3 to 4 days later open quietly and separate the frames away from this and then lift it out at once. The queen will most probably be found upon it. If not, then next search the combs on either side.

64. Another and more effective device is to put one selected frame with emerging brood into a hive body placed at a distance and fill up with frames with foundation or starters and shake in all the bees. The old bees will fly to the old stand and old brood. Next day the queen will be found on the selected comb. Alternatively, place the body with queen and bees over an excluder on the old body until the second day. The first method should only be used when nights are warm, as it involves risk to the brood.

Catching and Holding a Queen

65. A queen, not on the run, may be caught by approaching the forefinger from behind and placing it on the thorax. This will stop or slow her sufficiently so that she may be picked up at once between the thumb and second or first finger. So lifted, by the right hand she may be transferred, if desired, to the left hand, placing the left thumb under her thorax and forefinger on top. Her wings and abdomen will then project and a wing can be cut if required.

If the queen is shy and moves rapidly, as do many dark queens and most virgins, she is best caught by dropping a (well-aired) empty match box over her and sliding the cover over the box and gently under the queen. In a closed room near a window she can be allowed to walk out and is then readily picked up, the workers making for the window.

Always be most careful not to press the abdomen.

Clipping the Queen's Wing

66. This is sometimes done as a method of marking, but more generally to prevent the queen flying with a swarm. If the dipped queen follows the swarming bees she falls to the ground near the hive, where she will be found unless perchance destroyed by an enemy or unless she has crept back into the hive. The bees, finding themselves queenless, will return to the hive, and the bee-keeper being informed of the attempt to swarm, will take steps to deal with the matter (1121). Failing means of notification, the bee·-keeper must periodically examine all hives for preparations for swarming. The bees, however, are liable to take things into their own hands and supersede the queen, in which case the swarm will issue later and be lost. The use of this practice is fast diminishing, other methods of swarm control and of marking (69) being preferred.

67. To clip the wing, lift and hold the queen as in 65 above, then remove not more than half the area of wing on one side, cutting on the slant so as to avoid the thick part of the back rib of the wing. If too much is cut, the queen is likely to be superseded forthwith. Thus for marking only, it is best to remove only a small but noticeable portion from the tip of the wing. If all wings cut in one year are on the same side, say the left side, and those next year on the right and so on, one gets a rough indication of age.

68. Another method of holding the queen is to let her walk on to any convenient surface and hold her down by means of a small forked stick having fine elastic stretched across the fork. The queen is held under the elastic passing across the back of the thorax. Or, again, if the queen is caused to walk on a surface sticky with honey, she will go slowly and lift her wings in an effort to get free, and they may then be cut.

Marking the Queen

69. A distinguishing mark may be put on a queen so that she may be readily found. Wing-clipping is scarcely adequate for this purpose. In the races with yellow markings and even in hybrids from them, the queen usually has sufficiently large or prominent light markings on the abdomen to enable her to be readily detected. In the case of black bees, however, the queen is frequently hardly distinguishable by colour, and is also generally more shy. It is useful, therefore, to mark all black queens with a brightly coloured patch on the thorax. When coloured markings are used, age is indicated by using each colour for one year and employing two or three colours in rotation.

70. The queen must be held (65) for marking and long enough for the colour to become dry. I have seen worker bees remove gold paint, immediately after its application. The place to mark is the middle of the thorax.

71. Coloured sealing-wax is frequently used, dissolved in rectified spirit to the consistency of cream, but if ether is used as a solvent the paint will dry more quickly. Some of the enamels sold, made with a celluloid medium, the solvent being amyl acetate (smelling of pear drops), are also good and quick driers. Ordinary enamel paints are of no use, they dry too slowly, but some of the gold and silver paints are good, those known as " Ardenbrite," for example. Good results are also obtained with Canada balsam dissolved in xylol and coloured with powdered coloured chalk, bronze powders, or other convenient matter. Experiment first on a worker bee.

72. The German-Swiss bee-keepers employ a brilliant mark which can be seen even with the light that shines down between combs. This is obtained by fastening a piece of coloured bright metal foil to the queen's back. Shellac varnish made up with rectified spirit is used, both thorax and foil being painted and brought together when the varnish is tacky. The foil should first be moulded somewhat convex so that the edges may be readily secured without undue pressure. The bright convex surface catches the light and appears to shine like a small lamp. The "Eckhardt" queen-marking set has a supply of metal foils carrying small numbers, a supply of adhesive and a convenient punch adapted for cutting, moulding and applying the mark.

Supersedure of Marked Queens

73. Heavily marked queens and queens marked in full lay or in the spring, are liable to supersedure, thus partly defeating the owner's object and causing loss. The risk is less if they are marked as virgins or if the queen is first chloroformed or if only the wing-tip is cut, especially with the last-mentioned alternative, but one must not cut the wing of a virgin queen.

Loss of Queens and its Detection

74. The loss of the queen is a serious matter. Work in the hive is brought sometimes nearly to a standstill. Brood-rearing ceases. The population rapidly dwindles. The death of a queen may occur through: (a) Crushing during careless manipulation;
(b) Balling (see 56 to 58) after manipulation;
(c) Destruction or loss of direction during mating flight; (d) Destruction on returning to the wrong hive;
(e) Old age.
A queen failing of old age will, however, be superseded by the bees and then destroyed.

75. The internal signs of queenlessness are absence of eggs and young brood at a time when such would be expected to be present. Moreover, on opening or disturbing the hive, queenless bees make a distinctive "note" of distress which should be particularly observed.

Presence of eggs may, however, be due to the presence of laying workers, following queenlessness (107 to 110).

76. Other external and internal signs of loss of queens are the following:
(a) Bees running over the Right board and front of the hive seeking or awaiting the queen. This is seen during the mating Right or an hour or more after the removal of a queen by the apiarist, or on a swarm returning without .queen.
(b) Drones retained when they have been destroyed in other hives. This is one of the commonest external signs, but is not conclusive, as very strong stocks have been known to tolerate a few drones .out of season.
(c) Bees listless and idling when other stocks are busy. This, however, may be caused by disease. It is thus a sign calling for examination of the stock.
(d) An important sign is the storing of considerable pollen in the middle of the brood nest.
A careful bee-keeper will keep a look out for queenlessness, especially about swarming and mating time, and again, when closing down for the winter.

Test for Queenlessness

77. If the hive is thought to be queenless, the apiarist will take steps to remedy the defect, or if the stock is much depleted he will unite it with another. Following the above signs, the apiarist may be reasonably sure of queenlessness, but before introducing a valuable queen a final and conclusive test is desirable. A comb containing hatching eggs should be introduced from another hive, care being taken to avoid chill. If no queen is present, then, on examining this comb three days later, shaking off the bees, queen cells will be found started. The queen may then be introduced. The comb with cells started should be removed at the same time. Failure, however, to build cells in such a comb does not absolutely prove that a queen is present, as this failure may be due to the presence of laying workers (107-110) whose activities should be sought for. Do not, however, introduce a valuable queen unless queen cells are built, unless, of course, there is other proof of queenlessness, as, for example, when the bee-keeper himself has just removed the queen.

Re-queening

78. The queen is the most important thing in bee-keeping. What the bees will be and do depends upon what the queen is. Even in the best strains queens vary greatly. A queen that is doing badly from any cause should be replaced at once. In comparing queens, however, it is much crowded with bees may often be spotted at once on this account. The queen makes for the edge of the comb to get away from the light, so when the comb is lifted out look first all round the edges commencing with the bottom one, where she is most likely to be. On finishing the necessary to take into consideration the size of the colonies. A weak colony never develops as fast as a strong one (922).

79. When the performance of a stock is below the average of the year, the queen should not be retained for a second season. Some bee-keepers re-queen every year, but the best queens of 1 year's service will average better in the following year than the new ones, so the beekeeper may expect to keep one out of every four or five to advantage for a third year unless he has very prolific strains, and at least half his queens for a second year. With the dark races a greater proportion may be kept. In warm climates, however, where breeding is almost continuous throughout the year, it is best to re-queen every year.

80. It is not easy to raise good queens very early or late in the season, and late queens are apt to be thrown out in the spring. The best time for re-queening is generally several weeks before the honey flow ceases (see, however, Section XI). When there are two flows queens may be raised between them. Queens raised in nuclei during the late flow can be introduced by uniting the nucleus to the colony to be re-queened after the flow has ceased and stimulating the queen to continue laying (**1294-8**). Such queens will lay later than queens introduced in mid-summer or before and are less likely to swarm the next spring.

81. Nevertheless, with some good non-swarming, long-lived strains of even quality, superseding is left, to advantage, largely to the bees themselves. More superseding occurs than is known to beekeepers who do not mark their queens.

82. When destroying a queen some bee-keepers let the bees smell her, believing that they thus become more quickly aware of their state of queenlessness.

83. A normal queen raised by and emerging amongst the bees with which she is to stay is most likely to give satisfaction. Such a queen is sure to be accepted without supersedure or swarming, provided there was only the one cell and that she mates successfully. The risks are small if the queen cell is introduced from another stock, provided there are no other queen cells present or young brood or eggs. If a queen in full lay can be transferred to a well-fed stock to replace one just removed the prospects are also good, provided she is not of a very different strain. In all other cases there is some risk of supersedure even if there are no cells or young brood and eggs at the time of introduction.

Introduction of a Local Queen

84. Before introducing a new queen the old queen should be deposed. A new queen in lay may be introduced directly from another colony (stock or nucleus) without any precaution, if the receiving colony is in a prosperous state with food coming in and the transfer is made straight from the one colony to the other. It is best to place the newcomer on the same place on the same comb from which the previous queen was removed.

The most hopeful way of introducing a queen into a hive having laying workers is to introduce a queen in full lay straight off a comb from another hive (see **112**).

Introduction of Travelled Queen

85. As a rule, the queen to be introduced cannot be taken straight out of another colony, but has to be brought from a distance in a queen cage so that she is not ready to lay immediately and is liable to a mixed reception. To overcome this a number of precautions are necessary as detailed in the remainder of this section.

86. The first requirement is that the bees must be aware of their queenless condition. They are then in a favourable condition, and especially from 6 to 12 hours after removal of the old queen. After a while, however, they plan to raise queen cells and are then, on say the second day, less favourably disposed. After the cells are well started they are again in a favourable state, say from the third day, but the cells started must all be destroyed before introduction (**1364**). If, however, the bees have been left queenless for more than 10 days they may even have raised a new queen; a virgin may be at large in the hive. This state is quite unfavourable. If, however, the bees have been long without a queen or hope of raising one, this again is a relatively favourable state. It is well, however, to test for queenlessness (**77**) a colony suspected of being in this state, as they may have a defective queen, which would produce an unfavourable condition. A colony may be brought into a favourable state by the temporary removal of all young brood and eggs when the queen is removed. Brood removed may be put into a body stood on top of the hive and united later when the queen is well established. No flying bees need be lost. The brood frames should be examined and queen cells destroyed (**1364**).

87. The second requirement is that the bees should be in a prosperous condition, especially as far as food is concerned. Unless there is a honey flow on, the colony should be fed slowly for at least 2 days before introducing the queen, and feeding continued for several days afterwards in the spring until the honey is flowing again; in autumn, until laying is well established, say, 5 to 7 days. It is better not to examine for 5 to 7 days. In an out apiary, it is clearly desirable to re-queen during a honey flow.

88. Another important factor is the age of the bees. It is often not appreciated that emerging bees cannot feed a queen. Young bees do not normally produce royal jelly until the fifth day. A new queen is most readily accepted by young bees not more than 10 days old. The next most favourable age is over 30 days old, but there should be amongst them bees which, through queenlessness, did not exercise their nursing functions. It is the bees from 11 to 29 days old, bees well able to build queen cells and in no desperate need of a new queen, that cause the difficulties.

Introduction of a Virgin Queen

89. The introduction of a virgin queen presents especial difficulty. She suffers under all the disadvantages of being a stranger, and of making no immediate demand upon the sympathy or services of those who later would attend her. A virgin normally arrives in the hive as a newly hatched queen, and when introduced from without is most readily accepted if under 24 hours old, the younger the better. It is always easier to introduce a ripe queen cell, however, than a virgin queen, If the virgin is older than 24- hours .it is desirable to make up an artificial swarm by shaking bees of all ages in front of a hive containing empty combs and one of brood unsealed and either drop her amongst the bees as they are run into the hive, or alternatively, shake the bees off several combs into a well-ventilated body or box, with food and water; an empty hive with empty combs except for some honey and water, may be used; when the bees have roared for an hour, run in the virgin through a small hole. The bees are then hived as a swarm (**1115**) and after the queen has been fertilized (**276** *et seq*), and is laying, the lot may be united to the colony requiring the queen or built up as a new stock.

90.a Running a queen in at the entrance is still used. Anaesthetics are no longer used. JC

90. *A virgin queen may, however, be introduced into a colony having nearly ripe queen cells. She may be run into the entrance of such a colony, or into a newly formed nucleus, 24 hours old, from which the old bees will have returned to the parent hive, or an anaesthetic may be used for introduction (105).*

91. Home-raised queens are best allowed to emerge in nuclei, from which they are mated, and in which they are allowed to lay and prove themselves (**307**). Later, the nucleus may be united with a stock to be re-queened, or the nucleus may be built up into a stock.

Use of Nucleus in Queen Introduction

92. From the above it will be seen that a favourable condition for queen introduction can best be produced in a colony prepared for the purpose. A nucleus may be formed of three or four combs, including one mainly of emerging brood and one of unsealed stores and pollen, but without eggs or young brood. The nucleus is brought into a prosperous condition by slow feeding and warm packing. The old bees will return to the parent hive. If the colony is prosperous before the nucleus is formed, then the queen can be introduced a few hours after formation, feeding being continued as above. Alternatively, the nucleus can be fed for 3 days, all queen cells then destroyed (**1364**) and the queen introduced, but there should be no eggs or young larvae in the nucleus on this third day. In either case, when forming the nucleus, the reigning queen must be found and the comb on which she is must be set aside while forming the nucleus, and then returned to the parent hive. The nucleus method is the best for re-queening during a late honey flow, the nucleus being united with the parent stock later (Section XV).

Methods of Queen Introduction

93. Several methods of carrying out the introduction of a queen to a strange colony are described below, of which the first is in most common use, and the two first are generally to be preferred. The act is not difficult or risky provided the more important conditions above indicated have been observed. In purchasing a valuable breeding queen it is better to have her in the nucleus in which she emerged and either unite this with a colony (Section XV) or build it up into one. A valuable queen arriving in a cage may be safely introduced by the method described in paragraph **102**. It is better to introduce queens in the evening than during the more active period of the day.

Use of Travelling Queen Cage

94. The traveling cage in which queens are shipped with attendant bees is closed on one side with wire gauze and furnished with an exit, plugged with candy and covered with card or metal at the outer end. The usual plan, on receipt, is to remove the card or other cover off the candy and any cover over the wire gauze and to insert the cage on top of the frames in the hive, so that the bees have access to the queen through the wire gauze and access to the candy. In a day or two they will release the queen by obtaining access through removal of the candy. The Latham cage is improved by the addition of a second exit furnished with a shorter plug of candy and a queen excluder formed of two pins or pierced metal. The bees obtain access first through the short passage and can pass in and out only one at a time and thus come in contact gradually with the queen.

A better place for the cage is in the middle of the brood nest. It can be tied on its side to the bottom bar of a shallow frame inserted into the middle of the brood nest.

95. Travelled queens are too frequently superseded not long after introduction. This is in part due to unnatural conditions of reception and in part due to the fact that a caged queen, especially one long caged, takes some days to recover her laying power. Hence it is safer to introduce her first to a nucleus. It is becoming customary to change the attendants, and the "one-hour method" (**98**) provides this feature in a very convenient manner.

96. A good cage should have the candy chamber and exit coated with wax and the exit covered with a waxed card, as the bare wood absorbs moisture too readily from the candy and is apt to make it too hard (for Candy-making see **1316**).

97. Six attendant bees are enough to look after the queen. Some use more to act as a buffer and prevent injury to the queen in case the cage be dropped in transit. A queen in cage equipped as above can travel safely for as much as 10 days, but for such a journey, a laying queen should be put in a nucleus for not less than 24- hours to reduce her laying before she is put in the queen cage. Queens have been caged for as long as seven months and afterwards produced satisfactory offspring.

"One-hour Method" for Fertile or virgin Queens (Snelgrove)

98. Towards dusk take 20 to 30 bees in a well-aired match-box, from the hive into which the queen is to be introduced; add the queen to the bees in the box after an interval of 10 minutes. Wait 50 minutes or more, but not more than 90, then insert the box under the quilt or under a soft cover over the feed hole. Do this quietly, if possible timing the procedure for completion in the half-light after bees have ceased flying, opening the box on insertion so that the queen is free to walk out. Close the hive quietly without shock, and leave undisturbed for several days.

Before adding the queen to the bees she may be lifted off a comb, or caged temporarily for 10 to 15 minutes in a match-box. To assist in inserting her the author uses a strip of metal somewhat narrower than the box, with one end turned down to reach nearly to the bottom when the strip is laid on the box. Just open the box and insert the turned-down end, the long end lying on the top. The box may then be opened further without releasing any bees. The queen is then inserted, being retained with the thumb. The metal strip is then lifted out and the box immediately closed.

Unless the receiving stock was already queenless it should be dequeened at the time the box of workers is being secured. If the colony is a large one, look for signs that the bees are searching for the queen (**76**) before inserting the new one.

If the queen is a virgin, all queen cells (**1364**) should have been destroyed and there should be no eggs or unsealed brood present.

The match-box with bees in it may be carried in a warm pocket if made secure with a drawing-pin or rubber band.

Direct Introduction

99. This method, due to Simmins, gave excellent results even in difficult cases. The queen is separated from all attendants and food for 30 minutes before insertion, which should be done just after dusk. Put the queen in a clean odourless receptacle which has not been used before, such as a test-tube with ventilated cork or a tubular queen cage, without food, and in a warm place, such as the vest pocket. Approach the hive quietly and open without vibration. Use a minimum of light, shining across the top of hive. Fold back one corner of the quilt at a point, if possible, remote from the cluster, then immediately and quietly let the queen run in. Close down and leave for 3 or 4 days.

The author considers the "one-hour method" superior to Simmin's method.

Use of Water for Introduction

100. The queen is soused with water and dropped into the midst of the bees, who attend to her at once, and she acquires their scent during the process. The Snelgrove procedure is as follows: Put the queen in an empty match box. Fill with luke-warm water near the hive. Empty the box forthwith and tip the wet queen amongst the bees.

Use of Artificial Swarm

101. A queen may be added to an artificial swarm made up from a queenless stock. For example, shake most of the bees on to the alighting board or an extension of it as when hiving a swarm (**1398**). As soon as the bees commence to run in, drop the queen into the midst of them and she will run in with the bees.

Introduction Through Screen

102. The queen and attendants may be introduced to a comb of emerging bees placed between two combs of stores, including pollen, arranged over a wire screen over a strong colony which supplies the heat. After a few days the comb with queen laying and young attendants may be inserted in the colony below. When using this method, it is essential that the attendants be left with the queen, as the newly emerged bees cannot feed her. When the cage is in position and the covering ready to place over it, remove the end cover to make a hole in the candy so that the bees have a direct exit.

103. If the screen employed has a controlled orifice, preferably covered with excluder, this may be opened quietly next day to let the bees become acquainted with the queen above. If a double screen be used (**671**), the new queen may be placed as described over a colony the queen of which has to be replaced, the old queen continuing to lay until the new one is well established. The upper colony can be built into a new colony if it be helped out with combs of emerging bees from the colony below, or alternatively by bees from the colony below allowed to pass through queen excluder in the double screen.

Queen Run in at Entrance

104. A queen may be run in at the entrance of a·recently queenless stock in the spring during a honey flow, especially a queen in full lay.

Introduction by Use of Anaesthetics

105.a The use of chemicals to aid introduction is no longer recommended. The chemicals mentioned are not easily obtainable and are hazardous. See Appendix ? for more details. JC

Chloroform (Chloromethane).

H 302 Harmful if swallowed.

H315 Causes skin irritation.

H351 May cause cancer/genetic defects.

H373 STOT RE2 Prolonged exposure may cause damage to body organs.

Ethyl chloride (Chloroethane).

H220 Extremely flammable gas.

H351 May cause cancer/genetic defects.

H412 Harmful to aquatic life.

105. *On queen introduction the behaviour of the queen herself is of the first importance, as she is liable to be balled. If the queen is rendered insensible, or partly so, she cannot be intractable. This can be done without injury. Chloroform (862) may be used, but ethyl chloride (863), sold in a container emitting vapour or drops under control of a lever, is very convenient. Put the queen under a wine-glass or small tumbler inverted on a piece of paper on a plate. Drop 2 or 3 drops of the liquid on the paper beside the glass away from the queen and move the glass to cover it, or move the glass over the edge of the base to insert the vapourizer from below for, say, 10 seconds. When the queen is sufficiently incapable she may be dropped among the bees, but it is good to drop a few drops of the anaesthetic where the queen is to be dropped in, later inserting the queen quietly. It is good to effect the introduction after flying has ceased and darkness is approaching.*

Two Queens in One Colony

106. As a rule, a queen will kill any rival in the same colony unless prevented by the worker bees. Bees about to swarm will prevent the queen from killing a rival which may have emerged before the swarm has been able to leave. With some strains a failing queen may be found in the same colony as her daughter raised to supersede her, both laying.

For methods of working two queens in one hive, see **1177-80**.

Laying Workers

General

107. If some strains of bees are left queenless and without any means of raising a queen, for a period of 7 days or more, a number of the worker bees experience a development of their ovaries and in 7 to 15 days evidence may be found of workers laying in the brood nest. The darker races appear to be more disposed than the lighter races and the less domesticated than the more domesticated. Within a month, in a bad case, most of the workers will be found under dissection to show some development of the previously atrophied ovaries.

An individual laying worker may lay several eggs, not more. Generally a number of workers reach the laying stage about the same time and together may lay a considerable number of eggs.

108. Recent observations have shown that in a colony troubled with laying workers there are large numbers of worker bees having ovaries developed above the normal, though not to the laying stage. Further, it has been found that the ovarial development in worker bees begins to show as soon as brood production. passes its maximum or is seriously checked in other ways. That is to say, the development of prospective laying workers commences simultaneously with the reduction of the duties of the nurse bees. Further, it has been shown that in a colony showing a large percentage of abnormally developed workers, the provision of excess brood at once checks the development, and, if continued, results in the reduction of the number of bees showing such development, the reduction being apparently too fast to be accounted for by death and replacement, seeing that it is young bees that show the development in question.

109. A laying worker is indistinguishable from a normal worker in outward appearances and behaviour. Laying workers are, therefore, very difficult to find.

Laying workers show similar characteristics in their laying to those of a badly failing queen. They lay more or less irregularly and sometimes more than one egg in a cell and all the eggs laid produce drones only; these drones are undersized. Drones from eggs laid by laying workers in drone cells, however, are of normal size. The worker cells are found capped with the convex capping characteristic of drone cells (**326**).

110. The workers may even endeavour to raise queen cells from the eggs, thus further deceiving the bee-keeper, but the inmates remain sealed for 14 days after sealing instead of 6, and are often destroyed before the full 24 days are completed. These cells are generally tapered and pointed in shape, and of course no bee emerges until at least the twenty fourth day, and generally none at all.

Destruction of Laying Workers

111. Laying workers generally cease their attempts, or are stopped by the workers, when a laying queen has been accepted, but in a strain prone to the development of laying workers they may appear at swarming time. Furthermore, a colony having laying workers is most unfriendly to a strange queen, and especially to one of an alien race. Introduction is frequently unsuccessful, but the following procedures, if carefully followed, are generally successful. Choose the middle of the morning for the operation.

112. In a mild case, take a vigorous queen of the same race in full lay directly off the comb in which she is laying and putting her at once on a comb taken from the hive containing the laying worker or workers. The stock should first be fed for at least 24 hours unless honey is coming in, and well smoked. If the stock has been queenless for some time the condition of balance should be restored before giving a new queen by giving a comb of brood a few days in advance.

113. In a mild case in which young bees are still emerging, or were emerging within a day or two, feed the bees as before, then give a queen cell, or better still, a comb of emerging bees with queen cell. If a separate queen cell is given, it should be furnished with a wire guard. Insert the cell or comb between frames of young bees.

114. A colony found to have laying workers is generally already much diminished and hardly worth re-queening. It should be united with a strong colony by the newspaper method

(1412). Even if the colony is of good strength this procedure need not cause material loss, for, after successful uniting, frames of brood may be taken away and added to a queen-right nucleus, or the colony, when strong enough, may be divided, one half being re-queened.

Trouble attributed to laying workers may be due to a drone layer or to an undersized or scrub queen. Uniting may then lead to loss of the good queen. Before uniting, therefore, the bees should be caused to pass through a reliable queen-excluder. For example, remove the colony off its stand; put a framed excluder on the base with an empty brood box on it and arrange a board and sheet as for reception of a swarm. Shake all the bees on to the sheet, putting the combs into the box as soon as each is shaken. The colony will be ready for uniting the same evening, and the defective queen, if any, will be found later beneath the excluder.

115. If the colony is much diminished and consists of bees of flying age, feed them well with honey syrup, then break up the colony, shaking the bees off the combs in the height of the day and removing the hive. The bees will eventually find their way into other colonies where they will be welcomed as honey-bearers.

Races and Strains of Bees

General

116a Through numerous breeding programmes, it has been shown that all the various geographical races of Apis mellifera can be selectively bred to display characteristics favoured by the beekeeper. Temperament and swarminess are almost universally improved for reasons of basic manageability. That the races differ in their starting points does not mean that they cannot be improved to a broadly similar standard (see 118 – 123), but it may well affect their suitability to the target environment. Wedmore's point that "The average character across all races is much closer than the variation with any individual race" is important, and should be digested and borne in mind when reading the recorded merits and demerits of any particular race. Through a number of long-established breeding programmes, the average characteristics of some races may have been modified since Wedmore's time; this is particularly apparent when queens are sourced from professional breeders. DB

116. *The average Italian bee differs less from the average Carniolan, British, French or other leading race than does the worst bee in any of these races from the best of the same race. Experienced beekeepers write about French bees as evil tempered, Carniolans as inveterate swarmers, Dutch bees as grey, and so forth, yet others have French bees of the most tractable, Carniolan that do not swarm excessively, and Dutch bees banded with colour. Some indication is given below of the general experience with bees of different races, and particulars of noteworthy differences, but the purchaser must beware. The more important characteristics are, therefore, discussed before races, and the purchaser should seek what he wants by performance rather than by race.*

117. Bee-keepers are a friendly people; a little inquiry amongst those in your neighbourhood will show what can be bought to advantage. Bees should not be imported from a warm climate for use in a latitude of 50°, although there are bees in the Hebrides which can be wintered out of doors without heavy packing. On the other hand, those hardy bees adapted to the local conditions would make a bad showing in the south of Spain, where there is again a local strain which gives a good return. The return, after all, is the principal thing. Buy from a successful honey producer, or from one who supplies to successful honey producers. Fancy names, fancy colours and fancy pedigrees are unimportant.

Points of a Good Strain

118. The principal characteristic required of a strain of bees is the ability to make a good harvest over a period of years. This, then, is the supreme test of quality. It is frequently observed that the colony making the heaviest return is not the one with the most bees. The prolific races are most useful where there is a sure large harvest. In a region where the climate is hard and uncertain the strain required is one adapted to and proven under such conditions.

119. Stamina, power to resist unfavourable conditions of wind and weather, cold and damp, and especially disease, giving a long effective life, is a most important characteristic contributing to a good average harvest per colony. Stamina is transmitted by heredity, but may be injured by disease and generally by bad nurture. Long-lived bees are the best nurses for long-lived bees.

120. A characteristic of the first importance, both to the professional and the amateur bee-keeper, is manageability. The bees should be gentle and quiet and responsive to good management. Some are nervous and run over the combs, ball the queen, and sting the operator and other living things within a considerable distance. Others are almost stationary on the combs under examination; can be handled almost like flies and, when well managed, are indifferent to the presence of quiet strangers and animals. All bees require good management. Only some respond quickly and with reasonable certainty to the bee-keeper's aims and manipulations. Only some answer quickly to change of circumstances.

121. Bees which are untidy builders, produce irregular combs, much brace and burr comb, and employ hard or excessive propolis, are objectionable. Propolis is however, a natural sealing medium used in controlling ventilation and keeping out damp and enemies.

122. A tendency to produce laying workers is a nuisance, as is a tendency to rob or to submit to robbing.

123. Lastly, there is to be considered the swarming propensity. If considerable, it is bad in all conditions as with such bees it is bees, not honey, that is the main product of their activities. Bees showing but little tendency to swarm are now receiving increasing attention.

Racial Characteristics

124. Where bees have been located in certain regions for hundreds or thousands of years with little or no opportunity for inter-breeding with bees in neighbouring regions, owing to separation by water or mountain, ranges, they tend to exhibit uniform characteristics, and on the basis of the "survival of the fittest" they become adapted to the peculiar conditions of the region. It does not follow, however, that no other bees could do so well in the same region, for there are deep-seated differences established away back to the beginning over hundreds of thousands of years. Thus, while the British black bee was peculiarly adapted to prosper in the British Isles, there is no question that bees from some other stocks are better home cleaners, better resisters of disease, and easier to manipulate. Nevertheless, the characteristics of a particular strain, wherever it comes from, must be more or less especially adapted to the conditions of the climate and flora in the region where it was developed. In different parts even of a small island like Great Britain, different characteristics are required, hardier bees for the north, and adaptability to circumstances where the only harvest is late and of short duration, or both early and late with a long gap between.

125. Nowadays, races are becoming mixed all over the world through the interference of man, and crosses are being produced showing a mixture of good points, not, however, always stable in their qualities (**194**, *a*).

The evolution of the various races and their entomological significance is not dealt with here; we are concerned rather with the question of the peculiarities of the more useful races as now found, and their importance to the practical bee-keeper. Nevertheless, one or two broad observations may be made before dealing with individual races.

126. Bees belong to the old world. Those found in America and in Australia and New Zealand have been imported by colonizers in comparatively recent times. The wild black bees in Australia and New Zealand are mainly descended from the British black bee, while those in America are more akin to the German Heath bee.

127. In general, the bees in the colder regions are dark in colour and cap their combs with a white capping which allows room for expansion of the honey by moisture absorption without sweating. Those originating in warm regions in general have more or less yellow, orange or light brown in their colouring, and cap their honey without an air space.

128. Bee-keepers in most countries have been experimenting with mixed races for so many years and with improvement of stock, that the original racial characteristics as given below have been, and are, subject to modification.

Italian Bees

129a The Italian bees are now commonly referred to as 'Ligustica', or more formally as *Apis mellifera ligustica*. Two distinct strains have been described, the yellow southern bees and the northern tan-coloured bees, with differing behavioural traits. It is generally the paler southern lineage that is encountered, prolific but hungry in winter. DB

129. *Italian bees, or strains with a strongly marked Italian element, are more widely used for profit than any others, the representative racial characteristics being modified more or less by crossing with other races.*

130. The representative Italian bee is slightly smaller than the European black or brown bee, the difference being especially noticeable in the drones. The queens, being very prolific, may develop larger abdomens. The worker is characterized by the possession of yellow or tawny bands in the first three segments of the abdomen (the first segment in a bee's abdomen being small and next the waist). The underside of the abdomen is of a tawny shade. The so-called Golden Italians have been bred for colour and have more than three yellow bands. They may be suspected of Cyprian blood.

131. The segments of the abdomen are fringed with bands of short hairs tinged with yellow. The legs are brown. The queens run from almost pure golden or tawny or leather colour to coppery hues and dark shades, resembling the brown queens, but the latter have probably some dark blood. . Generally, more or less of the end of the abdomen is dark, and sometimes there are dark rings or spots elsewhere on the abdomen. The legs are yellow or tawny, as a rule, especially those of the queen.

132. These bees are very prolific and can build up rapidly under natural or forced stimulation, but are shorter lived than some of the dark races, and liable to continue breeding during a dearth of honey. They are not much disposed to swarming, answering readily to manipulations tending to check this activity. They do not choke the brood nest with honey. They are disposed to robbing (but so are some of the dark bees), and have developed a corresponding instinct of defence. They are also good house cleaners and cope with disease better than the dark races.

133. They are easy to manipulate, requiring but little smoke and remaining quietly on their combs if well handled. They are good comb builders tending to finish their work as they go, but poor cappers, although this latter feature is less noticeable in some strains and when there is a rapid flow.

They require better winter protection than races developed in colder regions.

134. Some strains showed a tendency to stop breeding sooner in the autumn and to start later in the spring than did black bees, necessitating autumn and spring stimulation to avoid dwindling. This feature still recurs and is especially noticeable with some highly prolific but short-lived strains.

Cyprian Bees

135. These are mentioned next because they have sometimes been interbred with Italian to produce a golden strain, although seldom used in the pure form on account of a tendency to vicious temper during a dearth of honey and to excessive building of queen cells.

136. The Cyprian much resembles the Italian bee, but is smaller and brighter, the abdomen being somewhat more tapered, and the orange colour extending over more than three segments. The hairs are longer and the thorax coloured, especially the forepart, altogether a very handsome bee. The underside is light in colour so that the bees are somewhat transparent. The modern Cyprians closely resemble the Italians but are somewhat smaller.

137. When well disposed they remain quiet on their combs under manipulation like the Italian. They are good comb builders but bad cappers, their combs being suitable only for extracting.

Carniolan Bees

138a *Apis mellifera carnica,* more commonly referred to as 'Carnica' or 'Carniolan', is native to central Europe. It was the subject of widespread adoption and improvement in Germany during the latter half of the 20th century. DB

138. These bees have much to recommend them, being very docile and hardier than the typical Italian bee, but in most places larger returns in honey are being obtained from strains with Italian blood. The Italian Carniolan cross is favoured by some, giving a hardy docile strain, building good sections. The cross generally recommended is one between the Italian drone and Carniolan queen. A Carniolan-Egyptian cross has given good results in Egypt.

139. The Carniolan is a large bee, larger than the Italian, generally greyish black, but those from some parts show a little colour. The hairs are longer and whiter than on the European brown or the Italian bees. The queens show colour in the legs and sometimes in rings on the abdomen, tending to brownish red.

140. The Carniolan bee is very prolific and must be assured of ample room if swarming is to be hindered. In their native country they are accustomed to a long season. Like most dark races they make fine white cappings. They breed late in the season without stimulation and start early. They hold well on the combs, being more difficult to shake than most.

Caucasian Bees

141 These are receiving attention on account of the exceptional length of their tongues (16), their exceptionally sweet temper and their hardy disposition. They are, however, predisposed to excessive use of propolis and inclined to robbing. Under scientific crossing they should prove useful hybridizers.

142. There are, in fact, several distinct races in the Caucasus, differing in colour and in more important characteristics. We may distinguish between the grey bees of the mountainous regions, the light yellow bees of the north and Abkhasian bees which have dark yellow or orange markings.

The grey bees are the largest, though small. They are prolific though moderate swarmers, inclined to give one large swarm. They are not disposed to robbing. They work well in bad weather and winter well. Their cappings are not white.

The light yellow bees, on the other hand, much like small Italians, are bad swarmers, white cappers, build much irregular and burr comb, and are good robbers. They will readily raise queens from fertile eggs in the presence of laying workers.

The Abkhasian bee, however, has commercial value, answers readily to manipulation and swarms in moderation, but is inclined to give trouble with laying workers.

British Black Bee

143. Although doubtless of the same stock as the brown or black bees of the adjacent Continent, this race had been established probably without intermixing for thousands of years before the nineteenth century, and was well adapted to the climate and methods of bee-keeping in vogue. It showed good results under improved methods of bee-keeping, but has now become merged with bees introduced from all parts of the world. Although much less prolific than the more generally favoured races, it made up for this by increased longevity and would gather more honey, especially in a bad year, than larger stocks of more prolific races. It failed, however, to withstand diseases, such as foul brood and acarine, was a poor house cleaner and defender and not so quiet under manipulation as the imported bees, but its comb capping was perfect, although even in comb building it was outdone by other races, being disinclined to work in the sections and leaving many unfinished.

The body was blacker than most of the so-called brown bees, the hairs being short. The queen had reddish legs and some indication of red· in the thorax and forepart of the abdomen. See Plate 1.

The above is written in the past tense, but there are many bees at least approximating to the original to be found still, especially in Scotland, and crossed strains are common.

French Bees

144. These appear to be of the European black or brown type, varying much in different parts. The more primitive bees are frequently vicious, but the best developed strains have much to commend them, being hardy, prolific, easy to manipulate, not excessive swarmers, and producers of beautiful sections, but inclined to run on the combs, making it almost essential to mark the queens. Those from the Gatinais district are amongst the best.

German Swiss Bees

145. This is a noteworthy strain produced under careful control for home consumption by co-operative bee-keepers. They are dark bees like the British black, long-lived, hardy and adaptable, the queens having a useful life of several years. They are greatly disinclined to swarm. It is only recently that they have been tried in other lands, but they have the habit of running on the combs and require more subduing than Italians and Carniolans.

Dutch Bees

146. These are of a grey colour, but those from some parts show some resemblance to the Italian. They are of medium size, hardy and active and· good comb cappers, but uncertain in temper and, above all, have a strong tendency to swarm, being prolific and having been kept for generations in skeps. These skeps have an entrance half-way up the side, and the bees do not always take willingly to a modern frame hive. Large numbers were imported into Great Britain during the First Great War to replace stocks lost through disease.

Spanish Bees

147. These are brown bees of a medium shade with many grey hairs, large and prolific, slight swarmers, somewhat aggressive, and doing better than imported bees under the condition of prolonged harvests and warm winters characteristic of the country. They might be worth trying in Northern Australia.

The Balearic Isles produce a very different bee of no commercial value, small and of a grizzly colour, builders of twisted and otherwise irregular combs.

German Bees

148a Widespread introductions of the Carniolan bee in the latter half of the 20th century replaced much of Germany's existing honey bee stocks, which were considered heavily hybridised and of poor character. It is likely that many 'German brown bees' were lost at this time; they shared observed characteristics of *Apis mellifera mellifera*. (See 138 – 140)

148. The German brown bee is generally similar to the British black, which see (143), but not so dark; but there are also the heath bees, a much less desirable race, being bad swarmers, poor house cleaners and disease resisters, running badly under manipulation, great drone breeders and inclined to fall off the combs, but gathering late and wintering well, and producing white cappings.

Egyptian Bees

149. These somewhat resemble the Cyprian but are of a lighter yellow and have strongly marked bands of white hair. Having been kept since time immemorial in small hives (of earthenware), they are inveterate swarmers. Some sting furiously and cannot be subdued by smoke, but carbolic is more effective; but there are now available some relatively, gentle strains. Fertile workers are of frequent occurrence and are tolerated in hives having laying queens.

African Bees

150. The prevalent race resembles the Egyptian, but the bees are of a more reddy-brown colour. In South Africa, however, there are bees less bad tempered, building relatively few queen cells and good combs, but inclined to migrate. They sometimes produce laying workers sufficiently developed to be partially fertilized. There is a black race in Togoland and in Madagascar. In Tunis the black Punic bees are found, good fighters, bad propolizers, using a watery capping and wintering badly.

150.a Eight geographic races of honey bees have been identified in Africa. Of these, the two north-western African races, *A. mellifera sahariensis* and *A. mellifera intermissa*, are considered more closely related to the European races than to their African peers. Of the African races, two have achieved greater general awareness. DB

150.b *Apis mellifera scutellata*, native to Central and Eastern Africa, derives its name from the distinctive yellow scutellum band across the dorsal surface of the thorax. Much of its native region benefits from forage and climate patterns which are considered to be amongst the most favourable worldwide. However, the presence of many predators has led to a very defensive temperament, and the prolonged dry season has resulted in a tendency for whole colonies to abscond and travel great distances in search of forage.

Its importation into Brazil as part of a breeding experiment, and subsequent accidental release into the wild, led to the hybridised 'Africanised honey bee', notorious for its bad temper and aggression. This hybrid is widespread in Latin America and in the southern United States, its spread aided not only by its inherent migratory prowess, but also by a number of behavioural mechanisms resulting in predominance of 'Africanised' drones and queens when honeybee colonies of Western European origins are encountered. It presents significant management challenges. DB

150.c *Apis mellifera capensis*, the Cape Honey Bee, is native to the southerly Cape Town region of South Africa. The workers posess unusually high numbers of ovarioles, and in a queenless colony a significant proportion of the brood produced by laying workers will develop as females. Investigation of these laying workers has never shown sperm stored in a spermatheca, so the eggs must remain unfertilised. The resulting viable female larvae may be raised as queens, through normal means of differential feeding, allowing what would otherwise be considered a "hopelessly queenless" colony to re-queen itself. Such behaviour has been noted as occurring extremely rarely in other species of *A. mellifera*, yet is commonplace only in *A. mellifera capensis*. DB

Syrian Bees

151. The native bees somewhat resemble Italians, but otherwise their characteristics are much like those of the Egyptians. They swarm excessively and many virgins leave with one swarm, living together until mated, when they may swarm again. They do not winter well.

Eastern Races

152. In India *Apis dorsata* and *Apis florea* are found difficult to manage in hives, absconding too frequently; the latter, though, is of good temper. Apis indica, however, is a good harvester, gentle, and does not use excessive propolis, but it is a poor defender against wax-moth, it is prone to swarm and rob, and absconds on danger of food shortage.

153. There are many races in Asia, some possibly forerunners of the European races, but having no special merit or other than local interest to the practical bee-keeper.

153.a Since Wedmore's time, one other Asian honey bee race, *Apis cerana*, has received much greater awareness in the latter half of the 20th century due to two negative associations. It is believed to be the originating host of the parasites Varroa spp. and *Nosema ceranae*. *Apis cerana* displays coping behaviours, assumed to be present through prolonged adaptation, that limit the population growth and impact of the Varroa mite. The absence (through lack of prior exposure) of such sophisticated behavioural traits in all races of the Western bee, *A. mellifera*, is arguably the principal cause of the problems precipitated by the presence of Varroa. Nosema ceranae was first identified in *Apis cerana* and named accordingly (ceranae = of cerana), and has been shown to be widespread worldwide since at least the early 1990's. DB

SECTION II

QUEEN AND DRONE RAISING

Introduction

General

154. Volumes have been written on this subject, it being one of the first importance to successful bee-keeping. The essential details of the numerous methods are here reproduced in such a way as to minimize repetition, the whole being brought into convenient compass by the omission of all that is mainly historical, of much argument and unnecessary description of details which can be seen in any Catalogue of Appliances, and by the orderly arrangement of the parts. Some repetition has been necessary, mainly for the sake of the worker on a small scale who has no occasion to master the details of methods employed by the professional. In connection with this, and indeed any, section of the work, the author will be glad to hear from anyone able to add information of importance in the practical art, furnishing any improvement or useful addition to what is here given.

Attention to Detail

155. It will be understood that in such a condensed account every detail matters. The bee-keeper who substitutes six days for three through some other engagement, or omits to feed where feeding is advised, or who in fact fails to observe any of the numerous details given, will deserve the results he will get, but they will not be the best, and no bee-keeper can afford to have any but the best queens.

Small Scale Working

156. At the same time, let not the small bee-keeper hesitate, if he so desires, to raise the few queens he requires, let that number be half a dozen or less, for he can raise them by simple measures in his own hives without special appliances, adding greatly to the interest of his hobby and his reasonable pride in his results. He has the advantage over the professional queen breeder that his queens do not have to suffer the common handicap of more or less detrimental internment in cages at a critical period, and transit by post.

The subject is first treated in a simple way for the small man, and is then developed in full detail.

Queen Raising Without Interference

General

157. Queens are raised by bees (*a*) under the swarming impulse; (*b*) to supersede failing queens; and (*c*) to replace queens lost from any cause.

Under Swarming Impulse

158. When the colony has expanded in the spring and drone .cells are reached or built, following the commencement of a spell of milder weather, drone eggs are laid. This is a necessary first step because the drone takes longer to mature than the queen (**4**). In latitude 50° to 55°, except in severe climates, this generally occurs in the beginning of April in the northern hemisphere, October in the southern hemisphere, and may occur even before this.

159. Queen cells are next started, and, in good weather, eggs may be laid in them when drone brood is well advanced, but this may be deferred any time up to about midsummer. This depends upon the weather, the strength of the hive and the swarming tendency of the strain. Swarming occurring later than this is due to bad conditions such as excessive heat and overcrowding. Swarming is most likely to occur in the normal course shortly after the maximum brood production is reached.

160. Queen cells built under the swarming impulse are built out on the sides or bottom edges of the expanding combs, but if frames full of comb are used, places may be made for them by the bees by cutting away comb on the face or edge. The cell is commenced as a cup resembling a small acorn cup, opening downwards and outwards, and the completed cell hangs downwards. Several are started in succession, and after a few days another batch is started. The egg is placed in position at some time after the cup has been formed. Say six to twelve cells are generally built, but some bees build fifty or more.

Under Supersedure Impulse

161. When a queen is failing through old age, or inadequate fertilization, or when, for any other reason, the workers decide to supersede her, they take steps to raise a new queen. The new queen may become mated and commence to lay before the old queen is discarded. Strains showing this habit are disinclined to swarm and will supersede without swarming, if not overcrowded, and frequently without being noticed. Swarming is nature's way of securing the propagation of colonies and continuance of the race, but the bee-keeper can see to this. Reduction of the swarming propensity is not found inconsistent with the retention of good harvesting habits. The production of a real non-swarming bee would be no more inconsistent with nature than the production of the pipless orange, and much more beneficial to man.

To supersede a laying queen, the bees raise only a few cells, from one to, say, five or six, all within a period of a few days. They are generally new cells, but may be built by enlarging the mouth of, and extending, a worker cell containing young larva of suitable age. The two methods are not used at the same time.

To Replace Lost Queens

162. A queen lost by accident, or through careless or ill-advised manipulation, will be replaced by the bees by one raised from young worker larvae, as in supersedure. While such queen bees are maturing no young brood or eggs will be found in the hive. Such queens are rarely raised in too much of a panic from larvae over the suitable age (**8**). The bee-keeper should, however, be suspicious of any queen in an undersized cell, or taking a day or two more than the usual 15 days to mature, or failing to show the characteristic small head, large thorax, relatively short wings, large abdomen and plain legs of a good queen (**6**).

163. To replace a lost queen the bees utilize a young larva by enlarging and extending the mouth of its cell, forming a cell which has much the external appearance of a normal new independent queen cell raised from a cup built on the comb. Such cells may also b~ found sometimes in queen right supersedure described below, and possibly because in the particular case the bees were not satisfied with the quality of the eggs last laid; they might prove to be or include drone eggs.

164. Bees don't swarm when raising a new queen on emerging unless the emergency happens at a time when the bees are already commencing preparations to swarm. The first queen to emerge will destroy any others coming forward.

Queenright Supersedure

165. If a portion of the brood becomes separated from the remainder where the queen happens to be, by a substantial barrier, such as combs of honey or a queen excluder, such portion including eggs or young larvae, the bees in that portion are likely to raise queens from young larvae. Such a colony is clearly not queenless. The queen may be in full lay and able to continue so. The apiarist discovering such a case must not be misled into concluding either that the queen is failing or that swarming is the impulse, although there is a risk of swarming occurring, as there is with normal supersedure.

How to Distinguish the Impulse

166. Queen cells raised under the swarming impulse are new cells with a thick acorn bottom formed generally on the margins of the combs, but some may, be started on the comb face; they are started in successive batches so that they are found in various periods of development. Drone brood is present, and may be increased. Egg-laying is reduced and comb-building ceases.

167. Supersedure cells are few in number, built as one batch, most usually raised on the face of the comb, and, in any but weak stocks, are heavily covered with wax so that the surface pittings may be deep enough to take on hexagonal form. Drones are usually flying before the cells are formed, and there is no increase of drone brood. Egg-laying of the laying queen is not reduced and comb-building does not cease.

168. Cells to replace lost queens are found in hives without eggs (other than those which may be laid by laying workers), with diminishing brood, and, of course, absence of queen.

169. Queenright supersedure cells are raised under conditions described in **165**.

170. If one of the queen cells is cut open and found to extend back to the mid-rib, retaining the original base of a worker cell, one may be sure that it is not a cell built for swarming.

Queen Raising to Plan on a Small Scale

General

171. For queen raising the apiarist may utilize queen cells raised under either impulse above mentioned, destroying those he does not require, but to do so requires as much supervision and involves as much disturbance as does any well-considered plan and is less certain in its results, save in the case of one who encourages swarming and relies on this, with his system of management, for the production of queens as well as the production of honey.

For such a one the main requirements are a prolific strain, regular but not excessive swarming, and disposition to make one large swarm. He will raise a new queen from the parent stock where required, and from his best stocks may take further queens by removal of cells. For further management of the removed cells see **251-8**.

Reducing the Swarming Instinct

172. The majority of bee-keepers, however, desire to suppress, or at least reduce, the swarming instinct, and such should not raise queens under the swarming impulse, because thereby they will eliminate, by breeding out, the very bees having, or tending to have, the very characteristic they desire. to retain and strengthen. They should utilize the supersedure impulse (**161**), or raise queens deliberately from selected stocks.

They should not await the manifestation of these impulses by the course of nature, but choose times and seasons profitable to themselves, and bring about what they desire by appropriate procedure.

Good queens may be raised under the queenright supersedure impulse during certain manipulations to prevent swarming (e.g. **1166-70**).

Time for Re-queening

173. In general, however, the best time for re-queening is in the late summer or autumn, not so late that the queen does not have a fine opportunity of mating and a chance of showing herself as a tested queen (see **307**) by the production of brood, but so that she may come through into the spring as a young queen able to develop a large brood nest of the finest bees without material risk of failure. A failing queen in late spring; is both a nuisance and a cause of serious loss.

174. Queens raised, however, earlier in the season and not required immediately for re-queening may be kept in nuclei during the late flow and, when the flow is over, united with the colonies to be re-queened.

A good time to start queen raising is five to six weeks before the expected end of a principal flow, or in general, in Great Britain, the beginning of June. See, however, **185-6**. A disappointing queen should be replaced when found, so it is a good plan to have one or more spare queens available, raised as opportunity occurs.

Importance of Parentage

175. To the worker on a small scale, from the standpoint of results, the question of parentage (**192**) of his queens is of as much importance as it is to the man in a big way, and in some ways more so, as he may ruin his strain by one mistake, whereas the professional will check out and reject the result of the mistake and not feel the loss.

The section on parentage may be studied with advantage, but the essence of the matter is to have drones as well as new queens raised from eggs laid by the mothers of the most successful stocks, noting that the drone should be judged by the performance of the workers of the stock from which his mother was raised as much as by the performance of her worker progeny.

176. Provided there are potent drones living in the colony from which the mating fight takes place, there is a high probability (some say 95 in 100) that the queen will mate with one of those drones. Never, therefore, let a queen fly for mating from a colony having drones of undesirable parentage. Either destroy all such drones and advanced drone brood before inserting a queen cell, shaking in some recently emerged drones of desirable origin. One avoids or minimizes the raising of undesirable drones by promptly replacing all unsatisfactory queens.

Drone Raising

177. Whether the queens are to be mated from stocks or nuclei, therefore, there should be suitable drones available therein, potent by the earliest date they will be needed. Assuming the queen may mate on the third day after emergence and that the drone may not be potent until the fourteenth day, the following notes give limiting dates for drone raising:

(a) *The drone egg should have been laid at least 20 days before the egg from which the queen is raised.*

(b) *The cells of drones so raised will be found sealed at least 10 days before the egg for queen should be laid.*

(c) *Such drones will have emerged before or by the time the queen larva is a day old, or at least four days before the queen cell is sealed.*

(d) *Such drones will be seen flying, in suitable weather, for four days before the queen emerges.*

From one or other of the above observations one may secure that suitable drones are available. It will be seen, for example, that if recently emerged drones are to be shaken into the mating hive so that they will make it their home, this should be done preferably before the queen cell is sealed.

If this critical matter does not receive attention there is no knowing with what kind of drone the queen will mate, but if it does receive attention, the large majority of queens will be mated as desired. The method is more sure than any attempt to flood your district with selected drones, unless your location is exceptionally isolated from neighbouring apiaries and wild stocks.

Raising a Few Queens

178. The methods described below are suited to the bee-keeper with only a few hives. The information on "emergence" and "fertilization" in **246-66** and **276-87** can be studied with advantage if further guidance is desired. Where nuclei are referred to, these may each be made from, say, three standard brood combs (**259** and **262**), but it will be found practicable and economical to use three or four shallow combs for the purpose, such, for example, as those used over a single brood chamber for wintering and into which the brood has spread in the spring. One can secure early drones by putting in the middle of such a super in the autumn, a few shallow frames having drone cells along the bottom edges. One such shallow chamber may furnish enough combs, food, brood and drones for three nuclei. If three nuclei are made up on a floorboard giving three flight holes (**672**) they will assist in keeping each other warm.

179. I t is true that a queen cell can be left in or given to a strong colony for queen raising, and if all goes well the queen is certain of a favourable reception, but in case of any mishap, even only delay in mating, the colony receives a serious set-back. Where, however, the bees are superseding a queen themselves, a cell should be left in the colony (**1366**) and the business left to the bees.

180. Mating from small colonies and nuclei is rather more certain than from large colonies, besides involving less loss if things don't go according to plan; moreover, nuclei furnish, not only a convenient store for spare queens, but provide for making increase with such queens, as they may be built up into colonies. With a little planning several queens in succession may be raised in one nucleus. A queen may be tested in a nucleus, but watch must be kept to avoid danger of swarming through overcrowding following raising too much brood.

181. *Method I.* Having formed some nuclei as above described, supersedure cells may be utilized as found, or cells raised from selected stocks during manipulations designed to check swarming, such as those described in **1226-62**. Such cells will be available at a favourable time from. the standpoint of weather, temperature and general prosperity. Note, however, that the low-swarming strains may omit to raise cells in an upper chamber, but may be caused to do so if temporarily cut off from access to those below.

Method II. A selected stock may be caused to raise several cells by temporary removal of the queen with a few combs to a nucleus hive. These cells may be transferred to nuclei after sealing and preferably when nearly ripe. If a stock be selected in which the queen has already given long service, opportunity may be taken to replace her by leaving one cell in the colony. In case of failure the queen in nucleus is united with the stock. If, however, the returned brood would emerge in time for a principal honey flow, it is better to return the queen and brood and requeen later in the season.

One may utilize any strong stock for raising queens from the offspring of a selected queen, by removing the queen from that stock, temporarily or permanently, with any queen cells, and destroying nine days later all cells then found., The stock is then in condition to receive a frame of eggs from the selected stock and will raise several cells. A larger number can be secured by following the procedure given in **205-9**.

Method III. When any of the methods of swarm control are used involving raising one brood chamber above a super (**1226-62**), queen cells raised in this chamber may be utilized. Insert a double screen board under the upper chamber with flight hole. If not inconvenient place the flight hole at back or one side, thus reducing the risk that the queen returning from her mating flight may enter the main entrance and be destroyed. The screen board should preferably have an excluder portion, which should be left open and the flight hole closed so that the bees can rearrange themselves, sufficient nurses coming up and the flying bees going below. Next day or later so long as there are eggs or very young brood still available, cover the excluder portion and open the flight hole. The bees finding themselves fully isolated are sure to raise queen cells, which they do not always do when separated only by a super and excluders.

If only the stock below is to be re-queened, the excluder portion may be uncovered at any time after queen cells have been started, for example when selecting a single cell to be left for utilization. Some prefer to exchange the top and bottom chambers before the mating flight. In either case it is practicable to have both queens laying at the same time, and the old one may be destroyed at convenience.

If several queens are required the upper lot of bees may be divided into nuclei, each with one cell, shortly before the queens are due to emerge.

When the upper lot of bees are to be isolated as above, it is important to make sure that they have ample honey and pollen.

Method IV. The more ambitious amateur may employ, on a small scale, grafting or other methods described later as used for raising large numbers of queens; but the queens so raised will not be better queens. Grafting involves no injury to combs, but the bee-keeper should be familiar with the procedure employed to secure repair of combs (**350-2**).

182. Whichever method is employed, drone raising must receive attention (see **176-7**).

Special Modifications

183. The details must be adjusted to the method of management employed, but it is convenient to recognize four types of district requiring different management.

The first or commonest is that in which there is an early flow, say from fruit blossom, and a late flow, say from clover and lime, and more or less of a quiet gap between. This condition is covered by the instructions given above.

184. The second type is that of a prolonged flow throughout the season, without any serious set-back, and sometimes even beginning early and finishing late. This type goes with continuous heavy breeding and makes a great demand upon the queen. It is best to combine queen breeding and non-swarming manipulation and re-queen every year (**1043** *et seq.*). This condition is not found in the British Isles.

185. The third type is that in which there is little or no early harvest. This is a trying type, as the swarming season is rendered uncertain and troublesome. The bees are apt to fritter away stores, breeding to no purpose. See **1033** for general management. Queens may be raised in the early summer, feeding being necessary. Aim to get the peak of brood development just before the harvest is due to commence. Use bees for breeding that suit the conditions.

186. This latter is still more important in the relatively rare fourth type in which there is little or no late harvest but a good early one. This type, however, may occur when type one is expected. In this case early breeding is essential and young queens should be inserted in the autumn and the stocks must go into winter quarters really strong. Long-lived bees (**18** and **19**) are desirable, and if the hives become strong after the profitable harvest is over, divide them, raising the necessary queens, and unite for winter under young queens.

Types 3 and 4 are not conducive to profit, and high skill in management will be required to secure good returns.

Queen Raising in or for Out-Apiaries

187. The condition here is that the operation must be carried out on infrequent periodic visits. If the out-apiary consists of only a few hives, as, for example, in an orchard, it should be re-queened from the home apiary, but the out-apiary may be a large one. Queens may then, be raised at home or in the out-apiary. Small out-apiaries may be treated as in **181**, *Method I*, due regard being paid to the timing of operations; or *Method II* may be used in out-apiaries, or by the week-end beekeeper, if timed as follows, drone raising receiving attention (**177**). The first method below suits also the weekend bee-keeper.

188. If larvae are raised as in **209**, they will be ready for cell starting in 7 days as in **226-38**, and can be timed for making provision for emergence (**246-73**) 7 days later, or up to 10 days, in the colonies in which they are to be used (**258**). A period of honey flow must be chosen for the operation.

189. Notwithstanding the adverse comments in **180**, it may be convenient in an out-apiary to insert cells raised on the spot, directly into the colonies requiring them, attending to any failures at a later date. Sufficient cells may be raised in one colony.

Where queens fail to emerge, mate and lay, stocks will have to be united on a later visit unless spare queens are to hand. To provide against this, the queen-raising colony may conveniently be divided temporarily into two or three nuclei, thus giving one or two spare queens; or any other convenient colony may be so divided temporarily, or used to provide a temporary nucleus.

Raising Queens in Nuclei

190. Good queens can be raised only in a colony able adequately to nourish the larva and maintain continuously the necessary high temperature. Good queen cells can be raised in the centre of a three-frame nucleus only if warm nights are assured, plenty of young bees and but little young brood. The attempt is risky in an uncertain climate and not to be encouraged.

Queen Raising to Plan on a Large Scale

Introduction

191. It is customary to describe detailed plans as practised by their authors. This involves much unnecessary repetition and in some cases description of details which have been superseded by better methods.

The business consists of a series of stages in most of which there are several possible alternative good methods of proceeding. The whole is covered by the diagram on the next page, which serves also as a key to the portions in which the various details will be found set forth.

Parentage

192. In raising new queens, parentage is by far the most important factor, as through the fertilized queen mother the new queen derives all her qualities. Her progeny will derive qualities at least equally from her drone mate. This aspect is dealt with under "Drone Breeding" (**292**).

193. A fertile queen used for the production of new queens is known as a breeding queen, and her suitability is decided mainly by examination of her behaviour and progeny (see also **153** and **307**). The estimation from examination of the progeny is by performance and requires that the queen should have been laying for at least 12 months, so that characteristics of the workers in all seasons and under most conditions may have been observed and noted. The points of a good strain are detailed in paragraph **118** and apply equally to a good stock. See also paragraph **43**.

194. In the future when the laws of heredity as applied to bees are better known,probably more attention will be given to pedigree even than performance, for no observations and conclusions of performance of parents can be complete even in 12 months, and the value of queens of the same parentage raised under identical conditions may vary greatly. The following may be provisionally noted:

(a) Variations among offspring are greatest the greater the difference between the bees mated. The sudden introduction of entirely new blood introduces so many possibilities good and bad that it should not be attempted except by breeders on a large scale who can afford to sort out the numerous bad elements and secure and develop the few good results.

(b) On the contrary, crossing of like with like tends to reduce variation and therefore increases the probability of daughter resembling mother. Inbreeding is practised for this reason and without deterioration, but it is well to avoid too frequent crossing of bees very closely related, by running several colonies of the same strain for breeding purposes. The introduction of new blood tends to hardiness and good health.

TABLE VIII

KEY DIAGRAM TO QUEEN RAISING
Parentage, **192**

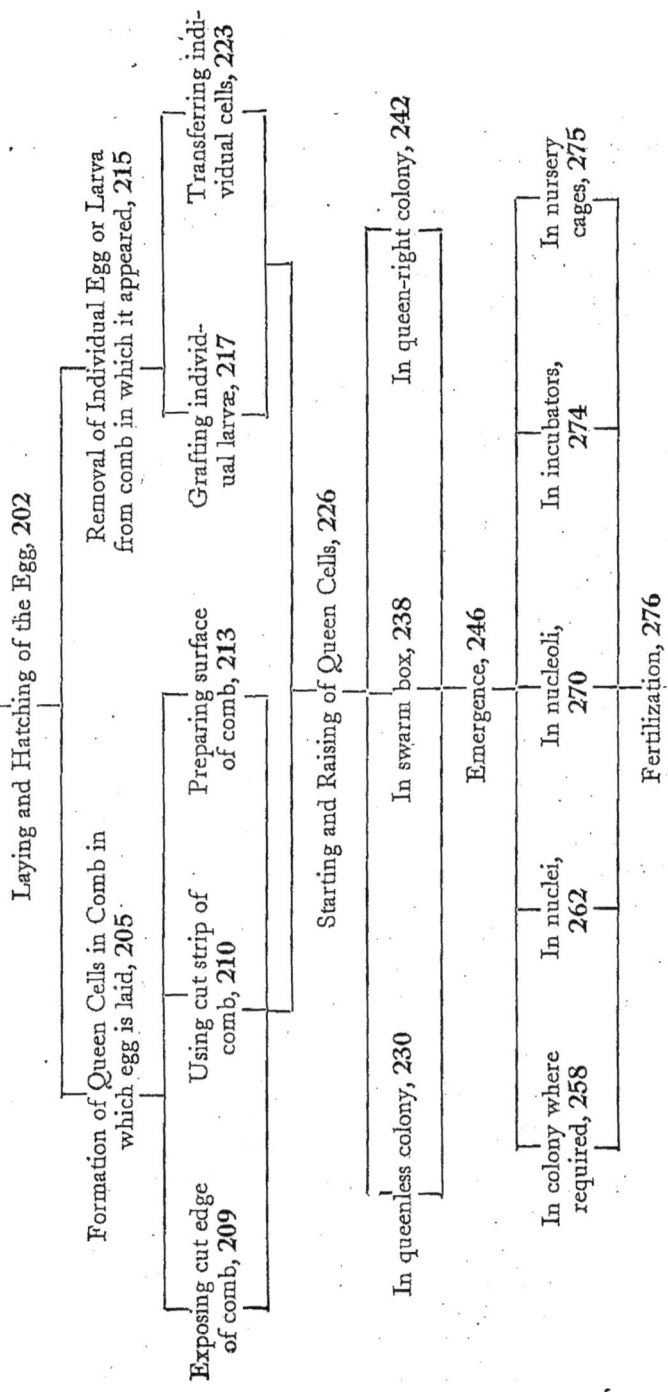

Laying and Hatching of the Egg, **202**

Formation of Queen Cells in Comb in which egg is laid, **205**

Removal of Individual Egg or Larva from comb in which it appeared, **215**

Exposing cut edge of comb, **209**

Using cut strip of comb, **210**

Preparing surface of comb, **213**

Grafting individual larvæ, **217**

Transferring individual cells, **223**

Starting and Raising of Queen Cells, **226**

In queenless colony, **230**

In swarm box, **238**

In queen-right colony, **242**

Emergence, **246**

In colony where required, **258**

In nuclei, **262**

In nucleoli, **270**

In incubators, **274**

In nursery cages, **275**

Fertilization, **276**

(c) While perpetual selection takes advantage of minor variations, undoubtedly certain characteristics depend upon the definite presence or absence of certain factors carried by the eggs and spermatozoa, and these characteristics may be definitely secured or definitely lost through an individual crossing. They do not proceed by progressive steps. The small breeder must therefore start with bees definitely possessing desirable characteristics in a high degree. The large breeder should study the work of Mendel, and researches on bee-breeding arising there from, or he may suffer deterioration of his breeding stock.

195. There is a third and equally important factor in selective breeding and that is the system of management to be employed and the conditions under which the bees are expected to work. Bees bred to give good results under a given system of management and a given set of local conditions of honey-flow and climate can hardly be the best for use under totally different conditions.

196. The breeder who raises his own queens tends to develop a strain which suits his methods and his locality. For example, it is undesirable that bees should build up too early when the honey flow is late, but they cannot build up too quickly where a good early flow is to be secured. For comb honey we must have queens whose workers make white cappings and do not leave many unfinished sections. Again, swarming may be controlled by suitable manipulation of bees that will stand it, or by the use of a strain which seldom swarms even in the queen's second year. The breeding of the latter type cannot be conducted except by using a strain in which the characteristic is already fixed or by selective breeding only from queens worked for two full seasons. A queen giving bees with a short working life may give better results where acarine disease is prevalent, but a long-lived bee may be expected to show greater economy under other conditions.

197. Curiously, if breeding out the swarming impulse be successfully pursued, the bee-keeper may be left with a strain in which the queen-raising instinct has been so diminished that the bees require a decided stimulus to raise queen cells. He can no longer rely on queenright supersedure. He must remove the queen and all brood from the queen cell-raising stock, to which the selected larvae are given, or he may follow an equivalent procedure by means of the "Shook Swarm" method (238). Some Swiss bee-keepers have carried the business so far that they can raise numerous queens and have them all emerge in the one hive without any fighting. A French bee-keeper, producing 30 to 50 tons of honey annually, has reduced swarming to not more than 1 per cent. by rejecting all swarm cells and swarming bees as breeding stock for thirty years, otherwise following normal procedure.

198. It has not yet been established what correlation there is between longevity of the queen and of her workers, but it is worth noticing, by the way, that some of the shortest-lived yellow bees come from strains raised where it is customary to re-queen every year, and long-lived blacks are the progeny of queens able to give several years' service. It is probable that a queen will give her best results in her prime, judging by experience with other animals, but it is certain that the more important characteristics other than the highest stamina are carried in the gene and will be transmitted in the egg as long as the queen is able to lay fertilized eggs. By keeping valuable queens in nuclei as breeding queens only, one secures also that the longest-lived of them have a chance to show their real quality.

199. It is essential to commence with bees having in the highest degree the characteristics the bee-keeper desires. He may then hope to maintain a high level of quality, and may even improve it if he exercises skill and care. To avoid excessive in-breeding he should occasionally purchase a queen from a breeder of bees having similar characteristics to those of his own strain.

200. Parentage is more important than impulse. Queens raised under the supersedure impulse from a strain addicted to swarming will produce bees that will swarm more than will queens raised under the swarming impulse from a strain little disposed to swarm. There is no proof that the impulse itself in a particular case has any effect, though it may do; nor is there evidence that the nurse bees have any influence on the tendencies of the queen they nurture, but they may have.

201. On the other hand, nurture is of comparable importance with nature; environment with heredity. Good bees are not raised by poor nurses or in inadequate surroundings.

Laying and Hatching of the Egg

202. Although queens can be raised from eggs laid in worker cells by control of the food supply, some hold that there is a difference between a larva fed from the commencement as a queen and a larva started in a worker cell. It is true that the term "royal jelly" covers a food variable in its content. For example, the food given to maturing queen larvae differs at least in the proportions of its constituents and in consistency from that given in the first few days.

It has been persistently claimed, without however the support of statistical evidence, that as good queens are raised from young larvae as from eggs, but in recent years an increasing number of breeders of high repute are using the egg.

203. However the egg be hatched, and wherever the larva be raised, it is most important that there should be a condition of prosperity and abundant fresh pollen.

204. The breeding queen having been selected, we may consider the formation of queen cells under two heads, thus:

(a) Formation of queen cells where the eggs are laid, commencingin **205**.

(b) Formation of queen cells from individual larvae or eggs removed from the comb in which they are laid, commencing in **215**.

Formation of Queen Cells where the Eggs are Laid

205. Queen cells raised for supersedure, or under the swarming impulse, may be used, preferably the former. In this case the cells are not disturbed until they are sealed and nearly ripe for hatching. The cells are then cut out and dealt with as in **252** and following.

On a larger scale it is, however, customary to cause the breeding queen to lay in a convenient comb, which is then prepared in either of three ways described below, particular note being made of the date of laying of the eggs or age of the larvae used (**8**) for guidance in removal of the sealed cells later (**252-8**).

206. In each case the operation must be carried out with precautions to ensure that the eggs and young larvae are not exposed to sunlight and that they are not chilled, especially the larvae. The most favourable temperature is at least 75° to 85° F. (25° to 30° C.), and it is best to transfer the comb in a warmed receptacle to a warm operating-room, and return it to the queen cell-raising hive with similar precautions, unless the shade temperature is above 75° F.

207. There are several methods of procedure as follows:

Method (1) due to Cowan and Miller. The cells are started on the prepared edge of a new comb (**209**).

Method (2) due to Alley. The cells are started on a new comb,a strip of which is removed prepared (**210**).

Method (3) due to Pechaczek and Hopkins. One face of a new comb is prepared and used generally in a horizontal position (**213**).

It will be observed that method (1) involves the least risk, but does not produce so many queens from one comb as do methods (2) and (3).

208. In methods (2) and (3) above, the cells are turned through an angle into a new position, and eggs are not so well received as young larvae. It has been suggested that the egg is hindered in final settling (**8**) by the change of position. If this is so, then day-old eggs should find better acceptance. If larvae are used they should be preferably less than 1½ days old. See Plate III*b*, facing p. 5.

A queenless stock is apt to destroy eggs. This has the effect of about halving the prospective nursing duties and prolonging the life of the existing bees, but further study is required.

Raising Queen Cells on Prepared Edge of Comb

209. Insert in the midst of the colony containing the breeding queen, a frame fitted with foundation extending only half-way down from the cross-bar. Leave for 6 to 7 days, seeing to it that the bees have plenty of open stores. When removed, the comb will be found filled with eggs and young larvae. With a sharp knife trim the bottom edge of the comb, cutting back so as to expose cells with eggs or preferably young larvae 1 ½ days old (**8**). This comb is given to the queen cell-raising colony (**226-45**). A larger number of cells per comb may be raised by using several V-shaped pieces of foundation hanging from the cross-bar with their bases about an inch apart and their angular ends reaching within 1½ inches of the bottom bar, but the completed cells are more readily cut out from the one-piece comb cut with a straight or wavy bottom edge.

The comb should not be out of the hive longer than necessary, 10 minutes at the outside, and must be protected from direct sunshine and from winds.

Raising Queen Cells on Prepared Strip of Comb

210. A frame fitted with foundation is inserted in the colony containing the breeding queen as in 209 above. When removed in 6 or 7 days strips of comb are cut from it. Each strip should contain larvae of not more than 1½ days old (**8**) in one continuous row of cells. Examine both sides of the comb and choose that showing the best strip. Cut the strip about 3/8-inch wide with the selected cells along the centre. Trim the mouths of the selected cells with a sharp, warm (but not too hot) knife, so as to leave the cells about 3/16 inch deep. With the head of a match or an old toothbrush destroy two out of three of the larvae, leaving every third one. With a conically ended stick open out a little the mouth of each cell containing a larva. All this should be done on a convenient flat surface.

211. The strip is now ready to be mounted in a frame for insertion in the cell-raising colony (**226**, etc.). The frame should be one with a horizontal cross-bar in the middle, with comb above it, which may contain sealed brood, but no larvae under 3 or 4 days old. This is prepared beforehand. The prepared strip is glued to the under-side of the crossbar with glue or melted wax, so that the selected cells hang downwards and avoiding heating the larvae. Alternatively, an empty comb in a standard frame may be taken and the bottom cut away, leaving a suitable straight or convex surface on which the strip is secured.

212. A strip containing eggs may be used instead of larvae, but see **208**.

Raising Queen Cells on the Prepared Surface of a Comb

213. A frame fitted with a full sheet of foundation is inserted in the colony containing the breeding queen. The queen should be laying well, as a full frame of eggs is desirable. Remove in 6 to 7 days. Choose the face most evenly covered with larvae of even age (**8**). With a sharp, warm knife or chisel cutaway the cells in horizontal rows down to the mid-rib, leaving every third or fourth row, then cut these down to 3/16inch (5 mm.) with a sharp, warm knife, then cut across, leaving every third or fourth cell standing, giving preference to any containing a larvae well developed as compared with its neighbours. Destroy with an old toothbrush or match-stick all larvae discarded. Finally, open out the mouths of the cells somewhat with a conically ended stick. The comb is now ready to be placed in the cell-raising colony (**226**, etc.), horizontally above the brood frames there in with prepared cells hanging downwards, the whole supported on a temporary wood surround, raising the mid-rib 1½ inches above the cross-bars. Cover over all, snug and warm, using at least an inch of soft packing above the frame.

This method is more economical of material and simpler in execution than that described under **209** above, where large numbers of cells are required.

214. A piece of comb containing eggs may be used, but see **208**.

Transfer of Individual Eggs or Larvae for Queen Raising

215. Individual larvae of suitable age raised in the colony containing the breeding queen may be removed from their cells into artificial cells conveniently arranged. This is known as grafting (see **217**). Alternatively, individual cells may be cut out each containing a single larva, or even an egg, and mounted in a way convenient for the formation of queen cells (see **223**).

The advantages of the procedure are that it is always possible to find a number of larvae of suitable age (**8**) in the colony containing the breeding queen, and the queen cells may be built directly on convenient individual supports facilitating handling them in later stages.

When the larva is removed in its cell there is no risk of direct injury to the larva, and its feeding is not interfered with. The operation of mounting is simple, though not quite so easily done, and there is very little cutting of combs.

216. The work should be done in a warm place in a temperature not below 75° F. or above 85° (25° to 30° C.) and with the air not too dry. Take care not to chill the larvae. The comb of selected brood should be removed under cover to the work-room in a warmed box, unless the shade temperature is up to 75° F., and the prepared cells handled in a similar manner. The larvae must not be exposed to too dry an atmosphere. A travelling box warmed by exposure to the sun, or by inserting one or two frame feeders filled with warm water, is very suitable for carriage of combs and cells. The larvae removed should be about 1½ days old (**8**) to give the best results, but if the cells are cut out younger larvae can be safely used. Eggs may be used, but see **208**. If the larvae differ much in age the bees may make a selection. To secure the maximum number of acceptances the larvae should all be about the same age. This facilitates later calculations also.

In a large area of brood of even age here and there a larva will be found somewhat larger than its neighbours. Such are stronger or better nourished and may be given preference.

Transfer of Individual Larvae-"Grafting"

217. The larvae are lifted from their native cells (which the beginner may first cut down somewhat) by the use of a quill or toothpick cut like a pen with the point turned up, or a fine camel-hair brush moistened in clean water and dried, or one made of a strip of soft wood. The end should be spoon-shaped. Quill should be dipped for a minute or two in boiling water for shaping. The end of a wood stick can be bitten until soft. The larva is scooped out, floating on the jelly in which it is immersed, and deposited, jelly and all, in the bottom of the proposed cell, the spoon being applied to the convex back of the larva. It used to be customary to prepare the cells by inserting in each a small amount of royal jelly, using the contents of one natural queen cell a few days old to supply fifty or more cups. Incidentally, this renders the safe deposit of the young larvae easier to the beginner. The bees, however, promptly remove this food and provide food of proper consistency, and for this reason it is now more usual to employ so-called dry grafting, i.e. the larva has only as much food as may be transferred with it.

To secure a good start with ample and suitable food some go to the length of double grafting, i.e. the cells are grafted with any young larvae for a day and these are then removed and replaced by the choice larvae. If, however, the receiving colony is well prepared, there is no risk of underfed larvae in dry grafting. The air in the operating-room as well as in any receptacles in which the larvae, combs or cells are placed, should be kept moist. It is advantageous to have the queen cells warmed up in a hive before grafting.

218. Queen cells once used are more readily accepted again than new artificial cells. For small-scale work old queen cells cut down in length have been employed, also old drone combs used once and cut down. Artificial cells are moulded on a stick of hard wood, with smooth rounded end 5/16 inch (8 mm.) diameter at the end and slightly tapered, say, to 3/8 inch (9 1/2 mm.) at 1 inch (25 mm.) from the end. The stick is wetted, then plunged 1/2 inch (12 mm.) into melted wax, using pure wax of light colour as obtained from cappings and heated only to the melting point. When the wax is set the stick is dipped again, and so on, about seven times in all, each time not quite so far as the last. Cups so formed may be stuck to wooden cross-bars with melted wax for insertion in the frames (Plate IV *b*, facing p. 150), or they may be stuck into wooden cups for easier handling.

219. Now the cell is generally moulded in the cup. For this purpose a cup is sold in the form of a wooden cylinder about 5/8 inch (16 mm.) long and 9/16-inch (14 mm.) diameter, bored about 13/32 inch (10 mm.) for a depth of 3/8 inch (10 mm.). These cups are filled to the brim with wax and allowed to set. They are then warmed uniformly until the wax is soft. The cells are then formed by pressing the wax moulding stick into the wax, leaving a projecting rim on removal of the stick.

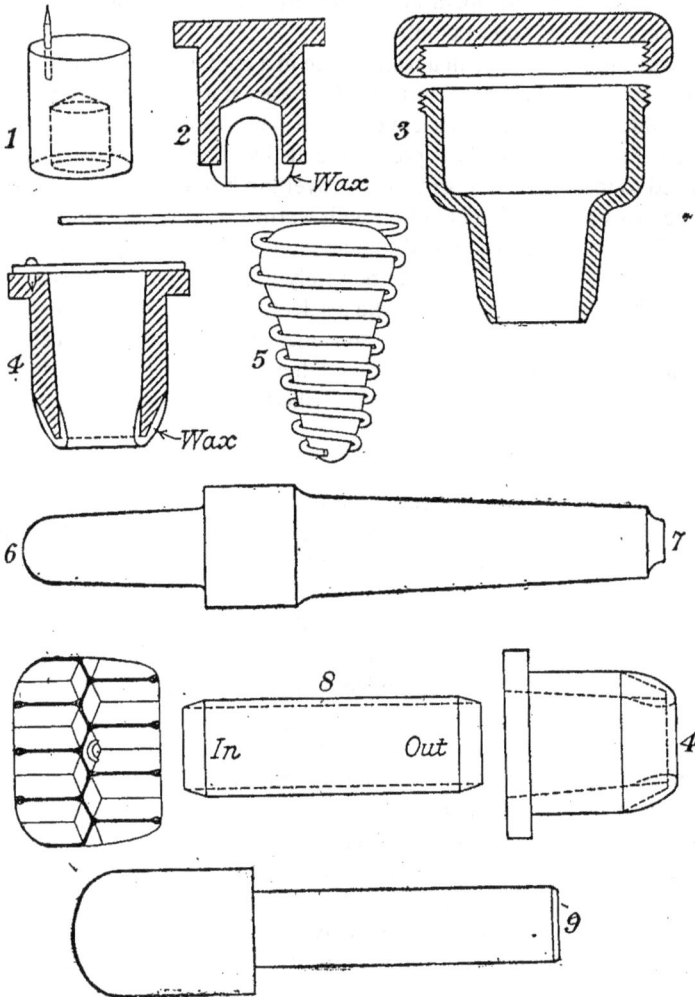

1. Doolittle cup.
2. Pratt cup.
3. Perret-Maisonneuve cup.
4. Cup for natural cell with larva, wax-dipped.
5. Cell protectors.
6. Cell-moulding tool.
7. Enlarger for modelling mouth of 3 and 4 and enlarging mouth of natural cell.
8. Tubular cutter for removing worker cell.
9. Extruder to push cell from 8 into 3 or 4

FIG. 2.—QUEEN-RAISING APPLIANCES.

The cups are sometimes furnished with projecting pins at the back for mounting, but the better form is that with a flanged end.

220. The more substantial cups designed by Perret-Maisonneuve, used in France and elsewhere and shown in Fig. 2, may also be employed as carriers for moulded cells, which are inserted in the mouths after moulding. Their form is such that plenty of spare wax can be furnished for the use of the bees in building up the cell, and very fine cells are thus produced.

221. The cups may be mounted in vertical or horizontal frames, the latter being the later, and probably the better, method.

For vertical mounting, frames are fitted with horizontal bars to take the cups, about 2½ inches apart, the bars being drilled with holes to receive flanged cups, which are inserted from the top so that each cup rests on its flange with the cell hanging downwards. Generally the upper part of the frame is occupied with comb, free from eggs or young larvae. The best results are got with not more than two rows of cells per frame. The lower bar should be at least 3 inches (75 mm.) from the bottom of the frame. The frames are inserted in cell-raising colonies as described in **226**, etc.

222. For horizontal mounting a shallow box is used about 1½ to 2 inches deep over all according to the cell mountings, which can be inverted over the top of the brood frames in the cell-raising colony. The inverted bottom is perforated all over to receive the cups, which are inserted from above and rest on their flanges as before. So placed, the bees have free access to them, the position is warm and easily ventilated by the bees, and a large percentage of acceptances and good cells is secured. Free access is obtained for removal of cells from the top. The whole is made snug and warm by packing with warm quilts or a cushion.

The box can completely cover the broad chamber or only a portion of it according to the number of cells it is to carry. Space may be left for a feeder, or a feeder can be placed on top, giving access by omitting a cell, or through a special feed hole.

Transfer of Individual Cells with their Larvae

223. This method can be applied to raising queens from eggs or larvae without removal from their cells, but see **208**.

The individual cells with the larvae they contain are removed from the comb by the use of a cylindrical punch in the form of a thin tube ½ inch (12½ mm.) internal diameter with sharpened edge, used at a temperature taken from the surrounding atmosphere or the vest pocket of the operator. The comb is placed on a flat surface and the selected cells punched out, the punch passing right through and being given a twist on the way and on the removal. Each cell as punched is removed from the tube by passing a rod trough it from the top end. The cell is then ready for mounting and trimming. It is best to avoid old comb in which a previous generation has been raise and essential to avoid cells having honey at the back of them.

224. The cells should be mounted in supports convenient for carriage in vertical or horizontal frames as in the preceding section, the supports protecting the back of the cell, the bottom of which should be not more than 1/8 inch (3 mm.) within the mouth and with wax around for the bees to mould in continuation of the queen cell. If too much cell is left or the mouth is not opened out, the bees may continue it as a worker cell.

Probably the most convenient form of mount is a cylindrical wooden cell with flange at back and narrow mouth, drilled through so that the wax cell may be inserted from the back and pushed through to the proper level. The back end of the wax cell is then packed and sealed with soft wax. The hole should be $7/_{16}$ inch (11 mm.) diameter in front and slightly larger at back, say, ½ inch (12½ mm.).

225. The Perret-Maisonneuve cup described in Fig. 2 is suitable where its size is not objected to. After inserting the cell containing the larvae, the projecting portion is slit up and opened out with a conical-ended stick. It is advantageous to coat the wood-cup well with wax beforehand. The bees will soon mould the cell and wax together and use any surplus wax in building up the cell, it not being their practice to build up queen cells from new wax as in the case of worker cells.

Starting and Raising Queen Cells

226. For successful raising of queen cells it is necessary to select a prosperous colony and keep it prosperous. The colony should be one disposed to raise queen cells, as in swarming-time with advanced drone brood present, or it should be forced to realize the necessity of raising queen cells as in the swarm box method described below, or to a lesser degree in the partition method, or in a combination of the two, as described later (see **244**). It is essential that there should be no queen cells existing where the required cells are to be raised (**1364**) and that there shall he no eggs or larvae on other combs in the same compartment. The surest way of avoiding the presence of eggs and young larvae is by removal or exclusion of the queen from the combs to be used, as described in greater detail below.

227. It is necessary to note that it takes from 7 hours to 7 days to prepare the queen cell-raising hive, according to the method employed, and steps must therefore be taken in advance of preparation of the prepared frames of cups or larvae which are described in previous sections. Further, although under the swarm box method a hive may be prepared in 7 hours, it is customary to transfer the cells to another colony the next day, and this must be arranged for in advance.

228. The queen-raising colony should occupy not less than eight frames, but a stronger colony crowded on to twelve large frames is none too good for raising, say, 100 cells on the horizontal plan (**213**).Do not make the mistake of assuming that newly hatched bees can nurse queen larvae. They are not able to until 5 or 6 days old. Nurse bees are at their best for queen raising when 1 to 3 weeks old. It is a good plan to feed the colony an hour before inserting the prepared frames.

229. The full procedure is described under three heads below as follows:

(a) Use of Colony Rendered Queenless: This is the commonest practice, but may not give satisfactory results with bees disinclined by circumstances or nature to raise queens freely (see **230**).

(b) Use of Swarm Box Method to Start Cells: This method is the most certain for ensuring acceptance of a number of cells, but they have to be finished in a queenless colony as in **238-41** or behind a partition as in **242-5**, below.

(c) Use of Queen-right Colony: The queen cells are raised behind or above a queen excluder partition. Fewer cells are raised than with the other method, but they receive good attention and the strength of the colony is maintained by the laying queen. Acceptance can be forced as described below, but the method is best adapted for work on a small scale, for which it is very good (see **242-5**).

Developing Queen Cells in Queenless Colony

230. Make a prosperous colony queenless 7 days before the prepared frame is to be inserted. Feed with1/2 lb. syrup or diluted honey every day. Destroy all queen cells raised (**1364**). Alternatively, remove queen and all eggs and brood under 4 days old at least 12. hours before inserting prepared frame, filling up with store combs at the outside with some unsealed honey, or with frames of emerging brood if more than one set of cells is to be raised. It is best to use fully built out combs. See that there is a supply of fresh pollen.

The conditions are particularly favourable if the stock from which the queen is removed is one making preparations for swarming.

231. Insert the prepared frame or place on top if horizontal. It is preferable to continue slow feeding. If the frame is horizontal a vertical frame feeder will be found convenient. If the frame is vertical a horizontal feeder on top may be used.

232. Instead of removing eggs and young brood, some prefer to remove all brood combs, which is safer as a few eggs may be missed. In that case, the removed brood combs are put in a temporary body or stand and replaced by empty frames, first ensuring that ample stores and pollen are present; then all bees are shaken back off the removed combs and the brood given to another colony to care for. Leave quiet for 2 or 3 hours before inserting the prepared frame of eggs.

233. Examine after the cells are due to have been sealed and destroy any badly formed or placed, taking note of the number of good ones, so that timely arrangements may be made for the emergence of the queens (**257**).

The cells must be removed a day or two before the queens are due to emerge and transferred to the colonies in which the queens are to emerge. If it is desired to use nurseries (**275**) the cells may be put into them any day after they are sealed and before emergence is due.

234. For continuous working the removal of the queen 7 days in advance gives the best start, as it ensures a supply of nursing bees for, say, 4 weeks after insertion of the first frame. A queenless stock is a liability, not an asset, save for starting cells; and if large numbers are to be started, it is desirable to use the stock only for this purpose, the cells being transferred to queen-right colonies for completion (**242**). A convenient arrangement is to put them in the top super of a Demareed colony (Section XV) or similarly in a body box placed above honey supers and containing young brood but no queen or queen cells, the queen being left below with comb containing emerging brood and empty combs. There should be honey and pollen in the top super. For emergence, however, see **246**, etc.

235. For continuous working in the queenless cell-starting colony, it is necessary to maintain the supply of bees and especially young bees. A comb of emerging bees may be added. Combs of emerging bees may be added at the rate of, say, two full combs per week; or, alternatively, well-fed bees may be shaken at the entrance, when the young ones will be accepted and the old ones will fly back home .

236. When the queenless colony is to be used only for one lot of cells it is more convenient to remove brood and queen and restore them 10 days later, uniting by the newspaper method. The queen-raising colony must be kept on the old stand so as to receive the flying bees and so be kept strong. Unless the weather is good and nights warm the removed brood and queen will require feeding and keeping warm, as, for example, by placing them over a perforated partition (**671**) over a prosperous colony.

237. Incidentally, a prosperous colony containing only a newly hatched queen and having no queen cells may be used for starting queen cells, but the cells should be removed before sealing and the colony must be kept prosperous.

Swarm Box Method of Starting Cells

238. A swarm box is used, taking five to six combs, because it is conveniently arranged for ample through ventilation and for closing the entrance, but any hive body similarly arranged may be employed.

Say the hive is arranged to take five combs, two combs, free from brood and without a single egg, are provided containing honey and pollen. Water is put in one of them. These are arranged one on either side of the body, leaving space for one at the centre. The hive is placed handy to a strong stock, which is to furnish the bees previously well fed for three days and having had access to ample pollen. At nine-thirty to ten in the morning the stock hive is opened up; the queen is secured, or the comb containing her put aside, then the bees from four frames are shaken into the swarm box. The box is closed, leaving ample through ventilation. The swarm box is then placed in a cool shady corner.

In a six-frame box, three combs may be inserted, spaced to receive two prepared·frames of cells.

239. In 6 to 8 hours after imprisonment of the bees they will be ready to receive the prepared frame or frames, and it or they may be inserted in the spaces left for them. All ventilators are then covered except one at the bottom or over the flight hole, and the bees left to the next day. Experience shows that under this procedure a vigorous start is made at raising queen cells. The cells are then transferred to cell-raising colonies, described under the preceding heading, for completion, or to partitioned queen-right colonies as under the heading following.

240. A rectangular funnel, made of metal or ply-wood, will be found handy for shaking in the bees. The narrow end of the funnel should be fitted into a wooden base, large enough to cover the swarm box. Give the box a jolt, to jar the bees so that they fall down, before removing the funnel to put the cover in place.

241. If a horizontal frame is to be used covering more than five or six combs, it is convenient to use an ordinary hive arranged with ventilating floor board and means to close the entrance. Remove all unsealed brood, replacing with empty combs, putting water in one of them, the removed brood to be given to another stock and the queen also placed elsewhere. Close down and put in the shade as before for 6 to 8 hours, then insert the horizontal frame, replacing the stock on the old stand. Next morning open the entrance and leave the colony to complete the cells.

Partition Method of Raising Cells in Queen-right Colony

242. Choose a colony with young queen, to reduce the risk of swarming. Divide off not less than four combs, one of the stores and three of brood, by means of excluder zinc so arranged that it is impossible for the queen to pass the division. It is better to have these combs end on to the exit, provided the exit portion is also fitted with excluder Leave for 3 days, then insert the prepared frame between two combs in the portion from which the queen is excluded, examining all combs in that portion for any queen cells before insertion of the frame and destroying them all (1364). By this also, if the operator desires to employ royal jelly in his cups (see 217) he has the opportunity to secure some on the day that it is wanted. Do not attempt more than twelve cells, and all may not be accepted.

Examine for any badly started lots on the third day. Do not disturb unnecessarily. Proceed later as in **233**.

243. For continuous work a further frame may be inserted between two brood combs when the cells previously inserted are sealed. It is desirable to have adjacent to-the comb with newly grafted cells a comb of emerging brood and a comb of honey and pollen.

Frames can be inserted, each between two brood combs every 2 or 3 days in a strong colony, but the number of acceptances is reduced and the conditions must be maintained, and stray queen cells regularly sought for and destroyed. For such rapid working the frames are removed to colonies rendered queenless to complete them.

Some divide the colony in half and transfer operations from side to side about every 7 days.

244. In a strong colony requiring more than one brood box, the upper box may be divided off by an excluder and worked as above. This gives better results, as the colony cannot be too strong. Alternatively the colony may be Demareed (see **1226-47**) and the upper brood chamber used, the frames being inserted after any queen cells formed therein have been destroyed.

Acceptance will be assisted by utilizing in part the idea of the swarm method. After the three days' pause the bees in the divided portion are confined by the insertion of wire gauze completely covering the excluder, 4 or 5 hours before inserting the prepared frame, leaving the gauze in position until the next day.

245. The success of the partition method is dependent upon a continuous flow of food. If the weather turns bad the bees may decide to destroy the queen cells even though artificial feeding is resorted to.

When using a horizontal queen excluder, opportunity for escape should be provided for the drones. For example, the top covering can be lifted for a few minutes during a sunny spell, at intervals of 4 to 7 days, using a little smoke if found necessary.

Emergence in General

246. The emergence of a queen from a queen cell is sometimes described as hatching, but the pupa is not an egg.

247. Sealed queen cells should not be turned upside down and are best retained roughly upright. Ripe cells can be laid on their sides when handling. If, however, a cell is placed and left in a horizontal position soon after sealing, the queen does not develop properly. Queen cells should never be jolted unless it is intended to destroy them. If there is the slightest doubt about requirements, do not shake bees off a comb that may have good cells that you may want, but use a feather to remove the bees (**875**).

248. It is important to keep queen cells warm while out of the hive. They should be protected from wind, but not left in the sun. It is convenient to have a small box, resembling an egg box, with felt-lined cavities and a felt flap cover; it should stand in a warm place before use.

249. For emergence the queen cells should be inserted in a colony prosperous according to its size, well fed with honey and pollen. The colony may be any size, from a large one in which the queen is to remain, to a small one containing even only a dozen bees, as detailed below. Except where nursery cages are used, the colony must be queenless and definitely in need of a queen. Ripe cells may be examined by holding them so that the sun shines through them, when the pupae may be seen and cells with dead larvae, if any, rejected. Do not expose them for long. A dark tube like an egg-tester is needed, and cells on new comb.

250. A queen at large in a colony will normally endeavour to destroy any other queen whether in a cell or not. It is essential, therefore, to secure that the cells raised are removed and distributed amongst the colonies in which the queens are to emerge, before a single queen hatches from anyone of them, and if two are put to emerge in one colony both must be covered temporarily with cages so arranged as to give the emerging queen access to both honey and pollen.

251. Queen cells raised in cups or on other individual supports are ready for transfer to the colonies in which they are to emerge without further preparation. Queen cells raised on combs require further preparation.

Preparation for Emergence of Cells Raised on Comb.

252. Unless one or two cells are to be given to a colony on the comb on which they were raised, each cell must be cut out individually by means of a sharp, warm knife, leaving ample wax round the base of the cell, so as to avoid all risk of injury to it, and remembering that the bases of some supersedure cells, depending from the face of a comb, run back to the mid-rib.

The best cells are of full shape but not stumpy. Care must be taken in handling queen cells that they are not jarred or chilled, but they can be laid on their sides temporarily without harm.

A temperature of 75° to 95° F. is most suitable. At lower temperatures there is serious risk of chilling, but work can be done at shade temperatures down to 55° F. (13° C.), in still air, avoiding all unnecessary exposure and using warmed packing.

253. Cells are sometimes cut out attached to a piece like a dovetail, to be dovetailed into a comb in the receiving hive. There is no need for this, however, and in fact the best place for the cell generally is not at the bottom or edge of a comb, but near the top where it is better protected.

254. If the cell is cut away leaving a wider portion at the top it may be supported by this portion when inserted between two combs, separated if necessary, from the top. So long as the cell is held temporarily in place, the bees will make all secure. Alternatively the cell may be secured by a stout pin or a piece of wire bent like a staple and thrust into the adjacent comb.

255. The emergence of the queen can be ascertained without disturbing her or the bees if the cell is at the top of the combs. If found by removal of the capped tip, all is well, but if found torn open at the side, the queen has been killed. A few days after emergence most of the cells will have been cut away.

Protection of Queen Cells

256. Queen cells require no protection from the bees unless the bees are not in a condition favourable to receive a queen. If eggs are present or any brood, even up to four days old, the bees are liable to have a preference for a queen of their own raising. Such a condition is unfavourable unless the colony is also queen-right as in **242-5**, but if the worker bees attack the cell, they will do so near the base, and this can be prevented by the use of a spiral wire cell protector of the type shown in Fig. 2, save that more of the tip of the cell should be exposed than is there shown. See also nest-boxes in **275**.

Preparation of Colonies for Emergence

257. Arrangements may be made for the queen to emerge in the colony in which she is
to remain (**258**), in a nucleus (**259**), or in a baby nucleus or nucleole (**270**). The first plan is obviously mainly suitable for occasional home use. The nucleus is most commonly used. Nuclei can be maintained with but little attention and require fewer workers. Standard combs are generally used in nuclei, so that they can be added to, on short notice, or built up into stocks at any time. A queen can be tested in a nucleus. Nucleoli (small nuclei) are suitable for use mainly in an apiary devoted to queen raising on a large scale, with climatic conditions favourable. They may be used on a small scale, however, by anyone who will master the essentials and not jeopardize results by omitting any. The procedure proper to each type is detailed below:

Emergence in Strong Colony

258. A queen cell may be given to a strong colony 24 hours after the colony has been made queenless, all cells existing in the colony being destroyed (**1364**) when the colony is made queenless.

If all goes well, such a queen makes a good start without loss of time. It is safer, however, to make the colony queenless seven or eight days before introduction of the cell, destroying all queen cells raised in the interim. No cell protector is necessary and the queen is sure of a favourable reception. If, however, she fails to mate, the colony is subjected to a serious set-back.

Emergence in Nuclei

259. In making up a nucleus from a stock it must be noted that most of the old bees on the combs removed will fly back home, so additional bees must be shaken in, first seeing that the old queen is safe. The return of the old bees can be prevented by putting the nucleus, with ample ventilation, in a cool cellar; or it may be largely reduced or prevented by stuffing the entrance with hay, which the bees must remove, again seeing to ventilation.

In making up several nuclei, it is frequently convenient to use a single strong stock for the purpose. The queen is removed 2 or more days before the nuclei are formed and 3 or more before they are wanted, the colony being fed before and after division. When dividing into nuclei, set·each lot in the shade and close the entrance, well ventilated, however, for 24 hours, when the nucleus may be set out where it is to stand. If the nuclei have to be carried some·distance they get well shaken and may be opened on arrival immediately after setting out in their places.

260. In case it is not practical to wait several days and destroy any queen cells then formed, the cells inserted must have cell protectors. It is more simple to wait, say, 7 days, before dividing, and then cells can be destroyed while the combs are being sorted out.

262. A nucleus may be made up of three Langstroth or British Standard frames, two well filled with brood and one with honey and pollen. A larger nucleus tends to build up, but may be used if increase is desired. A smaller nucleus requires care and attention to maintain it, arid should be used only if there is a shortage of combs and bees. It is desirable, however, to have room for at least a fourth frame in a 3-frame nucleus box.

263. Nuclei are frequently made up, two, three and more in the same body, having independent entrances well separated. A standard body can be used with a special floor board, having the necessary entrances cut in it on three or all four sides, and division boards reaching to the floor board. The spaces beneath the lugs on the division boards must also be closed with felt or other packing. Separate quilts or other covers should be used. Such a body and floor can be arranged for mounting above a strong stock which will help provide heat. The bees in adjacent nuclei also materially assist each other in maintaining a proper temperature, thus reducing food consumption and increasing efficiency.

264. It always pays to feed nuclei and to keep feeding them, but they must not become clogged with food so that there is no room for egg laying after fertilization. For fertilization see **276**.

A convenient shallow frame feeder for shallow frame nuclei is described in **887**.

265. If there are more queen cells available than nuclei, two may be inserted in one nucleus and covered with cages (**250**). The queens should be examined as soon as possible after emerging and the better one retained (**43**). An imperfect queen should, of course, be destroyed.

266. To prevent robbing when there is little or no honey flow a small entrance should be used. The entrance can be reduced to one bee space, but a floor-board ventilator is then desirable.

Use of Small Frames in Nuclei

267. The use of standard frames enables nuclei to be made up from frames in working colonies and working colonies to be made up from nuclei. This advantage should not be sacrificed without securing some definite compensating advantage. With the nucleoli described in the next section this compensation is found in the much reduced number of bees required with the very small frames employed.

There is, however, a useful intermediate stage, having certain advantages. Nuclei may be built up of special frames of about half the size of a standard frame and adapted to be used in pairs as or in a standard frame, so that they may be inter-changeable with standard frames.

268. Such half-size frames may be arranged in a double row in a modified hive body having two entrances available, both at the front and the back, so that not only is the hive divided lengthwise but each half may be further subdivided by inserting small division boards. Each side should carry six to ten half-frames and division board, and the walls, floor, etc., should be thick to conserve the heat.

Either half can be worked as a self-supporting nucleus, but on emergency the set can be converted into four small nuclei, mutually supporting and adequate for emergence and fertilization.

269. Provision should be made for closing completely, or with gauze or queen excluder and ventilating through the floor board and of course for the insertion of queen cells. When a queen is laying, the entrance of these small hives should be closed with excluder, as the bees may swarm if the queen needs more room.

If the centre division and division boards are pierced with several 1/8 inch (3mm.) holes near the top and bottom, the bees acquire the same scent, and there is no difficulty in rearranging the combs and divisions at any time.

Emergence in Nucleoli, Baby Nuclei

270. To prevent robbing and to control mating, nucleoli should be located in an apiary employed primarily for queen raising and at least three-quarters of a mile (1,250) metres from any other apiary. Entrances one-bee·space wide should be employed, and care exercised that the small frames employed, when not in use, are protected from wax moth (**357-69**).

For the mating of queens, frames as small as a standard section (4¼ X 4¼) have been employed and as few as a dozen bees, but for satisfactory use for emergence, mating and testing for fertilization, it is generally necessary to use a much larger number of bees, except where the night temperature is uniformly high.

The supply of bees must be kept up, but the greatest success in quick fertilization is obtained with a moderate number, not more than, say, five hundred.

271. Nucleoli must be fitted with frame feeders, and should each have one small comb with food, pollen and a little water to start with, and room for building a second. The bees are furnished by shaking from a frame into a box and shaking or shovelling a cupful into the nucleoli. For continuous working the supply of bees must be kept up. The nucleoli should be prepared the day before the queen cells are to be inserted and the entrance closed for 24 hours.

Nucleoli can only be used for emergence, fertilization and immediate sale of "untested" queens (see **307**).

272. As soon as may be after a queen commences to lay, she should be removed, when another cell may be inserted, preferably protected (or after 5 days a virgin queen will be accepted for fertilization).

273. Nucleoli must be well packed for warmth and should on no account be opened or disturbed when the shade temperature is below 55^0 F. (12^0 C.), indeed, much higher temperatures are desirable.

It will be seen that nucleoli require constant expert attention and are suited only to the needs of the breeder working on a large scale.

Use of Incubators

274. Incubators as used for chicken raising can be adapted for use for the emergence of queens. The temperature should be adjusted to a maximum of 98 F. ($36°$ C.), and the humidity should be about 52. The queen cells are inserted in separate cages furnished with Good's candy (**1316-8**) for use by the queen on emergence. The young queen should be introduced as a virgin (**89** and **272**) into a colony for fertilization as soon as possible after emergence.

Nursery Cages

275. These are designed to enable one colony to look after a number of emerging queens, but if uniform success is desired it is not good practice to keep a young queen caged longer than is absolutely necessary. Each cage should contain its own supply of Good's candy (**1316-8**), so that a queen may not be left without food, as she will have to feed herself. The use of such devices affords temptation to bad practices. There is a danger of cells becoming chilled in nursery cages unless well placed in very strong colonies. Some cages have perforation covered with excluder zinc or celluloid to admit attendants, but it is more convenient to have no attendants with the queen when the cage is removed for introducing her elsewhere.

Conditions Favourable

276. The mating flight normally takes place from the colony in which the queen has emerged, though on occasion a virgin hatched or held elsewhere may be inserted in a colony, large or small, for fertilization (**105**).

277. One of the most important requirements is to secure that the virgin queen returning from her mating flight shall find her way back to the colony she left, for if she enters the wrong hive she will be killed and her own colony will be left queenless.

To this end it is useful to have mating hives painted a variety of colours and arranged in groups in a distinctive manner. The queen is greatly helped by distinctive features in the neighbourhood of the front of the hive in the line of flight, and these should be provided where none exist naturally (**601-3**).

278. When several colonies are located in one hive body it is important that the entrances should be well separated, and where the flight board is common to two or more, they should be separated by vertical partitions extending from the hive body to the edge of the flight board. The surroundings of the entrance may be brightly coloured. Some make the entrance itself conspicuous with black paint. Distinctive forms, such as crosses, arrow-heads, circles, triangles, 3 or 4 inches in size and strongly outlined, are probably of more help than distinctive colours (**776**).

279. If several mating colonies are to be located in one hive body the divisions·are sometimes made with holes to allow interchange of air (and scent), either a number of holes 1/8 inch diameter (3 mm.) are used, or larger holes covered with wire gauze. This gives the bees a common scent so that combs of bees may be interchanged and rearranged freely as between compartment and compartment. The division boards are frequently made movable to allow of varying the capacity of individual compartments. With such a hive it is however found important to secure that all the cells furnished should be of the same age, so· that as far as possible the queens may be fertilized the same day. This furnishes evidence of the importance of the condition of the colony (see **281**).

280. The colony with bees and queen to be fertilized is sometimes·placed for warmth above a normal strong colony. This is best done by the use of a bottom board for the mating colony having a large hole in it covered both sides with wire gauze, with not less than 1/2 inch between gauzes, letting warm air through but not allowing the bees to pass. This board should have a flight hole let into it and a small extension to act as flight board. It will then stand on and receive standard hive bodies without any cutting of entrances, etc., in them.

281. The condition of the colony itself is an important factor in successful fertilization. Everything that helps to render the colony aware of the need of a fertile queen favours fertilization. This is evidenced by Pratt's success with minute colonies of old bees and by the fact that fertilization may be seriously hindered by maintaining a supply of young larvae. Nevertheless, a colony having decided to supersede its queen may raise a virgin and have her fertilized and laying before destroying the old queen (**106**).

Conditions favourable are a temperature above 64° F. (18° C.) and a supply of fresh pollen and thin honey or syrup. At such a temperature a supply of nectar will generally be coming in, but it is helpful, especially with small colonies, to supply a small quantity of stimulative food (**1294** et seq.) about 1½ hours before mating may be expected. Except during a good honey flow there is danger of robbing and small colonies should be stimulated with sugar syrup rather than with honey as it gives off less scent. Sugar syrup may be used, adding one measure of fresh milk to two of syrup.

282. At a suitable temperature drones should be flying freely, but it is advantageous to give slow stimulative feeding to any drone breeding colonies at the same time as to those containing the virgin queens. See, however, **176-7**, **192** and **290-4**.

Procedure and Period of Fertilization

283. A virgin queen will not fly for about 2 days after emergence and may conveniently be examined during this period to see that she shows no sign of abnormality, defect or deficiency.

It has been proved by observation of marked queens that an imperfectly fertilized queen may fly again for a second mating, returning with signs there of before settling down to lay. In these experiments, drones for mating did not fly from the same hive.

284. Mating flights generally occur between 11 a.m. and 4 p.m. (Summer Time), and most frequently between 1 and 2.30 p.m. The flight usually lasts 4 to 10 minutes and drones and queens take flight about the same time.

285. Drones are known to have flying grounds, or more properly, favourite exercising grounds, and for many years it has been stated that these are the mating fields, also that the queen flies high and is chased by many drones, being caught by the fastest and strongest. In contradiction, many observations of recent years have shown that the queen is frequently mated within a few yards of the hive. The drones in the hive from which she flies, if potent (**176-7**), will fly with her at the time, while drones of neighbouring hives do not have the same good opportunity and notice of her intended flight. If the hive is so placed that the queen is limited and directed in her flight, say by an avenue of trees, the prospect of her mating with a local drone is almost ensured.

286. If only old bees are present they are apt to swarm out with the queen on her mating flight. The remedy is to avoid this condition by supervision.

Queens Going Astray

287. Virgins, also queens just mated, have been known to fly to queenless stocks, where they have been welcomed, but a queen entering a queen-right colony will almost certainly be destroyed. A marked mated queen, even, has been known to fly after laying and to be received by a neighbouring stock which happened to be queenless. It is held that in the early spring a queen may take a flight, but this is a rare occurrence.

Egg-laying

288. The queen generally commences to lay about 2 to 3 days after fertilization. Eggs should generally be in evidence within 10 days of emergence of the queen, and they may appear on the seventh day. Very late in the season, if the honey flow ceases about the time of fertilization, laying may be deferred until the spring, but this is hazardous and may lead to loss of the queen.

289. As soon as laying commences in a nucleole or a small nucleus it is desirable to close the entrance with queen excluder as, if the queen lacks laying space, the bees are apt to swarm.

Drone Breeding

General

290. Although mating cannot yet be absolutely controlled, except by laboratory methods (**295**) not yet fit for use in the apiary, this affords no excuse for not giving equal care to the selection of drone, as of queen, parents.

291. Enquiry shows much difference of opinion as to the heredity of vicious bees, but in many instances they appear in the second generation after crossing different races, notably yellow queens with dark drones.

Selection of Drones

292. In selecting a colony for drone breeding it should be remembered that the male parent of the drones in that colony is not the drone that crossed with the queen mother therein, but the drone that fathered that queen. One must look not so much to the characteristics of the worker bees in that colony for evidence of the quality of its drones as to the workers produced by the mother of the queen. One may, however, also examine the drones themselves for such evidence as they may show of race and vigour.

293. If the strain is not uniform in quality, the drones will be found to have greater variation than the queens, and the chance that a new queen and her drone mate will reproduce in their offspring all the desired characteristics from their inheritance, or indeed any of them, is greatly reduced. Uniformity of quality is much to be desired in breeding stock and should be diligently sought after. The amateur especially, is prone to introduce variety, with results that are often disappointing.

294. Reference should be made to the information given in **176-7**, about the utilization of selected drones. The alternative course is to flood the neighbourhood with very large numbers of drones, produced in selected drone breeding stock, rigorously suppressing all others, and arriving at an understanding with neighbours who may have different ideas and different stock.

Control of Mating

295. The only method of absolutely controlling mating is by artificial insemination. Such a method not only requires special training and apparatus and considerable dexterity, but is likely to remain too hazardous for commercial use in fertilizing queens for sale. It should prove, however, of immense help to those engaged in unravelling the complicated laws of hereditary transmission and ultimately of value in establishing pedigree stocks of breeding queens and drones of various classes suited to the needs of different districts and methods of working.

296. Some success in the laboratory has also been obtained by the hand-mating of queens. It is possible that the technique of this method may be developed sufficiently to be of value at least to professional queen breeders, but further study of genetics is necessary before such a procedure would be justified for commercial work.

297. Another method of laboratory value in controlling the parentage for scientific purposes of individual queens or workers is that of Barratt and Quinn. They employed fresh laid drone eggs, hand fertilized. The orifice in the egg is quickly closed after laying by the drying of mucus, which, however, Barratt removed by wiping with alcohol. The sperm from the selected drone is immediately applied with a brush. The operation was performed at a temperature in the neighbourhood of 80° F. (27° C.), in a humid atmosphere. It is very desirable that further evidence of this possibility should be obtained, and development of technique.

298. Meanwhile much can be done by any working bee-keeper who will master what has been given herein, especially the instructions in **176-7**. Other steps may be taken in addition, and are detailed below.

Mating at Abnormal Hours

299. The weather being favourable, both queens and selected drones may be stimulated to fly abnormally early on a sunny day by stimulative feeding at 7 a.m. to 8 a.m., leaving other colonies severely alone or giving them extra shade overnight. Flying may be expected between 9 and 10 a.m. (G.M.T.).

300. Alternatively, and with greater certainty, the hives containing the queens and the selected drones may be closed before 10 a.m. and put in a cool place, free ventilation being provided through floor ventilators or the like. Feeders are placed in position and stimulative food given about half an hour before the drones from freely exposed colonies are known to cease flying. At the end of the half-hour the closed colonies are exposed on their original stands, which are best put in a position facing towards the sun late in the day, and opened. There will be great activity of the bees and mating generally occurs forthwith. If necessary, a second attempt may be made in a similar manner the next day.

301. Some close the entrances with queen excluder only, thus allowing the workers free flight and confining only the queens and drone. This is not so effective in securing flight of the queen at the desired time.

302. It is good to provide the hives containing the virgin queens with a supply of two or three dozen selected drones of suitable age (see **177**), thus avoiding the necessity of closing the hives in which the drones are bred.

Mating out of Season

303. This is based on a similar idea of securing mating when unselected drones are not flying, but is less certain as a wild colony may well be a prosperous and vigorous one, sending out early drones, whilst less desirable queenless colonies may send out drones late in the season. Furthermore, queen raising very early or late in the season is a hazardous business.

304. For mating early, the colonies selected for drone breeding must be built up early by adding combs of bees from other colonies, feeding and packing warmly and given drone comb towards the outside of the brood nest, noting **292-4**. Queen cell raising can be commenced with eggs laid not less than sixteen days after the drone eggs have been laid.

305. For mating late the drone-raising colony must be made queenless before the late honey flow ceases and drones are massacred. It should be kept queenless, but provided with any necessary brood and united with another colony after the drones are no longer required, or re-queened with a laying queen if worth keeping as an independent colony.

Mating in Enclosures

306. Many attempts have been made to control mating by confining the queens and drones in an enclosure, the least unsuccessful involving the use of large barn-like cages of wire screen. To obtain reasonable economy, the drone breeding and the mating colonies should open, not only into the enclosure, but also through excluder into the outer atmosphere, so that the workers can continue their important functions of providing food and water.

Incidentally such an arrangement would constitute an ideal swarm catcher, from the standpoint of reliability in securing the swarm.

Tested and Untested Queens

Classification and Definition

307. A *Virgin* Queen is one unmated, but should be sold as soon as possible after emergence (see **89**).

A *Selected Virgin* Queen is one selected by appearance (see **43**) and heredity and obviously cannot have been tested in any way.

An *Untested* Queen is a mated queen that has begun to lay.

A *Selected Untested* Queen is one that has been selected by appearance (see **43**) and heredity.

A *Tested* Queen is one that has been laying for more than 3 weeks in a satisfactory manner and whose progeny show no signs of incorrect mating.

A *Selected Tested* Queen is one selected by appearance (see **48**), heredity and performance.

A *Breeding* Queen is one whose progeny have shown excellent results and characteristics for at least a full year, and is one selected for the raising of new queens on the evidence so obtained and according to its heredity and appearance.

Notes

1. The above definitions have no authority behind them but will be recognized by any good breeder.

2. A queen should not be passed as fertilized until some of her brood has been sealed with the flat capping characteristic of worker brood, proving that she is no drone layer.

3. An untested queen from a reliable source is nearly as good an investment as a tested queen.

4. The buyer of an untested queen takes the chance of her having been badly mated, but the seller of a fertile queen is responsible for her being properly mated and should replace a queen proving actually defective.

5. A queen in full lay should not be sent by post. Her laying should first be reduced. A queen in full lay may be sent by rail or carrier on comb in a nucleus hive or other convenient travelling box.

COMB, WAX AND PROPOLIS
Comb Building

Method

308. Bees build their combs outwards from the surface on which they start and generally downwards from the more or less horizontal roof of the chamber they have for a home. To fill vacant spaces at a lower level, for example, between widely spaced crooked frames, they will build out from the side wall, even though vertical. If hindered from building in the normal way, they can build upwards.

309. Building is commenced by accumulating on the chosen spot a small body of wax, moistened with saliva, masticated and toughened. This wax is moulded from opposite sides, hollows being worked, which, when nearly meeting, bottom to bottom and side by side, form respectively the commencement of the mid-rib and cell walls. The first cell walls are built at right angles to the surface on which they are founded. Thus, comb depending from a horizontal surface, as it most commonly does, is found with its hexagon cells having two vertical walls. Comb built on a vertical face, or one sloping about 30° from the horizontal, will be found to have cells each with two horizontal faces. It frequently happens that through stretching, or some irregularity in building, a comb started with cells placed either vertical or horizontal way, undergoes a transformation during construction.

The bees have a natural preference for a substantially horizontal surface from which to depend their combs, but no preference for the vertical or horizontal cell wall *per se*. The comb is, in fact, equally able to carry the weight whichever way it be built. The mounting of foundation in frames to give vertical walls gives the bees a better start from the top bar, but a more difficult job at the sides. If a full sheet is mounted the other way, the bees will make a job of the top fixing and finish the sides more readily.

310. While the above is the normal procedure in building, the bees can, in fact, construct comb, starting by tracing hexagons in wax, on a vertical flat surface.

311.a Wax is naturally white and very pliable. Once the bees have completed comb building they treat it with propolis and gets stained by pollen. The propolis hardens the wax comb and is principally laid on the top surfaces of the comb. ID

311. *The finished wax comb is white, but is treated by the bees before use with a balsamic or resinous substance, giving it a yellow colour, the material being obtained possibly from pollen or from other vegetable source.*

Building Drone and Worker Comb

312. In nature a swarm starting a new colony first builds worker comb, but when a good brood nest is built up, drone comb is built on the outer margins and used for drone raising, early in the season, or honey storage, later in the season. Next spring drone eggs are laid, when, or soon after, the brood nest has expanded to the drone comb. This is not the whole story as drone comb may be built on the edges of the brood nest before full expansion of the nest. Moreover, a first swarm hived without foundation may build drone comb within 3 weeks of hiving, especially if there is but little honey coming in. A swarm on foundation will build worker cells for about 3 weeks, but the use of worker foundation gives no absolute guarantee that drone cells will not be built.

313. If the swarm is headed by a young queen of the same year, drone comb is seldom built until the next spring. Queenless bees, if they build, will build only drone comb, and are likely to do so especially if they have very little.

314. One must remember that drone comb is required mainly for raising drones and will be built whenever the bees desire to raise drones. Even with a young queen, the bees apparently like a few drones, against emergency. They do not work so well if they are allowed none. They will tear down worker comb to build drone comb if they desire it and there is no other place for it.

315. Where the bee-keeper has to contend with crooked combs, he can get good flat combs by having them drawn out between other flat combs, preferably between sealed store combs in a super. For economical wax production see **370-4**.

The best worker combs from foundation are built by young bees in a full brood nest or in an upper chamber, where two or more combs are used to give them a start, and the queen is confined below. Combs so built are far more frequently carried down to the bottom bar than where the bottom bar is next the floor.

Brace and Burr Comb

316. These terms are sometimes confused. A brace is a mechanical contrivance, a fixing between two parts to give additional mechanical support. Thus, after moving a hive some distance one frequently finds many pieces of brace comb built, from hive wall to comb and from comb to comb, to resist further shocks. If a comb threatens to break loose from any cause, or a frame shows a tendency to swing, the bees build brace comb to secure it. Brace comb generally takes the form of an extension of the cell walls at a junction between three of them, carried out to in adjacent surface to which it is secured, thus avoiding blocking any cell.

317. Burr comb is a term applied to any odd pieces, not part of the regular structure, generally found in gaps of more than a bee space, or as the commencement of uncompleted combs, on the hive wall, or on a division board.

Inclination and Disinclination to Build Comb

318. Comb building requires a high temperature (**29**) and honey sacs well filled with honey or sugar syrup (**370** and **1286**). Bees build combs willingly and most readily when there is a good flow on. A good flow presupposes a high temperature, but it also involves the handling and inversion of large amounts of nectar, and activity in the hive, all of which unavoidably brings about a high temperature and probably also the automatic production of a certain amount of wax. Hence the necessity for giving some opportunity for comb building (**1127**).

319. Notwithstanding that the bees are busily occupied with handling nectar during a flow, that is just the time that they are willing to build and to produce wax for capping, but if the flow suddenly ceases, building and capping is liable to be brought almost to a standstill. This supports the suggestion made in the previous paragraph.

320. In a hive with the combs arranged parallel to the entrance, the bees are less readily disposed to draw out foundation placed at the front than at the back.

321. The normal direction of extending a comb is downwards as the cluster of bees expands, the cluster conveniently overlapping the edge. Bees will expand upwards in case of necessity but object to breaking the cluster, as this involves loss of heat. For the same reason, in expanding at right angles to the comb surface, the normal procedure is to build the next comb immediately adjacent to the last so as to have no break in the cluster. This shows why the bees object so strongly to building in those little section boxes so beloved by the bee-keeper, and why they will make a better start if attracted by partly drawn comb, say by a shallow comb, adjacent to the sections.

322. For the same reason bees dislike entering a super fitted with foundation when that super is placed above the occupied combs, although they will readily enter the same chamber if placed below.

Nevertheless a stock, developing well and ready for another brood chamber (**935** and **1394**) at the start of an early flow, may be given one, placed on top and filled with foundation. The bees will draw out and occupy the combs on the flow, and build out every comb to the bottom.

323. As a corollary to 318 it is peculiarly difficult to persuade bees to build in cold weather unless artificially fed. Comb production under such circumstanses is uneconomical (373).

Cell Walls

324. The top and bottom walls of cells intended for breeding purposes are built out substantially horizontally and nearly at right angles to the plane of the mid-rib, but cells provided for honey storage are given a slight inclination upwards towards the mouth, which assists in retaining the nectar. Frequently, the inclination is very considerable, say 1 in 3. Rarely, the cells take the form of nearly vertical open pockets, presumably to accommodate a rapid flow of exceptionally thin nectar. When cells in the brood nest are extended to receive honey, some upward inclination is given.

325. The mouth of the cell is always thicker than the walls. As the cell is extended, material is first added around the mouth and then thinned out and extended by moulding from within. The thickening at the mouth serves to protect the thin wax from injury and also stiffens the whole structure, helping to prevent distortion, especially in the lower parts of combs, which parts may not be secured to the side walls. This thickening is much greater in old cells, the opening tending to a circular rather than hexagonal form (**342**).

Cell Cappings

326. The cappings of cells containing brood are porous. They owe their porosity to their structure being built up of superposed portions of wax. Pollen may be present in small quantities, but only by accident, the chance of which is cumulative with successive recapping in which old material is used again. Those of worker cells are practically flat, sunk portions being a sign of foul brood. The cappings of drone cells, or of worker cells containing drone brood (109), are markedly convex, almost hemispherical and are thus readily distinguishable.

327. Cappings of honey cells are built with wax, but many contain pollen grains also. They are slightly porous. The darker races produce a white capping with an air space beneath, which adds greatly to the appearance and makes some provision against expansion of the honey by water absorption during the winter. Italian bees and yellow, races in general from warmer climates, frequently fill up their cells to the capping, which is wetted thereby, but when honey is coming in quickly they also may leave an air space; some strains more so than others. Comb honey with white cappings, with an air space beneath, is preferred in the market for its better appearance. Thus the yellow races can be used for the best section comb production only during a rapid flow, and then only strains that show the desired characteristic.

328. Some bees, notably the French bees in certain districts, produce bright yellow cappings, the colouring matter, it is believed, coming from certain flowers, e.g. the dandelion. Wax made from willow herb nectar is exceptionally white; that from sainfoin is primrose yellow, and from ragwort a deep yellow.

Dimensions of Cells

329. The cell tends to a true hexagon, although many irregular cells are built. The cell bottoms are still more irregular, to accommodate differences in the two sides, due mainly to incidental curvature of the midrib. Such accommodation affects the cell form and dimensions. Sometimes patches of cells·are found practically square. It is rare, indeed, for natural comb to compare in regularity with that built on machine-made foundation. The size of a hexagon cell is generally measured across the flats. The diagonal dimension across the corners is larger in the ratio of 1 155 to 1·000. The length of the side is half the diagonal dimension. If the dimension across the flats = f, then the area = $0.866f^2$.

330. Foundation for worker brood is generally made to give comb with about 4¾ to 5 cells per inch run (measured across the flats), the latter more commonly, and 4 cells per inch for drone brood. The area of the hexagon of a cell then becomes such as to give 26 to 29 worker cells per

square inch and 18·5 drone cells, in each case counted on one side of the comb, and not allowing for stretching.

331. In nature the inside dimensions of worker cells across the flats may vary even as much as from, say, 0.195 to 0·235 inch with the same lot of bees, and will vary still more as between races having the smallest and largest bees. Similarly, in one lot of bees drone comb may run from 0.22 to 0.26 inch.

332. In new comb the cell wall in worker cells may be less than 0.004 inch thick and more than 0.006 inch in drone cells. There is not much data on this, but the average cell wall in used worker cells exceeds·0.004 inch. Owing to rounding out of the cells where the flats meet at a corner, the diagonal dimension within a used cell is reduced by about 0.015 inch, more or less.

Importance of Dimensions

333. If the cell diameter is such as to cramp the bee, the workers produced therein are undersized and particularly in tongue-length and size of honey sac. Foundation running 5 cells per inch is not the best for the larger strains and that running 4¾ per inch is not large enough to give the best results with the largest bees.

When the cells are already too small, it is stated that an increase of 5 per cent in diameter, i.e. 10 per cent. in area, may give an increase of ¹/₁₆ inch in tongue-length. Too large an increase in cell diameter involves increased size of brood chamber and some loss of economy in wintering, the cluster being less compact. Undoubtedly the bee-keeper needs to study foundation in relation to the size of his bees. Although larger cells produce larger bees, there is no evidence that they are better bees. They are of lighter build.

334. Finished new worker comb measures about ⁷/₈ to ¹⁵/₁₆ inch (23 to 25 mm.) from face to face and drone comb about 1¼ inch (31 mm.), the depths of the individual cells being rather more than half these figures. Old comb with thickened cell bottoms will be thicker overall in a corresponding degree, but see **342**.

Size of Combs and Method of Using

335. In hives the size of comb is determined by the size of frame. In nature it is determined in part by the size of the cavity in which the comb is built and in part by the ability of the wax walls to withstand the dependent weight. Where support can be obtained from the sides of a cavity, say 10 inches wide, as in a hollow tree, the length may run to several feet. Where the weight is considerable, the upper part of the comb is found strengthened by additions of hard propolis. The upper part used for honey storage is also wider from face to face than the base used for brood rearing. This increase also assists in obtaining security and strength to carry the weight below. Comb full of honey from face to face runs nearly 7½lb. per square foot per inch of thickness (4 kg. per 1,000 sq. cm. per 2.5 cm. thickness).

Bees will build odd pieces of comb in which to store honey, in any odd outlying parts adjacent to the brood and stores generally, but for brood raising they prefer to provide large continuous surfaces over which the queen may roam without hindrance.

336. The queen does not pass readily from one chamber to another above it, across an intervening gap, and across the edges of the frames, neither does she pass readily from one comb to another parallel comb if there are two or more combs of honey and pollen between. Barriers of this kind are apt to lead to the raising of supersedure cells (**1153**). Thus the bees prefer large continuous surfaces limited only by the inherent strength of a comb, built with due economy of costly wax. Thus while a bee space at the side of a frame is respected by the bees, a bee space between frames vertically one above another, and especially between brood frames, is disliked and a *more or less persistent endeavour is made by the bees to bridge the gap*. In extreme cases, where the combs are out of alignment, the lower part of the upper comb may even be reconstructed to overhang its frame and extended downward to join the top bar of the frame below. Where the combs in two bodies are in vertical alignment, the bees tend to build rows of cells between them and over the top bars and adjacent bottom bars (see use of Vaseline, **876**).

337. In the winter, when the bees are clustered closely, the cluster tends to the spherical form. In the spring, when the cluster first breaks, the brood is spread over a larger area, frequently of fewer combs, because with a higher temperature surrounding the brood nest, a relatively few bees can pack the edges of the spaces between combs to conserve the heat and the combs themselves serve as adequate packing in the other directions. We may at that time find, for example, three batches of brood 8 inches mean diameter, with eggs and young brood on the margins and old brood at the centres and small patches of old brood on two adjacent combs with no eggs, clearly showing that a transition of form is taking place. After this period, in hives with the largest frames a given amount of brood which would fill a large number of smaller frames will be found spread over a smaller number of the larger frames, showing the desire for large patches and consequently for large combs.

338. Combs larger than the M.D. are not very practicable for the beekeeper. The justification for the use of smaller combs, such as the Langstroth and British Standard, is their use in double brood chambers in a different system of management or incidentally with a less prolific strain of bee. (See **906-10**.)

339. As another example of the bees' objection to a comb not continued in one piece may be cited the difficulty of getting bees to take to a super provided only with starters, a difficulty often not appreciated by the apiarist (**323**).

Combs from starters are best built out placed each between two finished combs. The frames may be wired in advance with vertical wires, not too coarse.

340. In nature, having built a large enough brood nest, the bees build drone comb outside it and utilize this for drone raising and honey storage. Honey is stored also in worker comb, the upper part of which is built out for the purpose as the comb is extended downward. It is true, therefore, to say that the bees build drone comb for storage, but it is equally natural for them to store in worker comb.

It is not natural to have drone comb above the brood nest, and,unfortunately, the queen will go up into it, on occasion.

341. Honey can be extracted rather more easily from drone comb than from worker comb, but this advantage is more than offset by the risk of raising excessive drone brood at great cost, and the·consequent necessity of using an excluder. Moreover, the bees are apt to leave the central parts of the comb for the use of the queen even when she cannot get access to it. There is no material saving in wax by the use of drone foundation as the cell walls have to be made thicker to compensate for the increased spacing and provide the required strength.

Use of Old Combs

342. It is commonly taught that, with the raising of successive bees in the same cell, the cell walls become thickened more and more by the deposit of larval skins and cocoons, gradually reducing the diameter of the cell. On the supposition that the cells must soon become too small, many good tough combs are destroyed unnecessarily every year. The fact is, the bees limit the reduction of diameter by stripping the walls, letting the deposit accumulate, however, at the bottom of the cell as this is readily compensated for by extending the cell mouth. In time the mid-rib may thus be thickened to 1/4 inch or more and such combs are most excellent for the winter cluster.

The writer has seen old comb with thickened surrounds to the cell mouths and greatly thickened mid-rib, but is waiting to see one with overthick side walls. Dr. Miller wrote to the same effect.

343.a   The received advice these days is to replace brood combs at least every 2-3 years. This is because it has been shown that pathogens and poisons build up in the comb and can become a source of infection. However it is true that a brood cell can be reused many times without any significant reduction in the cell size. Brood combs that have been used many times are more rigid than new comb and is less likely to sag or distort when held in a horizontal plane. ID

343. *The author has examined and measured worker and drone combs in frames which have been in use for 7 to 18 years without having received attention other than that given by the bees, and has failed to find side walls of worker cells thickened by more than three thousandths of an inch, a negligible increase. He has particulars of an apiary where no frames have been touched for 21 years and the bees produced are still of normal size. Drone cells have somewhat thicker walls, but they also are not allowed to become too thick.*

It is a great mistake to discard old combs merely on account of age. Nevertheless, old combs of doubtful origin should be replaced at convenience as they may carry spores of foul brood disease.

It is much harder to work over an old comb than a recent one, hence, although old combs are preferred for wintering, new comb is preferred for rapid expansion of the brood nest.

344. a Stored comb may be 'freshened' by placing in a warm oven (at about 35° C) restoring the bright appearance. If stored comb has meildew on it it is preferable to melt the wax and recycle it. ID

344. *Stored combs of dull appearance may be freshened by exposing them to strong sunshine, or by passing the flame of an alcohol lamp over them, or by holding them for a few moments in a warm oven. If mildewed they may be brushed with methylated spirit. A strong colony, however, will restore such combs without help, but at some small cost.*

Cleaning Old Pollen-clogged Comb

345. The utility of combs well charged with good pollen is dealt with in **1276**. The bee-keeper will, however, occasionally have pollen-clogged combs he wishes to clear. If put in the middle of the brood nest of a strong stock or swarm, the bees will clear them out. Be careful, however, to watch that the bees do not raise unexpected queen cells in the brood nest on the side on which the queen is not laying (see **1153**).

346. If it be desired to remove the pollen by hand, the comb should be soaked for 24 hours in rain-water and then syringed or put in an extractor, or the comb may be placed in a warm dry place to shrink the pollen which may then be extracted in an extractor, or shaken out by holding the frame and striking the far edge gently against the edge of a horizontal wooden support, a table or box. When the pollen is consolidated with honey, a syringe may be used. The comb should then be well dried and the pollen shaken out. Again, a refrigerator may be used to loosen the pollen.

347. Pollen not covered with honey loses food value seriously after a few weeks and may be further ruined by fungi. Pollen exposed on the borders of the brood nest during brood raising is, however, rapidly consumed as gathered and is not in danger of deterioration at that time.

Weight of Combs and Foundation

348. The wax in new brood comb runs to about 650 square inches per pound. Thus a pound of wax will be found in:

British Standard Frames .... say 6 frames
British Deep Frames .... 4½ frames
Langstroth Frames 5.frames
Modified Dadant .... 4 frames

This is with cells running about 5 to the inch. The weight may be somewhat less or appreciably greater. The weight is increased by additions of propolis and later by additions of cocoons. The above corresponds to about 37,000 cells per pound. This figure may run from 30,000 to 50,000, the latter only with bees making rather small cells. The smaller cells take less wax per cell but more wax per square inch. The comb in a section weights about one-fortieth of a pound, more or less, or, say, one-thirty-sixth with cappings.

349. Medium weight foundation well worked out by the bees will provide about two-thirds of the wax required in building brood combs. If heavier weight foundation is used some bees frequently fail to thin down the bases and thus wax is wasted.

In melting down old combs, the weight of wax obtained will be somewhat less than the figures in **348**, even with the assistance of a wax press to follow the use of steam or boiling water **(396)**. If the combs exceed 1 inch from face to face, or the cell walls are unusually heavy, the weight will be increased. For weight of foundation, see also Table XX.

Repairing and Cutting Comb

350. If it be desired to secure repair of a worker comb, as for example after cutting out a queen cell or portions of drone comb, the comb should be given to young bees for re-building, late in the season, feeding them well at the time. To make more sure of a good repair made with worker cells it is, however, desirable to trim the damaged portion neatly and insert a portion of worker foundation accurately fitted. This is readily done if it be remembered that accurate placing is more important than securing contact at all points. The bees will make the junctions good.

In fitting the foundation it should be noted that the horizontal facet of the cell base forms the bottom of cells on one side of the comb and the top of those on the other side. The foundation must be cut so as to match.

351. If a large portion of the bottom edge of the comb has to be replaced, or if, as is frequently the case, one has to deal with a number of combs that have never been built out to the bottoms of the frames, they may be dealt with successfully in a very simple manner. They should be put in the middle of an empty body, filling up the sides with fully built combs, then fasten them all in place by screwing down thin strips of wood placed on the lugs. The box of combs is then inserted *upside down* under a colony in spring-time, ready to receive and occupy an additional chamber **(935)**. The bees will build every comb upwards to its bottom bar **(308)**, as they abhor waste space in the middle of the brood nest. When completed the chamber may be reversed, the clamping strips removed, and the whole re-stored. If the stock is ready and fine weather holds, this operation will take only a few days.

352. To cut comb cleanly dip the knife blade in methylated spirit. To prepare sections of comb to show cell walls, etc., embed the comb in plaster of paris, cut with a fine hack saw and finish with fine sandpaper. The plaster should be poured separately into each cell and the lot shaken down.

Use of Foreign Substances in Comb Building

353. The bees add a small amount of foreign material, for example, propolis, to their combs, so that commercial beeswax contains such. The bees will, however, use considerable quantities of foreign material which they can mould in with the wax if given the opportunity, such as bitumen of suitable consistency, harder waxes of the paraffin series, and others. Such adulteration should be avoided, however, as such additions are sure to find their way into the melting-pot sooner or later in smaller or larger quantities, and may spoil the market for pure beeswax, an article which already finds none too high a price having regard to the cost of producing it.

354. Some foundation is marketed containing a layer of tough vegetable wax of slightly higher melting-point than beeswax. Its use is subject to the objection that such wax cannot be again separated from the beeswax and the meltings ought to be sold only as a mixed wax suitable, for example, for part use in foundation building. For adulteration of beeswax in general, see **386-7**.

355. If beeswax is reduced to fine scrapings, the bees will gather and use them in comb building. Bees given cappings to clean up sometimes weld them into weird irregular forms, and in time they may carry the wax into occupied parts of the hive. They have been known to store the wax by greatly thickening the walls of drone cells on the border of the brood nest.

356. Bees will carry old propolis into the hive for use again, as well as wax.

Preservation of Combs (Wax Moth)

357. The principal danger to wax combs is the wax moth, of which there are two kinds. The large, *Galleria mellonella*, measures about 1 1/2 inches across the open wings, a soft-looking, pale drab-coloured moth with darker brown markings over the upper wings. The grub is about 3/4 inch long. The smaller, *Achroia grisella*, measures about 5/8inch across the wings, being light drab or fawn-coloured all over with but little marking.

358. The larvae are almost indistinguishable when young, but those of the larger moth run when disturbed, while those of the smaller moth feign death. They feed upon combs and cocoons and prefer old combs, but although not able to complete their development on wax alone, they will attack unused comb, securing cellulose by attacking the adjacent woodwork. Completion of development is greatly delayed if the diet is unbalanced, but follows the normal course where old comb is attacked.

359. The larger moth takes 4 weeks to reach maturity, dating from the egg, at a favourable temperature, say 75° to 85° F. (25-30° C.), and 20 to 30 weeks in the winter. The smaller insect matures rather more quickly in the summer. The larvae are killed in about 3 minutes at a temperature of 40°C. (104°F.), but there appears to be no record of what the eggs will withstand.

360. Eggs laid late in the season may not mature until the next spring. It is not safe to store combs for the winter on the assumption that they are free of wax moth unless stored where hard frost will reach them. The eggs are killed by severe frost (time and temperature conditions not established).

Remedies for Wax Moth

361. Bees which are good house cleaners will not tolerate wax moths. The moths get no hold, save in weak and diseased colonies, but they readily and rapidly attack combs in storage if care is not exercised to prevent this occurring.

The wax moth will not lay eggs in comb freely exposed to the light. If one side is exposed and the other is against a wall or another comb, the moth will choose the protected side. Light, however, does not destroy the larvae, and when the eggs are laid the mischief is begun.

362.a The perfect remedy should destroy both eggs and larvae. The larvae are destroyed by many substances and several have been described as egg destroyers which cannot be relied upon to bring this about unless development has already reached a certain stage. Eggs will not withstand hard frost however.

The difficulty is avoided by treating the combs twice with an interval to allow eggs to hatch. In warm weather, for example at extracting time, a fortnight may be allowed.

Freeze the combs in a deep freeze for a week and then store in moth proof bags. Alternately the combs may be treated with acetic acid but follow suggested procedures and wear full personal safety protective gear. (Safety goggles, disposable gloves and lab coat or boiler suit). JC

362. The perfect remedy should destroy both eggs and larvae. The larvae are destroyed by many substances and several have been described as egg destroyers which cannot be relied upon to bring this about unless development has already reached a certain stage. Eggs will not withstand hard frost however.

363.a The difficulty is avoided by treating the combs twice with an interval to allow eggs to hatch. In warm weather, for example at extracting time, a fortnight may be allowed.

p-dichlorobenzene (PDB or 'mothballs').

H319 Causes serious eye irritation.
H351 May cause cancer/genetic defects.
H400 Acute toxicity to aquatic life.
H410 Very toxic to aquatic life with long term effects.

This chemical is no longer available. JC

363. *One of the most convenient and economical substances to use is paradichlorobenzine (sometimes referred to as p.d.b.), which, in the form of white crystals, is easy to handle. The fumes are heavier than air, so the crystals should be put at the top of the pile, in an empty body. Use a saucer or spread the crystals on a piece of paper. Allow not less than the following quantities for each treatment:*

Langstroth and British Standard Brood Chambers.	$3/8$ *ounce each.*
Langstroth and British Shallow Brood Chambers.	$1/4$ *ounce each.*
Modified Dadant Standard Brood Chambers.	$5/8$ *ounce each.*
Modified Dadant hallow Brood Chambers.	$5/16$ *ounce each.*

364.a Burning sulphur is not recommended. The sulphur dioxide produced is toxic and there is the danger of fire. JC

Sulphur dioxide gas.

H314 Causes severe skin burns and eye damage.
H331 Toxic if inhaled.

364. *Sulphur is easy to get and frequently employed. It produces fumes on burning which rise in the air. The combs are treated in the bodies or chamber in which they are to be stored, placing an empty body or two at the bottom and a cover on top. Unless all parts fit well, the cracks should be sealed with strips of paper. The gummed strip sold in rolls for sealing parcels is very convenient for the purpose. The sulphur is burned at the bottom of the chamber, using 2 oz. for every 10 cubic feet, or about two-thirds of the quantities scheduled in the above paragraph. The empty bodies at the bottom keep the combs from too close contact with the burning sulphur.*

365.a Do not use these chemicals. They are difficult to handle and extremely dangerous. JC

Carbon disulphide.

H225 Highly flammable liquid and vapour.
H315 Causes skin irritation.
H319 Causes serious eye irritation.
H361 May cause damage to fertility or the unborn child.
H372 STOT RE1 Prolonged exposure may cause damage to body organs.

Ethylene oxide (oxirane).

H220 Extremely flammable gas.
H315 Causes skin irritation.
H319 Causes serious eye irritation.
H331 Toxic if inhaled.
H335 May cause respiratory irritation.
H340 May cause genetic defects to germ cells.
H350 May cause cancer.

Calcium cyanide.

H300 Fatal if swallowed.
H400 Acute toxicity to aquatic life.
H410 Very toxic to aquatic life with long term effects.

365. Many other substances have been -successfully used in a similar manner, as follows:
Carbon disulphide   2 1/2 ounces per 10 cubic feet.
Ethylene oxide   1/3 ounce per 10 cubic feet.
Calcium cyanide   1/3 ounce per 10 cubic feet.
Carbon disulphide has a most unpleasant odour, and calcium cyanide is highly poisonous. Of the three the second is to be commended. It is a liquid and inexpensive.

366.a Do not use this chemical.

Methyl bromide (bromomethane).

H301 Toxic if swallowed
H315 Causes skin irritation
H319 Causes serious eye irritation
H331 Toxic if inhaled
H335 May cause respiratory irritation
H341 May cause genetic defects
H373 STOT RE2 Prolonged exposure may cause damage to body organs
H400 Acute toxicity to aquatic life
H420 Harmful to atmosphere (ozone)

Another chemical that is out of bounds to the non chemist. JC

366. Methyl bromide also is being employed when there are large numbers of hives, justifying the provision of suitable chambers for treatment of the combs. It is a gas sold compressed to liquid form in sealed tins. On puncturing the tin the liquid quickly evaporates. A volume of 1,000 cubic feet requires a 1lb. tin, and at this concentration it is held that both larvae and eggs will be killed in 8 hours. The gas is poisonous and should be used only under skilled supervision.

367.a Sounds a good idea for a method of biological control but is there evidence for this? JC

367. It is interesting- to note that spiders will attack the moth and certain centipedes will attack the larvae, and may even be used for clean up.

368. In hot climates, where wax moths may be a great nuisance, bees have been known to make a habit of tearing down unoccupied combs so as not to give the enemy a chance of establishing a footing.

Storage of Treated Combs

369. After treatment the combs should be protected by newspaper, which the moths dislike. If the bodies carrying the combs fit well, lay a sheet of newspaper between each when piling them up. Where they do not fit closely, use two sheets, turning the ends of one up and the other down and tying in place, or use a strip of tape or gummed paper.

Beeswax

Production

370. Beeswax is produced by the bees as a secretion in the form of scales extruded on the segments of the abdomen on the under side. These are masticated and saliva added during the process of comb building.

Comb as first built is quite white. Young bees then work over it, polishing the surface, and the comb then has more of a yellow tint, very yellow in some cases, and it would appear that the bees have made some addition (**311**). The bees produce wax on a honey diet, and even more abundantly on a sugar diet. The requirements are a full honey sac and high temperature (**29**).

371. The bees form a compact hanging cluster for wax production, which takes about 24 hours. This is a voluntary act, but it is thought that the process is involuntary if the circumstances are favourable to wax production. Wax is required for capping wherever brood is being capped and some new is produced, although mostly old stuff is used. If the abdomens of young workers are examined even in February one may detect some transparent scales of wax.

Bees have been known to build comb in candy boxes in early spring, which supports the idea of the automatic production of a certain amount of wax when sugar is being inverted. The provision of space for deposit of wax by comb building is reckoned amongst the hindrances to swarming. Bees ready to swarm are at any rate loaded with wax.

372. It is estimated that in the production of 1 lb. of wax the bees will consume from 6 to 20 or more pounds of honey and 10 or 11 is generally taken as an average; so wax production is a costly business to the bee-keeper. The higher figure named comes from laboratory experiments by Liebig, made under unnatural conditions such that the bees must have had to consume considerable food to maintain the necessary high temperature. The lower figure was given by De Layens, based on certain comparative tests made with two groups of colonies, one of them being given combs, whereas the other had to make them, an experiment worth repeating under better scientific control. No wax should be wasted (see Wax Rendering, **390**, *et seq.*).

373. Nevertheless De Layens claimed that a certain amount of wax could be produced each year without loss of honey. He claimed that colonies given a few sheets of foundation in the supers early during a heavy flow in hot weather showed as good an average harvest as other similar colonies provided with drawn combs throughout. The experiment requires repeating with a larger number of stocks. The statement involves either that some wax is produced and wasted if the bees have no combs to build, or that the production of some wax is indirectly helpful to the bees. It may be true. At any rate, it is easy to get good combs made under the conditions stated, and it is probably the cheapest way. Foundation is also drawn out quickly while syrup feeding if inserted between combs of brood one sheet at a time.

374. The production of wax for profit may be undertaken where labour is cheap and a good flow of inferior honey is available. Frames with starters should be inserted between store combs in supers and withdrawn when half or two-thirds completed, any honey being fed back to the bees.

Physical Properties

375. Beeswax has remarkably high resistance to the passage of heat; two or three times as much as other good non-porous heat insulators such as the resins, varnishes, bitumen and rubber.

376. Wax cooled rapidly is quite brittle at 60° F. (15 ½ C.), but after rolling at 100° F. (38° C.) it becomes quite pliable and remains tough when cooled. When melted and then cooled rapidly, the density of the solid wax depends upon the temperature to which the wax was raised when melting. When preparing sheets for foundation making, it is desirable not to raise the temperature above 175° F. When heavily rolled as in the "Weed" process, the wax becomes transparent. At hive temperatures approaching 103° F. (40° C.) beeswax retains sufficient strength to support a considerable weight of dependent comb, yet at 97° to 98° F. (36° to 37° C.) it can be moulded by pressure of the bees' jaws.

377. It is difficult to see how bees could have experimented with secretions until by some physiological process they arrived at a material having not only the desired extraordinary mechanical properties but also the remarkable thermal property above referred to. An analysis of the wax scales as first produced would show whether cerotin is a constituent of the original product or is a later addition. The latter would be a solution of the problem much easier to understand, as the bees can and do experiment with additions.

378. The specific gravity of commercial beeswax is about 0.96 to 0.97, i.e. it is a little lighter than water and weighs about 62 lb. per foot cube.

379. Beeswax becomes friable at 120° F. (49 C.) and melts at about 147° F. (64° C.).

The melting-point of white wax from new comb or from white cappings should be as high as 150° F. (65° C.). Figures as low as 140° F. (60° C.) have been got for commercial beeswax, due to admixture of resin from propolis added by the bees. The melting-point rises somewhat with age, two or three degrees Fahrenheit, within a few months, and bleaching by exposure to the sun is said to raise the melting-point several degrees.

380. Beeswax should always be melted over a water bath to prevent overheating. It loses its nature at about 250° F. (120° C.) and more so as the temperature is raised further. It does not boil with further heating, but decomposes, giving off smoke. Apparent boiling at lower temperatures is due to the evaporation of the water content.

381. The strength of beeswax depends upon the treatment it has received (376), but tests on cast bars have shown a tensile strength of 21 lb, per square inch (say 1 ½ kilograms per sq. cm.), at a temperature of 95° F. (35° C.), and compressive strength of about 33 lb. (2,350 grams per sq. cm.). These figures are roughly halved if the temperature is raised to 107° F. (42° C.). The tensile strength of wrought wax as found in the comb probably exceeds these figures.

Chemical Properties

382.a   Beeswax is a combination of hydrocarbons and long chain esters and fatty acids. Analysis has shown that there are over 200 separate components. Myricin and Cerotic acid were terms used to identify the non soluble and soluble components of beeswax when dissolved in hot alcohol. ID

382. Beeswax is frequently stated to consist mainly of myricin, with about one-seventh part crude cerotic acid. Small quantities of myricyl alcohol and certain esters are found also, but it is also held that commercial beeswax normally contains, say, 10 to 14 per cent. unspecified hydrocarbons.

The cerotic acid can be dissolved out in hot alcohol and precipitates in fine crystals. Cheshire gave 172° F. as the melting-point of cerotic acid and 127° F. for the myricin, but the latter should be 162° F., suggesting that Cheshire's myricin contained resins commonly added by the bees.

383.a   However the key characteristic is the melting point of fresh beeswax. This is set at 63.4°C +_ 0.5°C. ID

383.   Under chemical tests, beeswax (as taken from the hives) shows characteristic properties as follows:
Acid value ....16 ½ to 22
Saponification value ....86½, to 103½
Iodine value ....6 ½ to 12
the values depending upon the source.

384.a   Sulphuric ether is now commonly known as diethyl ether and is a powerful anesthetic. The components mentioned in 384 are considered dangerous to health and are generally not available to purchase. ID

384.   Water and cold alcohol have no effect upon beeswax, but alcohol dissolves the resinous matter in propolis quite readily. Hot alcohol dissolves also the cerotic acid.

Virgin wax scales dissolve freely in turpentine; white virgin comb dissolves less freely, but comb which has been worked over assuming a yellow colour resists turpentine but dissolves ultimately.

The yellow colouring is removed by cold sulphuric ether, which breaks up the comb. When virgin comb is broken up in sulphuric ether, it is partially dissolved. The yellow colour disappears under treatment by nitrous oxide.

Beeswax is saponified by caustic alkali. Propolis resists caustic alkali, i.e. it does not dissolve, but the removal of the wax causes the propolis to crumble readily, hence it can be

removed from metal fittings by alkalis.

Beeswax readily dissolves in carbon tetrachloride. It will dissolve in hot benzine, but a large part separates out in cold benzine.

385. Wax extracted from the refuse of the wax press (slum gum) by chemical solvents shows an altered nature, the acid value being materially increased and the iodine value by several hundred per cent., and suffers by loss of aroma.

Tests for Adulteration

386.a and 387.a The advice given here is sensible, but I would leave the Sulphuric acid test for resins to the professionals.

If necessary I would use the alcohol density test and a simple melting point test when judging.

In the laboratory I have also used Thin Layer Chromatography to detect adulterants and this was confirmed using a Mass Spectrograph and Gas Chromatography in a University Research Department. JC

Alcohol 'spirit' (Ethanol or Industrial Meths).

H225 Highly flammable liquid and vapour.

Sulphuric acid (10%).

H290 May cause metals to corrode.
H314 Causes severe skin burns and eye damage.
H318 Causes serious eye damage.
H335 May cause respiratory irritation.

---

386. Testing wax for impurities is a difficult matter, even for the specialist. Any material addition of single adulterants is readily detected, but by a judicious combination the properties for which the chemist usually tests can be reproduced by a mixture of adulterants, some tending to give higher figures and others lower than normal on test. The author suggests that a careful study of the mechanical properties would be of considerable assistance in detecting expert adulteration.

387. It is useful, in making most of the following tests, to test at the same time a piece of wax of known purity for comparison.

Beeswax when fractured shows a fine-grained dull surface; when cut with a knife it shows a glossy surface; when chewed it does not stick to the teeth. When handled until plastic it moulds readily and pieces readily weld together, showing no cracks on further working.

It has a characteristic aroma and taste.

A material addition of resins gives a glossy sheen to the fracture. Resins can be dissolved out in cold alcohol.

The presence of resins is detected also by melting wax in a little water containing, say, a tenth part of sulphuric acid. The resin produces a red colouring with the acid, which is stronger the more resin is present.

Hard fats reduce the density, which may be detected by comparison with a good sample.

Float the good sample in a small quantity of water and add spirit slowly until the sample just sinks. A sample of lighter density will then float in the same mixture.

388.a Carbon tetrachloride has been banned as a dry cleaner as it has ozone destroying properties. The two common solvents for beeswax (carbon tetrachloride and chloroform) are both very difficult to obtain. JC

Carbon tetrachloride (tetrachloromethane).

H301 Toxic if swallowed.
H311 Toxic in contact with the skin.
H331 Toxic if inhaled.
H351 May cause cancer/genetic defects.
H372 STOT RE1 Prolonged exposure may cause damage to body organs.
H412 Harmful to aquatic life.
H420 Harmful to atmosphere (ozone).

Turpentine.

H226 Flammable liquid and vapour.
H 302 Harmful if swallowed.
H304 May be fatal if swallowed and enters airways.
H312 Harmful in contact with the skin.
H315 Causes skin irritation.
H317 May cause allergic skin reaction.
H319 Causes serious eye irritation.
H332 Harmful if inhaled.
H411 Toxic to aquatic life with long term effects.

Alcohol (Ethanol or Industrial Meths).

H225 Highly flammable liquid and vapour.

388. *Wax dissolves completely in carbon tetrachloride, which is therefore a good cleanser for wax-soiled articles. When dissolved any solid impurities will be found in the clear liquid.*

Wax dissolves completely also, but slowly, in turpentine, leaving no deposit.

Solid impurities, flour, colouring substances, etc., will separate out if the wax is melted in a water bath and allowed to cool slowly.

A little wax dropped on a hot shovel so that it fumes, gives off the characteristic smell of sulphuric acid if acid has been improperly used in its preparation, or the characteristic smell of burning fat (acreoline) if it contains fats.

Dissolve about 24 grains of wax in 1 oz. of warm alcohol and filter into a 2-oz. vessel. Add an equal quantity of water to the filtrate and insert a piece of blue litmus paper. It should take about ¼ hour to turn red, and the liquid should become opalescent.

Full instructions for the detection of carnauba wax are given in the American Bee Journal for March, 1930.

Bleaching Wax

389.a Most beekeepers stick with using sunlight to bleach beeswax. The earliest methods involved laying out sheets of wax in the sun. The method suggested here (pouring a stream of wax into cold water so that small pieces are formed) is a good one. The small pieces of wax are then dried and hung up in a plastic bag in the sun. Fresh wax is brought to the surface by hitting the bag as you go by! Vinegar can still be used in hard water areas to 'soften' the water.

Most chemical methods of bleaching destroys the aroma as well as the colour. JC

Chloramine.

H315 Causes skin irritation.
H319 Causes serious eye irritation.

Vinegar. (contains acetic acid)
H320 May cause eye irritation.
Hydrogen peroxide (depends on % strength).

H271 May cause fire or explosion: strong oxidiser.
H 302 Harmful if swallowed.
H314 Causes severe skin burns and eye damage.
H332 Harmful if inhaled.

389. Beeswax is bleached by exposure to the sun. To expose a large surface, the wax is melted and poured as a fine stream into a large vessel of cold water so that it solidifies as it flows. Unless rainwater is available, a little vinegar should be added to prevent the lime in the water combining with the cerotic acid in the wax. Add a teaspoonful to the pint or 1 oz. per gallon.

Wax may also be bleached with chloramine, which is powerfully disinfectant, without detriment to its use in making up foundation.

Hydrogen peroxide, animal charcoal and ozonized air are also used in wax bleaching, as well as more powerful and less harmless chemicals.

Wax Rendering

General

390.a_Sulphuric acid should not be used as it changes the nature of beeswax. In a steam extractor the water does not come into direct contact with wax so hard water may be used. JC

Sulphuric acid (10%).

H290 May cause metals to corrode.
H314 Causes severe skin burns and eye damage.
H318 Causes serious eye damage.
H335 May cause respiratory irritation.

---

390. *Water is used in wax rendering and unless clean rainwater, or other soft water, is obtainable, it is necessary to add acid to prevent the lime in the hard water from combining with the cerotic acid in the wax, which would lead to waste of wax and of labour in separating the spoiled portion. For this purpose vinegar is convenient, and at the rate of 1 oz. per gallon. Sulphuric acid is sometimes employed, at the rate of ¼ to 1 oz. per gallon, but is said to affect the properties of the wax to disadvantage. When the water is comparatively free from lime less acid may be used. The water should be so treated both for use in the vessel in which the wax is melted or separated and any into which it is allowed to pour to solidify. Where the steam alone comes in contact with wax, hard water may be used without acid, as the lime is not carried over in the steam.*

391, 392 and 393.a Avoid using all metals except stainless steel.

Use pyrex glassware and high melting plastic. Use a water bath wherever possible for heating wax to control the temperature. Learn how to use a steam extractor if rendering cappings and old comb – but keep them separate. Avoid heating your wax above 75 deg.C as it will discolour.

Keep a fire extinguisher handy when heating wax. It is highly flammable. JC

391. *To keep the colour good it is necessary to use vessels of stainless steel, aluminium or nickel. Subject to the comments below copper can be used, or tin, provided with the latter that the tin coating is intact throughout as iron rust is peculiarly bad in its effects.*

Porcelain, glass, glazed earthenware, enamelled ware and wetted wooden vessels may be used to receive the hot wax. If acid is used in tin-coated vessels the tin should first be rubbed with mutton fat.

Zinc is attacked by hot wax. White wax may even be discoloured by dark honey.

392. *Wax acquires a blackish brown colour in contact with iron rust. Zinc and galvanized ware cause an olive brown stain, brass a greenish colour, and even copper some orange discoloration, but this latter can be reduced if the wax is re-melted and granulated by pouring into cold water.*

393. *Where much wax is handled, it should be sorted, as the lighter sorts fetch more money. Clean wax of excellent quality is obtainable from cappings, from virgin comb and from recent comb in which no brood has been raised.*

Wax Extractors

394. A solar wax extractor will serve for the small bee-keeper and for saving odd scraps in a large apiary. It consists essentially (see Catalogues) of a wooden box with double-glazed cover, the two sheets of glass being separated by an air space. The box is provided with a metal tray for combs, generally of tinned iron, which slopes to let the wax run down. A grid is furnished to stop the dirt and comb following the wax and a vessel in which the dripping wax accumulates. A screen is put over the vessel to catch the dirt. All odd bits of comb are thrown in as found.

To improve the colour the cake of wax may be put through the solar wax extractor two or three times, removing of course all debris and cleaning the container between each operation.

395. For commercial work on a large scale, large steam extractors are employed, also large presses, and even chemical solvents to secure the last remains, the solvents being evaporated and recovered by condensation; but there are several devices and methods suitable for general use on a moderate scale.

396. A hot wax-press will remove nearly all the wax from old, dark, dry combs, but a steam extractor may leave nearly a third, the cocoons, etc., in the old comb holding back a considerable quantity.

Before rendering combs, break them up and soak for 1 or 2 days in soft water. This will saturate the cocoons, pollen, etc., and much less wax will be left in the debris. The crushed combs may then be placed in a weighted canvas bag and boiled in water, the wax rising to the top. The debris from bag or steam extractor will still contain wax, some of which may be extracted in a press if there is a large quantity to be dealt with, but if the combs have been well soaked before heating there will not be so much waste. A steam extractor makes the most complete job, with a press to follow if the quantity justifies this.

397. If it be desired to save a wired frame complete with its wiring, the comb may be melted by pouring boiling water over it. The bee-keeper will readily devise a wire mesh tray to hold the comb and allow the wax and water to return to the boiling vessel. A large spoon or other dipper is used to apply the boiling water, and a guard keeps one corner of the vessel clear of wax for dipping.

398. The rough cake first formed may have debris on its under side. This should be scraped off and accumulated for re-melting and straining, but if rain or acidulated water be used, there will not be so much to remove. Sometimes some of the wax on the under side gets water-sodden and granular and is mistaken for refuse. It may be nearly pure wax and should be re-melted with the next lot of combs. A scum sometimes forms on the melted wax which will sink if blown to the side of the vessel.

399. Where a steam-heated wax press is not available a press as used for heather honey may be employed (see maker's Catalogue for apparatus). In applying canvas, use it so that the part near the bottom plate or exit plate is free from folds and all is well wetted so that the canvas will not absorb much wax.

Wax Cappings

400. One to two pounds of wax cappings will be obtained from every 100 lb. of honey uncapped. Even if the cappings are stirred and well drained, say one per cent. of the honey will remain on the cappings.

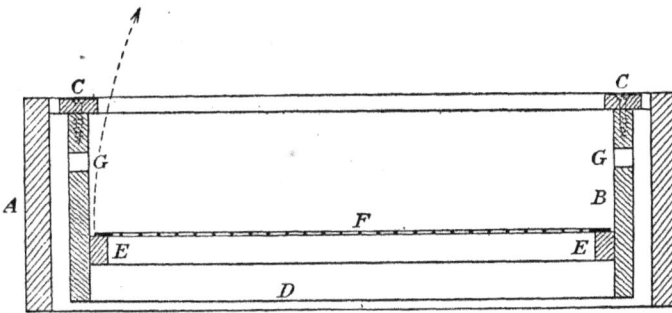

FIG. 3.—CAPPINGS BASKET.

A is a super, receiving the box B, 15 ½ X 14 X 5", carried by cross pieces C, C, 17 X 1 X 3/8 resting on the frame supports in A. The bottom D is of metal with bee-space below. On an internal rim, E, E, rests a framed wire screen, F, of 1/2" mesh wire-netting. The bees enter through the holes G, G, of which there are several on either side. The wire frame is lifted out as indicated.

The bee-keeper with only a few hives may give the cappings to the bees to clean up, putting them in a super in some kind of basket with coarse gauze bottom and a tray below to catch the wax particles as the bees break them up. A convenient form is shown in Fig. 3. A bee-space is provided below the tray and the bees gain access to the basket through the sides. A bee-tight cover must be provided, and the cappings should be removed after, say, two days.

Alternatively the drained cappings may be washed in warm water, the washings being used for making syrup. Again the broken cappings may be put in fine gauze baskets in the extractor, if in quantity.

For large scale operations the cappings may be pressed, or again melted over a water-bath or steam bath, the honey collecting below, but the honey suffers some loss of flavour and may be suitable only for feeding back unless the honey is quickly cooled. Special apparatus may be purchased, arranged for continuous working.

Wax Moulds

401. Moulds of convenient size are made both of metal and of hard wood. The latter should be soaked in soft or acidulated water (**390**) before use. The former should be lubricated to prevent the wax sticking. A little soft soap may be used, or glycerine, or flour paste.

Never overheat the wax and always let moulds cool slowly, preferably by standing in hot water, well covered over.

Wax for Exhibition

402. To produce an exhibition article the whitest cappings alone should be used. The wax cake may be improved in colour by passing through the solar wax extractor. For moulding, the cake should be melted finally, avoiding overheating, and the wax passed through a filter of white flannel or other fine white fabric, which may be secured over an aluminium funnel. The mould should have sides nearly vertical so as to avoid making a brittle edge, and should be lubricated with yellow soap, rubbed off to a polish. The wax should flow in gently so as to avoid making air bubbles. Any air bubbles appearing should be broken while soft by holding the heated bowl of a spoon near them. The mould should be supported in a large vessel of warm water, hot enough to melt the wax, and when the wax has cooled so as to set on top, water may be scooped up and poured in gently, then the whole covered and allowed to cool slowly. The wax will float up when cold. All water used should be clean rain-water or acidulated with white vinegar. The cake may be kept in light-coloured honey to retain the aroma until required for exhibitions. Needless to say, exhibition wax is not a commercial article.

Judging Wax

403. The quality of wax on exhibition is judged mainly by appearance and colour, also by aroma and texture. The following points have been used in the places named:

TABLE IX

POINTS IN JUDGING BEESWAX

Quality.	Australia.	Irish Free State.
Colour	25	20
Appearance	20	20
Aroma	10	10
Texture	–	10
Transparency	25	–
Tenacity	20	–

Propolis

Use of Propolis

404. Propolis is the name given to a resinous substance used by the bees as a lute for sealing cracks and preventing air leakage and the ingress of damp through porous parts of the hive and for smoothing rough places. It is used also to strengthen combs both at the point of attachment and in the corners of the cells. Carnolian bees use mainly wax as a stopping, a habit inherited by some hybrids. Some bees seem to prefer very hard and highly coloured resins.

Source and Colour

405. Propolis is collected by the bees from trees, shrubs, etc., exuding a resinous substance and is a cause of coloration of beeswax. That from poplar imparts a bright yellow, that from flax a rich red brown. From the ngais tree in New Zealand, propolis almost black in colour is obtained, colouring the wax red. Other common sources are the elder, birch, chestnut, pine and willow, but these are but a small selection of the sources visited. Other colours are scarlet, green and brown, and there are many shades between. Bees will also gather and utilize resinous, gummy and balsamic matter from many other sources; turpentine, pitch, bitumen, varnish, boiled oil, etc., as opportunity occurs.

Propolis is carried in pellets in the pollen baskets on the hind legs in a similar manner to pollen. Some races gather large quantities, others use but little (see **116** *et seq.*).

Propolis and Wax

406. Propolis as collected from bee-hives, by scraping frames, hive-bodies, etc., always carries with it a certain amount of beeswax, say 15 to 60 per cent. The propolis proper consists mainly of resins with melting-points from, say, 140° F. (60° C.) or less to 212° F. (100° C.) or more, the higher figures generally predominating.

On melting together beeswax and propolis they combine to a certain extent. If allowed to cool slowly some separation occurs, but the waxy portion is found to have changed in character, being tenacious and adhesive, but more plastic than pure beeswax.

407. In wax rendering by steam from old combs, the inclusion of a certain amount of propolis in the wax is inevitable, and common commercial beeswax always shows a certain amount. The effect is to reduce the melting-point from, say, 150° F. (67° C.) to 147° or 145° F. (64° or 63° C.).

The lowering of the melting-point is of no importance in itself as under working conditions the highest temperatures are 40° to 50° F. below the melting-point, but loss of strength at

temperatures of 95° to 105° F. is a more serious matter. This aspect appears to have received very little attention, save that in commercial wax rendering for foundation making, steps are taken to remove excessive resin.

Removing Propolis

408. Propolis is readily soluble in both alcohol and petrol, and these may be used for cleaning purposes. When handling frames, much propolized, the finger-tips should first be lightly vaselined, or talc powder may be employed. Lard may also be used to soften propolis before washing the hands with soap and water. (See also **384** and **870**).

78

SECTION IV
HONEY-PART I

Crops and Honey Gathering

Times, Distances and Temperatures of Flights

409. Bees engaged on a particular duty (**23-5**) such as nectar from a particular source frequently pursue that duty for many days. The following are representative times of journeys from exit to time of return to the hive:

Water-gathering	say 5 minutes
Pollen-gathering	say 12-16 minutes
Honey-gathering	say 35-45 minutes

After return, the bees spend some time in the hive. Longer times are taken under unfavourable conditions, for example, if there is no suitable water source near the apiary; there are generally long waits in the hive between journeys, sometimes lasting for some hours. On the other hand, during a heavy flow, harvesters may remain at rest in the hive for only a few minutes while unloading between flights. In fine weather also bees will sometimes rest outside the hive, being absent for 2 or 3 hours per trip.

410. Bees will fly about 15 to 25 miles per hour, but if there is a wind exceeding 15 miles per hour, few bees will fly. Bees have been observed, however, chasing swallows, which shows what they can do when roused.

411. Bees have been known to fly 6 to 8 miles to an isolated but very favourable source of honey or pollen, but they will not, as a rule, fly more than about 2 miles. The economical distance for honeygathering is about 1 to 1½ miles, that is to say, unless they can secure their supplies within an average radius of about a mile, undue wear and tear and risk is involved in gathering and the returns will begin to suffer. The economical radius for pollen-gathering is very much less, as is that for gathering nectar in the spring. Bees will carry pollen long distances, but in placing hives in orchards it is found necessary to furnish a hive per acre to secure the most effective pollination. In the spring, bees will not fly more than about 50 yards if they can get pollen nearer home, and if hives are placed not more than 70 yards apart each way, this will give a maximum radius of 50 yards to the centres of the squares so formed.

Some hardy bees, notably the dark races, fly at lower temperatures than do those from warm climates. Flying may be seen at shade temperatures of 45° to 50° F. (7° to 10° C.). At such temperatures, if the bee lands in the shade and reaches shade temperature, it is likely to be chilled so that it cannot return. In cool weather, bees return rapidly to the hive when the sun becomes covered with cloud, even though there be no threat of rain (see Feeding, **1265** et seq.)

Bees have actually been observed flying and returning with pollen in shade temperatures down to 42° F. (4° C.) and for water down to 33° F. (½° C.) but under exceptional conditions.

412. Nectar can be gathered at any temperature at which the bees can fly, but there is very little to gather at the lower temperatures. A normal colony will not increase in weight in the spring until the maximum shade temperature is, on an average, in the neighbourhood of 63° F., or, say, 65° F. in the summer. Above this limit, the daily increase is greater the greater the excess temperature, but the rate, of course, is dependent upon the presence of honey-bearing plants in season. In very bleak regions, the limiting temperature is lower. Probably, however, a day temperature of about 80° F. with cool nights after a rainy spell gives the heaviest takings, especially on a light soil. On a clay soil a high night temperature is helpful.

A good flow is obtained from fruit trees at a much lower temperature than is necessary with clover and heather. Sainfoin gives good results at lower temperatures than white clover and the brassicas at still lower temperatures. Clover needs about 73° F., and 83° to 86° F. for a really good flow (23° C. and 28° to 30° C.). Humidity is more important than temperature; if it is low, not only is the flow greatly reduced, but in very dry weather the nectar in certain sources, e.g. lime, becomes granulated and the bees are hindered from gathering it. Sugar collection by the bees is more efficient if the sugar content of the nectar is neither too low nor too high. Bees are not much interested in sugar contents below 20 per cent. and it is likely that the optimum value is in the neighbourhood of 56-60, depending in part upon the distance of the source and increasing with distance. A good harvest is taken when there is sufficient temperature and a wide range of temperatures during the day and night. The latter by varying the strength of the nectar secretion helps to secure that the sources are brought within the range corresponding to efficient collection for some part of the time. In summer and early autumn, temperatures are usually high enough, and it is this varying humidity that is most important. In the spring both temperature and hours of bright sunlight are liable to be inadequate.

In general high humidity increases nectar flow, and the daily increase in hive weight, but it is likely that the amount of sugar brought in, i.e, the ultimate amount of honey stored, is not on the whole increased by high humidity unless the sugar content of the nectar is already above optimum. Where the nectar is too thin low humidity of the air, during part of the day, must help.

Daily, Climatic and Seasonal Changes in Nectar Flow

413. The principal nectar flows occur between, say, 10 a.m. and 1 o'clock and between 3 and 6 ("summer" time one hour earlier). These flows overlap during a good flow. The time varies with the source and humidity as well as with temperature. Rain is bad as it washes away the nectar. Wind is bad if it hinders flights (see **410.**) (For Temperatures, see **411**).

414. Seasonal changes are of great importance considered in relation with the normal flowering periods of different crops. The bee-keeper should ascertain by inquiry and observation the principal sources and their seasons in his locality **(455)**. Crops and weather reports should be watched and the situation judged. There is no fixed order of appearance of flowers. For example, in an early year, apple is followed by hawthorn in Great Britain, and that by sycamore, whereas, with a retarded spring, the honey flow from these sources may be almost simultaneous. In one year, fruit and clover may give a large return, but hawthorn, sycamore and lime give hardly any owing to the vagaries of the weather; or vice versa. The next year, apple, hawthorn and sycamore may give a heavy yield nearly together, the apple and hawthorn blend having an almond-like flavour.

415. Between crops, colonies built up rapidly on the earlier crop may starve before the later one comes. The flow of pollen is equally important for building up stock and maintaining stamina. There is always some pollen with the honey flow, but in some districts early pollen is hard to come by, and it may be necessary to plant specially to secure it (for sources of pollen, see **455**)

416. In studying the weather and weather forecasts, it is useful to note that a blanket of clouds acts somewhat differently in summer and winter. In the cold months, it is conducive to warmth, and a falling barometer ushers in not only rain but warm air which has travelled at a low altitude. A fine sky then represents low temperatures. A clear sky with dry air portends a frosty night. In the warm months, on the other hand, a blanket of clouds reduces the action of the sun and leads to a general reduction of temperature, at the same time hindering radiation at night and keeping the temperature more uniform. Thus signs·of rain are welcome in the early spring, but not in the summer.

Weight of Honey Taken

417. During a good flow, hives should increase in weight several pounds a day. The gross increase in one day may be much more than this but a large part of the water taken in, in the nectar, will be evaporated within the first 36 hours. Based on hive weight alone an increase of 35 lb. in a day has been recorded, from purple sage, in U.S.A. In Great Britain, Grant, of Cheltenham, has noted 17 lb. Doolittle recorded taking 66 lb. in 3 days, which cannot be readily compared with the figures just given. Increases of 10 lb. per day are not uncommon, weighing at the same time each day, and thus allowing for what is evaporated at night. Having regard to the heavy takes of honey in **419-21** much larger daily figures must have been reached.

418. Small bee-keepers able to show an average take from season to season of 50 lb. per colony extracted honey, or 30 sections, are doing well, but for commercial work it is essential to choose good districts. With good management considerably higher averages can be reached from year to year, and colonies may well average 100 to 170 lb., according to the district, except in extreme northern regions. In Vancouver, 80 lb. would be good, and less in the Kootenays.

419. Individual bee-keepers have recorded some extraordinary takes from individual single colonies, that is, colonies having a single queen and receiving no brood or bees from other colonies. This last condition is not easy to ensure, as the end hive of a row frequently receives a considerable accession of bees from other colonies. However, R. A. Morgan, of Vremillion, South Dakota, claimed 803 sections from one colony (Caucasian) in 1898. In 1925, he had a record of 616 sections.

In 1921, C. B. Hamilton, of Michigan, reported taking 577 finished sections from one colony of American Italians, and in 1929, F. Marquette, of N. Dakota, reported 593 lb. honey from one colony of Caucasian bees. W. B. Wright took 496½.

420. Brother Adam, of St. Mary's Abbey, Buckfast, S. Devon, England, made a remarkable record for Great Britain in 1925. He had 39 stocks. A heavy flow of clover was on and a stock was relieved of about 1 cwt. and fresh supers of drawn comb piled on. The flow continued for 5 days and then the weather broke and it commenced to rain and the stock was not examined until the tenth day, when 135 lb. of sealed honey was taken in the supers, and it is reasonable to suppose that substantially the whole of this was taken in 5 days, or at the average rate of 27 lb. per day. It is noteworthy that out of the 39 stocks, seven or eight others did nearly as well.

421. Australian records, however, put all others in the shade, enormous crops being obtainable from the honey-bearing forest trees. An *average* weight of 610 lb. per hive has been obtained in a good season in an apiary of 66 hives.

In Victoria, one colony has been known to collect 824 lb. in the Eucalyptus region and another 807 in an apiary showing 480 lb. average. In Australia, however, bees have been seen flying by moonlight.

Weight of Honey in Comb

422. This depends in part on the density, but may be taken roughly at 7 1/3 lb., per square foot per 1 inch thickness face to face. On this basis and allowing bee-spaces between, a super of combs full of honey would show the average maximum weights in Column A below, the weight of honey alone being, say, 4 per cent. less:

TABLE X

Weight of Combs of Honey

Make of Frame.	Col. A.	Col. B.
	lb.	lb.
British Standard Frame, metal ends	6¾	5½
British Deep Frame, metal ends	9½	7½
British Shallow Frame, metal ends	4¼	3½
British Shallow Frame, wide ends	5½	4½
Langstroth Deep Frame, 1³/₈ spacing	8¼	6½
Langstroth Shallow Frame, 1³/₈ spacing	5¼	4¼
M.D. Deep, 1½ spacing	11¼	8½

but such combs are seldom seen, and the average weight of honey taken would be about 80 per cent. as in Column B above.

Yield per Acre

423. An apiary in a fair district may be built up to, say, 100 hives before the takings per hive appreciably suffer. In a good season at 100 lb. per hive and 1½ miles radius, this represents an average annual surplus yield in the district of 9 lb. per acre, and probably a total yield of, say, double this, allowing for what is used in brood production. There are no very reliable figures. According to an American bee-keeper, yields of 10 lb. per day per acre of clover or heather are not uncommon, however. Thus a bee-keeper can afford to offer free seed to a farmer to plant sweet or white clover near his apiary.

424. The following figures are taken from French sources. The yields from various crops depend upon the climate, the soil and the individual season:

TABLE XI

Yield of Honey per Acre

Plant.	Yield, lb. per Acre.
Buckwheat	95-150
Sainfoin	125
Lucerne	18-22
Trifolium inc	14-22
White Mustard	10-13
Alsike	9

Yield from Use of Drawn Combs and Foundation

425. It is estimated by experienced bee-keepers that for every 100 lb. surplus obtained from bees building their own combs from starters, a further 30 lb. may be expected if foundation is used in all supers or 60 lb. if drawn combs are used throughout. Put the other way round, a bee-keeper able to take 100 lb. surplus with drawn combs will lose about 20 lb. if he has to furnish foundation throughout, and nearly 40 lb. if the bees have to make their own combs as required. This does not mean that in gathering and storing 60 lb. of honey, another 40 lb. is used in wax production. Some of this 40 lb. is honey not gathered, and most is the cost of producing and maintaining a large number of bees which might be gathering, but are occupied in comb building during the honey flow. In 100 lb. of honey in the comb, there are, say, 3 to 4 lb. of wax by weight, which at 10 lb. honey per pound of wax produced represents another 30-40 lb. of honey to be consumed for its making. In using drawn comb to receive it probably 5 lb. of honey is enough for repairs and sealing, while with foundation, say 25 lb. would be wanted and 40 lb. with starters only. These figures are indicative only, but highly suggestive, as they clearly show that it must pay well to provide foundation and to save all drawn combs most carefully, also to have new combs built only under the most economical conditions **(372)**.

426.a It is very difficult to estimate exact yields and statistics on honey consumption due to geographical and climatic variables. An up to date estimate of some numbers in Para 425 - 430 may be calculated by the fact that in 2010

£0 0s 6d from 1945 is worth

£0.84 using the retail price index and £2.75 using average earnings

Currency inflation can be calculated from the web at

http://www.measuringworth.com/ppoweruk/
or http://www.nationalarchives.gov.uk/currency/
and for exchange rates www.xe.com/currencyconverter

These should be used together with the catalogue prices from a major beekeeping supplier. JJB

426. *For every six pennyworth (12 cents worth) of honey stored in a drawn comb the wax in the comb has cost two pence (4 cents). An additional farthing will cover the cost of wax for sealing. This shows the high value of drawn combs. If the comb is to be sold with the honey the value of the wax has to be added to that of the honey. This partly accounts for the difference in value between section and extracted honey.*

427. The honey sold has to pay for all costs before there is a profit, but when the profit stage is reached, any further honey sold has only to bear the cost of handling and sale. The use of foundation adds to this profitable margin, and if the honey is being sold retail by the bee-keeper at 1s 8d. per lb., foundation is certainly worth 10s. per lb. to him, or if sold in large quantities wholesale at, say, five pence (10 cents), then the foundation is still probably worth 4s. (a dollar) per lb. Foundation for sections is worth considerably more than this.

Statistics of Honey Production

428.a So large an amount of honey is produced by private beekeepers for home and local consumption, the amount of which is never recorded, and the methods of compiling statistics are so varied, that it is difficult to give any accurate picture of honey production. There has been an increase of honey production during the Great War. The following pre-war figures are likely to be exceeded in peace-time

It is safe to say that the annual world production of honey by bee-keepers employing modern methods exceeded 3,510.97 million lbs per annum and probably reaches £17.5 billions in value. There were probably more than 15.7 million beehives in the EU alone worked by around 631,230 beekeepers together owning equipment worth more than £7.85 billion. and of wax was about £8 wax figures still to be confirmed. The UK's National Bee Unit in York says there are 272,000 hives on the UK – a figures the Bee Farmers Association would challenge. Our members account for around 45,000 hives. MG

428. *So large an amount of honey is produced by private beekeepers for home and local consumption, the amount of which is never recorded, and the methods of compiling statistics are so varied, that it is difficult to give any accurate picture of honey production. There has been an increase of honey production during the Great War. The following pre-war figures are likely to be exceeded in peace-time*

It is safe to say that the annual production of honey by bee-keepers employing modern methods exceeded 400 million lb. per annum and probably reaches £10,000,000 in value (say $50,000,000). There were probably more than 7 million movable frame hives in use by more than 1 ½ million bee-keepers, together owning equipment worth more than £20,000,000. Wax production by these bee-keepers probably exceeded 10 million lb. per annum, not including waste, and fetched about £850,000. In addition, very large quantities of wax are obtained from wild bees and from bees kept by primitive methods, and some honey also, so that it is likely that the figure given for wax should be doubled to obtain the production from all sources and, say, 10 per cent. added to the figure for honey. The average market value of good honey was under £3 per cwt. and of wax was about £8.

429.a Figures produced by FAO and reported by the European Commission in 2014 show the major honey producers are as follows China 452,000 tonnes, The European Union 188,000 (Austria, France, Germany, Great Britain, Spain, and Sweden each between 2 to 4 million lbs) Turkey 88000, Argentina 76000, Ukraine 70000, America 67000, Russian Federation 65,000, India 61000, and Mexico 59000. MG

429. *America headed the list with an annual honey production of probably 200 million lb., Russia was stated to produce nearly 100 million lb., Canada came next with over 30 million, Australia 9 million, New Zealand 6 million, Italy 5 million, and Austria, France, Germany, Great Britain, Japan, Spain, Sweden, Switzerland 2 to 4 million lb. each, and others not stated.*

430.a In Great Britain the local, market value of good honey was nearer £6 per pound locally, the highest price being obtained for fine Scottish heather honey in the comb. The EU imported 149,641tons in 2012. MG

430. *In Great Britain the local, market value of good honey was nearer £6 per cwt., and the best honeys fetched from 1s 6d. to 4s. per lb. locally, the highest price being obtained for fine Scottish heather honey in the comb. Great Britain imported nearly 8 million lb. annually and exported about one-fifth of this quantity, against a home production of, say, 3 million lb.*

Composition and Properties of Honey

Water Content and Density of Nectar

431. Bees gather nectar from flowers, nectar being a watery solution of sugars with small quantities of salts, mainly phosphates, fats, albuminous matter, aromatic oils, etc. **(442)**, varying much in density in the neighbourhood of 1.12 to, say, 1.30 at normal atmospheric temperature, but sometimes as low as 1.05 from a rapid flow during damp weather, and as high as 1.40 in very dry weather. These figures represent nectar gathered. Nectars may be found showing a wider range of sugar content than here indicated.

432.a   The sugar content of nectar is primarily a combination of sucrose, fructose and glucose. Bees add enzymes to nectar to process it. Invertase is added to process sucrose into glucose and fructose, diastase in the nectar will starches into sugars and glucose oxidase is added so that during the processing of nectar any yeasts that might be present are killed. Glucose oxidase in the presence of glucose and water will produce hydrogen peroxide, a powerful antisceptic. Bees will reduce the water content to less than 50% by 'stropping', that is regurgitating the nectar onto their mandibles to evaporate some of the water content. The nectar is then spread over the top of vacant cells and the natural low humidity and high temperature in the hive will futher reduce the water conent to below 20%. At this stage the bees will pack the honey into cells and, once filled, seal the cells with a thin wax capping. ID

432. *The sugar content is mainly sucrose (or cane sugar) but is "inverted", i.e. converted into dextrose (grape sugar) and levulose (fruit sugar), by the bees by certain additions made to the nectar, the change being aided by heat. The modified nectar is spread over a large area of comb to expose a large surface for evaporation and is concentrated mainly by evaporation of the water assisted by vigorous ventilation (732). The nectar is moved into a smaller number of cells during concentration and during moving further additions are made. After a period of some days, the liquid (now honey) is sufficiently concentrated and is sealed over.*

433. If concentrated from, say, 1.12 density the volume is reduced in the ratio of about 4 to 1, whereas with a nectar density of 1 30 the reduction would be less than 3 to 2 in volume to bring the density up to that normal for honey. The change of volume is commonly 2 or 3 to 1.

434. The original moisture content varies with the source and the atmospheric conditions. The final moisture content varies to some extent with the source, but to a small extent with the atmospheric conditions also. Honey from fruit-trees and lime-trees (bass), for example, gathered quickly, in a moist atmosphere, is likely to be sealed with an excess of water, introducing a risk of fermentation.

Water Content, Refractive Index and Density of Honey

435. It is difficult to find reliable data on the moisture content of honey in relation to its density and composition. Some of the water requires a temperature considerably above the boiling-point of pure water to drive it off. The basis of figures and even the density scale used and temperature in question are seldom given by writers, but it is common practice to use "sugar" tables for the purpose of estimation. The figures may be corrected according to the proportions of dextrose and levulose present, but in fact there is frequently present a material quantity of salts and other matter sufficient to introduce errors into the figures. The author would welcome reliable detailed information from his readers. Meanwhile, the following approximate information may be given.

436. The moisture content of sealed honey, ungranulated, varies from, say, 14 to 22 per cent., but fruit honey gathered rapidly under high humidity may be sealed with as much as 27 per cent. water content. During granulation, water is first given up by the dextrose and mixes with the levulose content, which remains liquid, and this liquid portion, therefore, loses density and is in danger of fermentation (see **507** and **518**), hence the importance of seeing that honey is properly ripened, and especially honey that is expected both to granulate and to keep. Average densities do not have much meaning without reference to the honeys averaged.

A certain amount of honey sold, the world over; is liable to ferment unless retailed and consumed promptly. Honey taken reasonably ripened, and intended to granulate should not have more than 20 per cent. moisture, at 60° F. when extracted. This corresponds to a density of 1.415. In Great Britain honey bottles are designed to give correct weight at this density. Densities up to 1.440 and even 1.450 are reached, but even the latter would correspond only to 2½ per cent. overweight in a British bottle filled to the mark. Note, however, the light-weight honeys referred to in the first paragraph and check the weight for such.

In America lower densities have been allowed, and are satisfactory if the honey has been heat-treated to prevent granulation as has been customary in America and some other places. In recent years and with better public assurance against the use of substitutes, fine granulated honey is coming into increasing favour.

437. Honey exposed to contact with moist air readily absorbs water. A typical ripe honey will retain its normal content in contact with air having a relative humidity of 58 to 60 at normal room temperatures. At lower humidities (i.e. with drier air) it gives up water. At higher humidities it absorbs water. In Great Britain the average humidity in living rooms is 65, and water will be absorbed by honey from the air.

On the other hand ling honey will withstand a very high humidity without absorbing moisture and if exposed in a normal atmosphere seems to become drier. The *thixotropic* (jellifying) property of ling honey has been proved to be due to the presence of about 1½ per cent. of a certain protein which, transferred to clover honey, will make it also *thixotropic*.

438. The water content of honey may also be determined from the refractive index, using the following relationship:

$$\text{Water content per cent.} = \frac{1.538 - \text{refractive index}}{0\cdot0025}$$

The temperature coefficient of the refractive index is said to be $0\cdot00023$ per $1°$ C., the refractive index falling with increase of temperature.

Changes with Composition and Temperature

439. Most of the sugar in honey (see Analyses, **444**) is in the form of dextrose and levulose, the latter having the higher density. Honey from fruit-blossoms generally has a lower density than clover honey, but for the same water content, it has a higher density as it contains a larger proportion of levulose and smaller proportion of dextrose than does clover honey. For the same water content representative honeys rich in levulose run to densities about 0.003 higher than are given in the following table, while those rich in dextrose run to 0.003 lower. Comparison of the densities given for 20° C. and 15 ½ ° C. show a change of approximately $0\cdot003$ for 4½° C. or 8° F. It will be seen, therefore, that for accurate work in estimating water content, it is necessary to control the temperature of the honey and to know something of its analysis. Within the limits of the table a change of 1 per cent. of water corresponds to a change of 0.006 to 0.007 in density, thus if water content be estimated from density in an ordinary room without regard to temperature or source, it is easy to make a mistake of + or - 1 per cent. figured on the total weight, or 3½ to 7 per cent. on the water weight. If, however, the temperature be noted a correction may be made directly on the water content, of 1 per cent. for every 9° C. or 16° F., remembering that the density decreases as the temperature increases. By attending to temperature alone, neglecting source, the error is likely to be halved at least.

440. In the following table, columns are included to show readings given by the Baume hydrometer. Here again caution is necessary, as there is no universal standard for these instruments. The tables are based on a modulus of 145 now widely used, and standard in America for some years, but these instruments have been calibrated with a modulus as low as 144 and high as 146.8, giving proportionally lower and higher readings.

TABLE XII

WATER CONTENT AND AVERAGE DENSITIES OF HONEYS

Water Per cent.	Density, grams per c.c. at 20° C.	Density at 60° F. (15 ½° C.).	Baume at 20° C. M = 145.	Baume at 15 ½° C. (60° F.). M = 145.
15	1.445	1.448	44.9	45.1
16	1.439	1.442	44.4	44.6
17	1.432	1.435	43.9	44.1
18	1.425	1.428	43.4	43.6
19	1.418	4.421	42.9	43.2
20	1.412	1.414	42.5	42.7
21	1.405	1.407	42.0	42.2
22	1.398	1.401	41.5	41.7
23	1.392	1.395	41.0	41.2
24	1.385	1.388	40.5	40.7
25	1.379	1.382	40.0	40.2
26	1.372	1.375	39.5	39.7
27	1.366	1.369	39.0	39.3
28	1.360	1.363	38.5	38.8
29	1.353	1.356	38.1	38.3
30	1.347	1.350	37.6	37.8

Correction + .003 maximum for levulos in excess.
Correction − .003 maximum for dextrose in excess.

Use of Hydrometer
441. In using a hydrometer not calibrated to read density directly it is necessary to have a reliable table for converting readings to density. As seen by comparison of the last two columns above, the change of density with temperature within the range of the table is a decrease of approximately 0.003 for 4 C. rise, or 0.001 per 1½ degrees. Thus if a hydrometer scaled for use at 20° C. is used at 23° C., the rise of 3 degrees necessitates a correction of 0.002 which must be subtracted. If used at 17° C., 0.002 must be added to the reading. With the Fahrenheit scale, the correction is approximately 0.003 for 8 degrees.
The following short table gives an idea of the densities of diluted honey solutions:

TABLE XIII
DENSITIES OF HONEY SOLUTIONS AT 20° C.

Water. Per cent.	Sugar. Per cent.	Density.
10	90	1.480
20	80	1.412
30	70	1.347
40	60	1.286
50	50	1.230
60	40	1.176
70	30	1.127
80	20	1.081
90	10	1.038

Chemical Analysis

442. Published analyses more frequently give the source of the honey analysed by country of origin than by the botanical name, so there is a lack of information on the differences between honeys from different flowers. In giving values for the other constituents a uniform water content of 20 per cent. is here assumed for convenience. The figures given are representative and indicative. The figures for content other than water necessarily depend upon the water content. If this is reduced or increased, while the relative proportions of the sugars, etc., amongst themselves will not be affected, their percentage values will be affected to an extent corresponding to the difference in water content. Analyses of numbers of commercial honeys show average water contents below 18 per cent., although a figure as high as 27 per cent. may be reached, if not exceeded, in honey sealed over by the bees.

443.a Sucrose (cane sugar) is found in honey in small quantities, but tends to disappear with time, being converted to levulose (fruit sugar) and dextrose (grape sugar) by the action of heat in the presence of acids and found in the honey. From ½ to 2 per cent, of cane sugar is not uncommon in honey, but larger quantities are permissible. The pure food laws of the U.S.A. set a limit of 8 per cent. which, in fact, is high enough to allow for accidental adulteration, but probably low enough to secure against deliberate and serious adulteration. A limit is set also of ¼ per cent. for ash. KF

443. *Sucrose (cane sugar) is found in honey in small quantities, but tends to disappear with time, being converted to levulose (fruit sugar) and dextrose (grape sugar) by the action of heat in the presence of acids and found in the honey. From ½ to 2 per cent, of cane sugar is not uncommon in honey, but larger quantities are permissible. The pure food laws of the U.S.A. set a limit of 8 per cent. which, in fact, is high enough to allow for accidental adulteration, but probably low enough to secure against deliberate and serious adulteration. A limit is set also of ¼ per cent. for ash.*

444.a Levulose and dextrose are more commonly known as fructose and glucose. Higher concentrations of glucose in honey tend to make the honey granulate more readily and to produce finer crystals of sugar. The colour, aroma and flavour of honey is more due to trace compounds gathered from the nectar sources. ID

444. *The principal constituents of honey are the inverted sugars, levulose and dextrose. Levulose is generally present in the greater proportion, the amount generally lying between 33 to 40 per cent., with dextrose lying between 39 and 32 per cent., the increase in levulose being largely balanced by a decrease in dextrose. The darker honeys generally contain the more levulose. They flow more freely when of the same density, levulose solutions being less viscous than dextrose, and both are less viscous than cane sugar solutions.*

Equally important are the aromatic constituents, oils, gums, and other scented matters obtained from the flowers, of complex composition, which give the honey its aroma and distinctive flavours. Then, again, valuable assimilable salts of iron, phosphorus, lime, sodium, potassium, sulphur and manganese account for perhaps one-fifth or more of 1 per cent. Less important contents such as dextrine, albumin, waxes, fats, formic and malic acids, substantially make up the balance by weight, but complex digestive enzymes are present in minute quantities capable of converting sucrose to dextrose and levulose, starch to malts, etc.

Food Value and Medicinal Value

445. The protein found in honey is associated with the pollen content. The quantity is negligible from the standpoint of food value.

446. The vitamins in honey are also associated with the pollen content and are not sufficient to be of any importance in assessing the food value.

447.a   It is unwise to attribute and medicinal benefits to honey and this section is better considered as apocryphal or anecdotal evidence. However there is a small possibility that infants can suffer from infant botulism if fed honey and the advice is that honey should NOT be fed to infants who are less than 12 months old.

However, honey has been shown to be beneficial if applied topically (on the surface of the body). There is clear evidence that cuts and grazes heal faster and more cleanly if dressings containing honey are applied to the wound. It is believed that the natural antiseptic properties of honey aid healing and that honey placed directly on w wound help the natural scouring action to remove any foreign bodies in the wound. Some honeys claim better antiseptic properties due to naturally occurring antiseptic agents found in the nectar source. It is also clear that honey that has had minimal heating during processing from hive to honey jar will also be more effective as an antiseptic. ID

447. *The dextrose in honey can be absorbed directly into the human blood stream without further digestion, and the levulose follows with but slight modification, hence the great value of honey in stomach complaints such as ulcers and in malnutrition, especially with children, and in typhoid fever. Honey is a great producer of hemoglobin in the blood. It is useful in liver and heart troubles and does not produce gas. It is good wherever maltose or sugar of milk are indicated It may be used in baby feeding, for example, to advantage, for a nine months' child a teaspoonful being added to every 8 oz. of whole milk.*

As a stimulant use honey in hot water, adding orange juice.

448. One pound of honey of average composition has a heat value (when the honey is consumed and converted mainly to carbon dioxide and water) of about 1,450 calories per lb.

449. Honey from certain less important sources, such as the horse chestnut and dandelion, especially the first mentioned (a valuable source of pollen), is found to contain a much smaller proportion of sugar and more proteins. They are good bee foods for spring use, but poor as a source of surplus. When nectar contains as little as about 8 per cent. of sugars it is neglected by the bees as it is apparently not worth the labour of gathering, hence the serious check to harvesting caused by rain.

Honey is much used in toilet preparations for the skin and hair, owing to its value as a skin food.

Acidity

450.a   All sugars are naturally acidic and the primary reason for the acidity of honey is the three most common sugars, glucose, fructose and sucrose found in honey . The *p*H can vary slightly and this is due to the relative concentrations of the three sugars in the honey which is, in turn, determined by the nectar source used by the bees. ID

450. *The acidity of honey is commonly attributed to formic acid, but malic and allied acids are present also, varying in amount, and there is no formic acid in the poison of the bee's sting (865). The available information is vague and the subject still one for the scientist. Expressed in terms of pH value the clovers run about 3.45 to 3.55, the fruit honeys generally higher and sage honey, for example, lower.*

Honeydew

Sources

451.   Honeydew is a sweet substance derived from two entirely distinct sources which must not be confused. Many trees and plants -for example, fir trees; willows, balsams, vetches, and laurels- exude a sweet substance generally around the petioles where the leaves join the stems, which attracts insects. Where the tree is damaged by insects, a similar discharge may occur, notably with the ash and the eucalypti. This substance, dark in colour, but otherwise not unpleasant, is comparatively free from objection in small quantities, although undesirable as a winter food. Its addition frequently spoils the flavour of good honey, but it should be noted that the defect sometimes disappears if the honey is kept for a few months.

The other source of honeydew is the aphis, an insect which exudes a sticky sweet substance from its body, much beloved by ants, which latter keep and tend the aphidae as cows. This secretion is unpleasant and dangerous as a likely cause of dysentery amongst bees.

It would be good for bee-keepers to reach agreement with botanists as to distinguishing names for the two honeydews, viz. that obtained from extra floral nectaries and that from the aphis.

Characteristics

452.   Honeydew can be detected by its dark-generally dark green or blackish-colour. Once seen, it is unmistakable. It can be detected in the comb by holding up the combs before a good light, as when grading before extracting. Combs containing any appreciable quantity should be set aside, when extracting, being extracted separately. Such honey can be used for cooking or fed back to the bees when required for brood expansion, but not for wintering, i.e. use it during a cold spell.

453.   Honeydew is, in general, collected by bees only when there is a dearth of honey. This, however, frequently occurs in some districts between crops, and in dry seasons. If taken for immediate consumption, no harm need be anticipated, but if sealed, it is best disposed of by destroying the sealing by scratching or scraping in late August when the bees will use as food the honeydew so exposed. Some of the dark races of bees in certain districts are addicted to gathering honeydew in quantities. Careful supervision is necessary in such districts as honeydew is a bad food for wintering. It also causes discoloration of otherwise good extracted honey and of sections.

Sources of Honey and Pollen

454. All honey-producing plants produce some pollen and most pollen-bearing plants some honey. For practical purposes, however, one schedules the two separately according to their chief use. As a rule, there is no dearth of pollen while there are honey plants. Particular pollen-bearing plants are important, more especially for spring breeding. It is much easier to find a honey substitute than a pollen substitute. Early pollen sources are essential to provide fresh pollen for brood-raising, when stores of honey carried over from the winter can be used to supplement what little nectar is coming in. The dandelion, for example, gives a stimulating flow of honey, but is valued mainly for the pollen it supplies at a critical period. Plants are scheduled below separately as honey and pollen producers, according to their chief use, but where they have value for both purposes, from the practical stand-point, this is indicated.

Principal Sources of Honey and Pollen

455. A surplus may be regularly secured in certain districts from some plant or plants grown there in quantity, but in order to make a living from bee-keeping, it is essential that there should be within reach (**411**) plants giving ample supplies of nectar. The plants in List I are of that class, although some of the most important of them are not too reliable. It is generally desirable to have one or more of these sources in abundance within easy reach.

List II contains plants which are useful sources of honey, giving occasional surplus, some of considerable local importance, and most of common occurrence. Some less common sources have been included owing to their importance in filling the gap in honey flow experienced in June (northern hemisphere) so frequently.

List III, which might be greatly extended without difficulty, contains plants mainly of secondary importance unless occurring in abundance. Very many wild and cultivated flowers might have been added to this list, including most of the aromatic herbs, such as lavender, rosemary, thyme and the like.

List IV gives important sources of pollen.

456. In each list are given the most common names, also the generic name and the species known to be important. The species given are not exclusive, but are either important or indicative.

The author would be glad of reliable data for improvement of these lists in a future edition, and particularly data relative to time of bearing, nature and colour of honey, and territorial value.

On an average the difference in season between regions in the Northern and Southern Hemispheres is 6 months, but where more precise information derived by other means is available as to any particular source of nectar or of pollen, it is given in the detailed lists following.

457. Honey generally contains grains of pollen from the same source, but frequently contains grains from other sources. If, however, bees are known to be working a particular source and honey in the hive is found to contain pollen mainly from that source, it is good evidence that that honey came from that source. For the identification of pollen grains in honey it is necessary to compare them with grains from known sources which have also been immersed for some time in honey, as such immersion affects the appearance.

The colour of pollen grains must be determined by examining the pollen itself. The stamens may show a different colour at the tip, say, red or brown, while the pollen itself is pale yellow.

458. Honeys of too strong or otherwise undesirable flavour are not to be despised where they are or can be blended with milder or other appropriately flavoured honeys (see Blending, **551.**)

459. There are, however, a few sources to be avoided. List V contains a few such, but is doubtless capable of considerable extension when the subject of the properties of honey from particular sources has received more attention. Lists associated with particular countries follow List V.

LIST I—HONEY SOURCES

(For Explanation, see **455 - 6***)*

	Northern Hemisphere.	Southern Hemisphere.
Acacia. Wattle, Catclaw, Mimosa, Huajilla, Havardia.		
Acacia greggie, wrightii, mollis, etc., etc. (Tree)		
Honey very pale and very slow to granulate. Good for pollen.		
Important in Australia, California, Arizona and further south.	May to July.	November to January.

	Northern Hemisphere.	Southern Hemisphere.
Alfalfa. Lucerne, Spanish trefoil. *Medicago sativa, denticulata, lupulina.* (Perennial) Honey light colour, mild and mintlike flavour; granulates rapidly.		
Of almost universal importance. In U.S.A., west of the Mississippi; in British Columbia, the dry belt.	End June to August.	End December to February.
Alsike. A white, pink-tipped, or pink, clover, much branched. *Trifolium hybridum.* (Perennial) Honey pale of good flavour. Of universal importance.	May to October, especially June. Later in Minesota and B.C.	November to April, especially December and January.
Apple. No alternative names. *Pyrus malus.* Honey light amber of superb flavour. Pollen pale greenish yellow.		
Period depends upon the district and season and variety growing.	April to June.	
Box. Red, yellow, etc. (See Eucalyptus) *Buckwheat.* No alternative names. *Fagopyrum esculentum.* (Annual) Honey strong in flavour and dark in colour, even of a purple brown sometimes. Flow occurs early in the day. Important in Europe and the Northern States of U.S.A. and Canada. In the colder regions the flow ceases in August.		
Not much in evidence in the Southern Hemisphere or in Great Britain.	June to September.	December to March.
Citrus. Orange, Lime, Lemon, Grape Fruit, etc. *Citrus oruntium,* etc. (Trees and Shrubs) Honey delicious, pale and dense. Pollen pale yellow.		
Important in warmer climates, S. California, New Zealand, Haiti (December to April).	March to April.	September to October.
Clover, Sweet. Bokhara clover. Branches with long slender racemes of flowers. *Melilotus alba* (white), *officinalis* (yellow). (Biennial) *Melilotus indica* (yellow). (Annual) Honey pale, greenish yellow, slight cinnamon flavour.		
Of universal importance and of increasing importance in the Northern States and Canada. B.C., dry belt.	June to September.	December to March.
Clover, White. Dutch clover. Small leaves and short stems.		

	Northern Hemisphere.	Southern Hemisphere.

An important source in Scotland and other parts of Great Britain and Ireland. The Heaths, *Erica cinera*, etc., are sometimes wrongly described as heather. They add to the flow. — July to September, mainly August.

Ironbark. Ironwood. (See Eucalyptus)

Lemon. (See Citrus)

Lime. Basswood, Linden, Bee Tree.

Tilia Americana, Europoea, heterophylla, cordata, vulgaris. (Tree)

Honey of pronounced flavour somewhat minty, light amber or somewhat greenish, density apt to be low. Pollen greenish yellow.

Of universal importance. The flow generally lasts two to three weeks and is somewhat uncertain. There are early and late species. In Europe *T. cordata* is the most reliable. Flow in June in N. Dakota and July in Minesota and Great Britain. — June to July or August. — December to January or February.

Lucerne. (See Alfalfa)

Orange. (See Citrus)

Sainfoin. No alternative names.

Onobrychis sativa.

Honey lemon yellow. Pollen brown.

Universal, related to the clovers, gives nectar at a low temperature. — Late June to August. — December to February.

Sycamore. Allied to the Maples.

Acer pseudoplatanus. (Tree)

Honey pale with sometimes a greenish tinge, probably from admixture of honeydew; indifferent flavour.

Important in Great Britain, giving a short flow, somewhat uncertain, following the apple. — April to May.

Thistle. Canada thistle, Star thistle.

Carduus arvensis, etc.

Honey pale and of good flavour.

Universal in uncultivated areas. Important in Australia, Canada, California, and many other parts. — June to August. — December to February.

Tulip Tree. Tulip poplar, yellow poplar, white wood, cucumber tree.

Liriodendron tulipifera. (Tree)

Honey dark amber of a pronounced flavour. A reliable source.

Important in Eastern States of U.S.A., Ohio Valley, Appalachian Mountains. — May to June.

Tupelo. Sour Gum.

Nyassa aquatica, sylvatica, biflora.

Honey amber, fine, rarely granulates.

SOURCES OF HONEY AND POLLEN

	Northern Hemisphere.	Southern Hemisphere.
Important in Eastern States, U.S.A., and in the South.	April to June.	
Wattle. (See Acacia)		
Willow Herb. Fireweed, Rose Bay, Blooming Sally.		
Epilobium angustifolium and *montanum.* (Perennial)		
Honey white and of good flavour.		
Favours a dry soil and flourishes in ground burnt over.	July to	January to
Of universal importance.	Frost.	Frost.

LIST II—HONEY SOURCES

(For Explanation, see **455 - 6***)*

	Northern Hemisphere.	Southern Hemisphere.
Aster. Michaelmas Daisy, etc., etc.		
Aster ericoides, novae-anglica and innumerable others. (Perennial)		
Honey amber, rather pronounced flavour.	July to	January to
The later species are the more valuable.	October.	April.
Bearberry. Kinikiwi Manzanita.		
Arctostaphylos uva-ursi. (Tree or Shrub)		
California, Southern.	May.	
Bergamot. (See Horsemint)		
Blackhead. Knapweed.		
Centaurea nigra.		
Honey golden, thin and sharp flavoured.	June to	
Important in Ireland.	September.	
Black Mangrove. No alternative name.		
Avicennia nitida. (Tree)		
Said to be the heaviest known honeybearer.		
Important in Florida and Porto Rico.		
Boneset. Thoroughwort.	August,	
Eupatorium perfoliatum. (Perennial)	September.	
Borage. No alternative name.		
Borago officinalis. (Annual)	June to	
Important where in occurs in quantity.	Frost.	
Buckthorn. Coffee Berry, Cascara Segrada.		
Rhamnus cathartica, frangula, purshiana, californica. (Bush)		
Honey dark, slow to granulate.	May to	
Important in Canada and Florida.	September.	December.
Bugloss, Vipers. Salvation Jane, Patterson's Curse, Blue Weed.		
Echium vulgare, violaceum. (Annual)		
Universal, important in Ontario, New South Wales, etc.	July to September.	January to March.
Campanula. Bell flower.		
Convolvulacae, Ipomoea, etc. Numerous species.		

	Northern Hemisphere.	Southern Hemisphere.
Honey white and of fine flavour.	July to	
Important in tropical regions, Cuba.	August.	
Cape Weed. No alternative name (?).		
Honey golden.		
An important source in Australia.		
Cascara. (See Buckthorn)		
Century. No alternative name.		
Agave americana, etc.	April to	
Useful in semi-arid tropical regions.	May.	
Charlock. No alternative name.	June to	
Sinapis arvensis. (Annual)	August.	
Cherry. No alternative name.		
Prunus cerasus. (Tree)		
Useful minor source of honey and early pollen.	April to May.	October to November.
Clematis, Wild. No alternative name.		
Clematis ligusticifolia. (Perennial)		
Honey light amber, very dense, said to have a flavour resembling butterscotch.	June to	
Important in B.C. and California.	July.	
Clover, Crimson. A branched form.		
Trifolium incarnatum. (Annual)		
Honey similar to that of white or Dutch clover.		
Universal at least in Northern Hemisphere.	Early June.	
Currant. Flowering red and golden, cultivated red and black.		
Ribes rubra, aureum, anguineum, nigrum. (Bush)		
Useful between main spring and summer crops.	May to June.	November to December.
Universal.		
Dogbane. No alternative name.		
Apocynum androsoemifolium. (Perennial)		
Canada, B.C., West Kootenay.		
Eucalyptus, Grey Box. Candry wood, white gum.		
Eucalyptus homiphleia. (Tree)		
Honey nitrogenous, therefore, poor for wintering.		October to January.
Important in Australia.		
Eucalyptus—Stringy Bark. Red gum.		
Eucalyptus amygdalina. (Tree)		
Honey nitrogenous, therefore, poor for wintering.		October to January
Important in Australia.		
Fuchsia. No alternative name.		
Fuchsia, species. (Shrub)		
A rainproof source of honey useful in Great Britain and important in Ireland. Pollen primrose colour.	July to September.	

SOURCES OF HONEY AND POLLEN

	Northern Hemisphere.	Southern Hemisphere.
Golden Rod.		
Solidago squarrosa and other late kinds. (Perennial)		
Honey pale, of good flavour, granulates early.	July to Frost.	February to Frost.
Universal, important in U.S.A.		
Gooseberry.		
Ribes grossularia. (Bush)		
Honey good and pollen important.		
Useful in early spring when grown in quantity.	April and May.	October and November.
Hawthorn. May.		
Crataegus oxyacantha. (Shrub)		
Honey of fine nutty flavour.		
Important sometimes in Great Britain and Ireland.	May to June.	
Heartsease. Lady's Thumb, Smartweed, not the Viola.		
Polygonum persicaria. (Annual)		
Honey light and dark amber of good but delicate flavour, granulating readily.	August to October.	
Important in parts of U.S.A. and Canada.		
Holly. Gallberry.		
Ilex glabra, opaca, aquifolium, etc. (Shrub or Tree)		
Honey pale and of fine quality.		
Important in Southern States and other warm climates.	April to July.	
Elsewhere important for pollen.		
Horsemint. Bergamot. Bee Balm.		
Monarda punctata, fistulosa, didyma. Mentha citrata. (Perennial)		
Honey of good flavour, minty.		
Important in Southern States of U.S.A., also in Canada, June to September.	April to September.	
Knapweed. (See Blackhead)		
Labrador Tea. No alternative name.		
Ledum groenlandiensis, palustre. (Shrub)		
Important in C.B. wet belt.		
Limnanthis Douglasii. Butter and eggs.		
Limnanthes Douglasii. (Annual)	May and June.	
Important in Great Britain.		
Locust. (See Robinia)		
Logwood. No alternative name.		
Haematoxylum campechianum. (Tree)		
Honey pale and of superb quality.	October to February.	
Important in Tropics.		
Maple. No alternative name.		
Acer campestra, etc. (Tree)		
Honey pronounced flavour. Good source of pollen.		

	Northern Hemisphere.	Southern Hemisphere.
Important in many parts, Canada, U.S.A., etc.	April to June.	
Michaelmas Daisy. (See Aster)		
Milkweed. Silk weed.		
Asclepia tuberosa, etc. (Perennial)		
Important in Canada.	July.	
Mint. No alternative name.		
Mentha sativa, vividis, piperita, rulegium. (Perennial)	August to September.	
Universal.		
Needle Bush. Silky oak.		
Hakea leucoptera, Grevilla robusta. (Shrub)		
Important in Australia.		
Oleaster. Silver Berry, Buffalo berry.		
Elaeagnus argentea, olea. (Shrub)		
Important in B.C., East Kootenay and N. Dakota.	April.	
Peach. No alternative name.		
Prunus persica. (Tree)		
Honey good, important early source in warmer climates, including Northern New Zealand.	April and May.	
Pennyroyal, Wild. No alternative name.		
Hedeoma puligroides, Mentha pulegium. (Annual)		
Honey dark red.		
New Zealand, Southern States, U.S.A., Florida.	July to September.	January to February.
Phacelia. No alternative name.		
P. tenacetifolia, hispida, etc. (Annual)		
Honey flows freely, granulates quickly. Important in Europe and California.	June to August.	
Plum. No alternative name.		
Prunus domestica, etc. (Tree)		
Minor source, valuable for early pollen as well.	April to June.	October to December.
Radish, Wild. No alternative name.		
Raphanus raphanistrium, sativus. (Annual)		
Nova Scotia, etc.	July.	
Rape. No alternative name.	May to August, especially July to August.	November to February, especially January, February.
Brassica nigra, napus. (Annual)		
Honey good, granulates rapidly, important where grown in quantity.		
Raspberry. Wild raspberry.		
Rubus strigosus, idacus, etc. (Bush)		
Honey of fine flavour.		November to January.
Good intermediate crop where abundant.	May to July.	
Rata Tree. No alternative name.		
Honey white. (Tree)		
Important in New Zealand.		

	Northern Hemisphere.	Southern Hemisphere.
Robinia. Locust.		
R. *pseudacacia.* (Tree)		November to
Honey white, dense, of fine flavour, granu-	May to June.	December.
lating slowly.		
Sage. Black ball or button, white, purple.		
Ramona stachyoides, polystachya, nivea.		
(Perennial)		
Salvia officinalis, mellifera, apiana.		
Honey white, slow to granulate. Pollen		
greenish.		
Of wide importance in U.S.A. and parts	April to	October to
of Great Britain.	July.	January.
Sallal. Winter green, sallal.		
Gaultheria shallon. (Shrub)		
Important in B.C. wet belt.		
Siberian Pea Tree. No alternative name.		
Caragana.		
Important hedge plant in Canada.		
Silkweed. (See Milkweed)		
Silver Berry. (See Oleaster)		
Smartweed. (See Heartsease)		
Snowberry. No alternative name.		
Symphoricarpus racemosus. (Shrub)	June to	
Great Britain and Canada, etc.	August.	
Sourwood. Sorrel tree.		
Oxydendrum arboreum. (Shrub)		
Honey light in colour, granulating slowly.	June.	
Spanish Needle. Tick, Burr marigold,		
coreopsis.		
Bidens involucrata, aristosa. Correopsis		
species. (Annual and Perennial)		
Honey amber, dense, strong in flavour.	August,	
Important in warm, moist soils in U.S.A.	September.	
Spikeweed. No alternative name.		
Centromadia pungens. (Perennial)	August and	
Important in California.	September.	
Sumac. Sumach, Poison ivy.		
Rhus glabra, radicans, typhina. (Tree)		
Honey pale amber of fine flavour.		
Important in Ontario and where found in	June to	
quantity.	August.	
Tea Tree. Manuka.		
Leptospermum lanigerum, scoparium, Mela-		
luca. (Tree)		
Brown honey of such consistency that a		
press must be used.		
Important in New Zealand.		
Vetch. No alternative name.		
Visia villosa, faba, cracca, etc. (Perennial)		
Honey resembles clover honey.		
Important in B.C. wet belt, parts of Great	June to	
Britain and elsewhere, if widespread.	August.	

	Northern Hemisphere.	Southern Hemisphere.

Vitex. No alternative name.
    Vitex (Chinese species).   (Tree)   Early spring
    A reliable, continuous and heavy yielder.   to autumn.

LIST III—HONEY SOURCES

(For Explanation, see **455 - 6***)*

Aconite. Friar's Cap.
    Aconitum napellus.   (Perennial) January to
    Useful in Great Britain in early spring.   April.
Ailanthus. Tree of Heaven.
    Ailanthus glandulosa.   (Shrub)
    Honey of poor flavour.   August to
    California and Eastern States.   September.
Alfileria. Pin Clover.
    Erodium circutarium, moschatum. (Annual) April to
    Europe and U.S.A., especially California.   September.
Asparagus. No alternative name.
    Asparagus officinalis.   (Perennial)
    Honey amber.
Aubretia. No alternative name.
    Aubretia, species.   (Perennial) March to
    Great Britain, useful in early spring.   July.
Banksia. No alternative name.
    Banksia, numerous species.
    Amber honey.
    Useful in Australia.
Blueberry. Whortleberry, Wimberry.
    Vaccinium, species.   (Perennial)
    Important in East Canada, and in parts
      of Great Britain in the summer.   May.
Blue Vervain. No alternative name.
    Verbena officinalis.   (Perennial) June to
    Canada.   Frost.
Box. (European, not Eucalyptus, which
    see)
    Buxus sempervirens.   (Bush)
    Useful in Great Britain in early spring, March to
      pale yellow pollen.   May.
Box Elder (related to Maple and Sycamore).
    Negundo californicum, etc.   (Tree)
    U.S.A., N. Dakota.   Late April.
Broom. No alternative name.
    Citisus scoparius.   (Bush)
    Useful in Great Britain.   May to June.
Buffalo Berry. (See Oleaster)
Button Bush. No alternative name.
    Cephalanthus occidentalis.   (Bush)
    Honey mild in flavour, light in colour.
    Canada, swampy regions.

SOURCES OF HONEY AND POLLEN

	Northern Hemisphere.	Southern Hemisphere.
Cabbage. No alternative name.		
Brassica oleracea. (Annual)	May to July.	November to
Useful universally where cultivated.		January.
Catmint. No alternative name.		
Calamintha officinalis. (Perennial)		
Piquant flavour, good addition to mild honeys, granulates smoothly.		
Great Britain and elsewhere.		
Ceanothus. Indian currant, Brick Brush, White lilac.		
Ceanothus cuneatus. Symphoricarpos, racemosus. (Bush)	February to May.	
California.		
Chicory. No alternative name.		
Cichorium intylus. (Perennial)	July to	
Important in the Eastern States of U.S.A.	October.	
China Tree. Pride of India.		
Melia azedarach. (Tree)		
Important in Texas.		
Christmas Berry.		
Californian N. Coast and Bay Region.		
Cleome. Spider flower.		
Cleome serrulata, spinosa. (Perennial)		
Illinois, Rocky Mountains and sub-Tropics.		
Coconut Palm, and Thrinax palm and others.		
Cocos nucifera.		
Important in West Indies and Florida.		
Amber honey.	All the year.	
Cornflower. No alternative name.		
Lichnis githago, Centaurea cyanus. (Annual)		
Pollen pale yellow.		
Useful where growing extensively.		
Cotoneaster. No alternative name.		
Cotoneaster vulgaris.	March to	
Useful in Great Britain.	May.	
Cotton. No alternative name.		
Gossypium hirsutum.		
Imported in Southern States and parts of California.	June to August.	
Figwort. Simpson's Honey Plant.	July to	
Scrophularia, species.	August.	
Geranium family, Cranesbill, Herb Robert, etc., etc.		
Geranium and Erodium, pratense, Robertianum, phœum. (Perennial)	May to July.	
Hop. No alternative name.		
Humulus lupulus.		
Honey pale, pollen good.	July to	
Australia and where cultivated.	August.	

	Northern Hemisphere.	Southern Hemisphere.

Horehound.   No alternative name.
  Marrubium vulgare.   (Perennial)
  Dark amber honey of strong flavour.
  Universal, important in California.
Hound's Tongue.   No alternative name.
  Cynoglossum officinale.   (Perennial)
  U.S.A., Ontario.
Jamaica Dogwood.   No alternative name.
  Cornus florida.
  Important in Florida.
Juneberry.   Service berry.   March to
  Amelanchier canadensis.   (Tree)   May.
Lima Bean.   No alternative name.
  Phaecolus lunatus.
  California, Southern.
Meadow Sweet.   No alternative name.
  Spiraea latifolia.   (Perennial)   July to
  Useful in Great Britain and U.S.A.   August.
Mesquite.   No alternative name.
  Prosopias glandulosa, juliflora.   (Tree)
  Light amber honey, granulates quickly.   April to
  Useful in Southern States.   July.
Metopium.   No alternative name.
  Important in Florida.   (Shrub)   February.
Mistletoe.   No alternative name.
  Phoradendron, uiscana album.
  Useful in Southern States.   December to
  Early source of pollen and some nectar.   January.
Moca.   Cabbage tree.
  Geoffroea jamaicensis.
  Important in West Indies and Tropics
     generally.
Motherwort.   No alternative name.
  Leonurus Cardiaca.
  Canada.
Mustard.   Wild mustard.
  Brassica arvensis.   (Annual)
  Honey pale, inclined to granulate quickly.
     Pollen yellow.   June to   December to
  Important when grown in quantity.   August.   February.
Pear.   No alternative name.
  Pyrus communis.   (Tree)
  Honey good.   Important for spring
     pollen.   Pollen is greenish yellow not   April to   October to
     crimson.   May.   November.
Pin Clover.   (See Alfileria)
Pyrus Japonica.   No alternative name.
  Some use in Great Britain for pollen and   April to
     honey.   June.
Ragwort.   No alternative name.
  Senecio jacobaea.   (Perennial)   June to
  Honey of inferior flavour and colour.   September.

SOURCES OF HONEY AND POLLEN

	Northern Hemisphere.	Southern Hemisphere.

Red Bay. No alternative name.
   Persea borbonia. (Shrub) April to
   U.S.A., South and East. June.
Royal Palm and many others.
   Roystonea, etc. (Tree)
   Amber honey.
   Important in Tropics, W. Indies, etc.
Saw Palmetto. No alternative name.
   Sabal megacarpa.
   Honey light amber, dense.
   Important in Florida.
Sea Grape. No alternative name.
   Coccoloba uvifera. January to
   Florida. February?
Spider Flower. (See Cleome)
Sunflower. No alternative name.
   Heliunthus, tuberosus, annuus, etc.
        (Perennial and Annual) July to
   Australia and Florida. Frost.
Sweet Pepper. No alternative name.
   Clethra alsufolia. (Bush)
   Honey light amber, dense.
   New England, New Jersey, swampy July to
      woods. August.
Ti-ti. Leatherwood, Buckwheat.
   Cyrilla racemiflora. Cliftonia monophylla.
        (Shrub)
   Honey red and of strong flavour, lost in
    rain.
   In Florida, February to March. May to July.
Tobacco. No alternative name. July to
   Nicotiana tabacum. (Annual) August.
Wallflower, cultivated and wild.
   Cheiranthus, species. March to September to
   Pollen greenish. May. November.
Whortleberry. (See Blueberry)

LIST IV—POLLEN SOURCES
(*For Explanation, see* 455 - 6)

Alder. Black Alder.
   Alnus incana. (Tree)
   A universal early source.
Anemone. No alternative name.
   Anemone quinquifolia, etc. (Perennial)
   Universal.
Arabis. No alternative name.
   Arabis, species. (Perennial)
   Pollen green.
   Useful early source where cultivated in April to
    quantity. June.

	Northern Hemisphere.	Southern Hemisphere.

Ash. No alternative name.
    Fraxinus excelsior. (Tree)
Banana. No alternative name.
    Musa sapientum. (Tree)
    Useful in Tropics and sub-Tropics.
Barberry. No alternative name.
    Berberis japonica, vulgaris, trifoliata, etc. March to
    (Bush) May.
Beech. No alternative name.
    Fagus sylvatica. (Tree)
Birch. No alternative name.
    Betula alba, etc. (Tree)
    Gives a yellow honey, also pollen pale March to
    yellow. April.
    Universal, important in New Zealand.
Blackberry. Bramble.
    Rubus fruticosus. (Bush)
    Pollen greenish white. Honey of coarse
    flavour. June to December to
    Valued in Great Britain and Canada. September. March.
Buttercup. No alternative name.
    Ranunculus bulbosus, etc. (Perennial)
    Pollen orange. April.
Catnip. No alternative name.
    Nepeta cataria. (Perennial)
    Canada.
Celandine. No alternative name.
    Ranunculus ficaria. (Perennial)
    Pollen dark orange.
    Important common wild flower. April.
Chestnut. Sweet Chestnut, Spanish Chest-
    nut.
    Castanea dentata. (Tree)
    Gives a yellow honey also. May.
Clematis. (See List IV)
Coltsfoot. No alternative name.
    Tussilago farfara. (Perennial)
    Pollen orange colour. March to
    Important early source. May.
Cornflower. No alternative name.
    Lichnis githago, Centaurea cyanus.
    (Annual)
    Pollen pale yellow. July to
    Locally a useful source of honey also. August.
Crocus. No alternative name.
    Crocus sativus, etc. (Bulb)
    Pollen bright orange. February to August to
    Important wherever cultivated in gardens. April. October.
Elm. No alternative name.
    Ulnus campestris (Tree)
    Pollen yellowish green.
    Universal, early source. February. August.

SOURCES OF HONEY AND POLLEN

	Northern Hemisphere.	Southern Hemisphere.
Gorse. Furze bush.		
Ulex Europeus. (Bush)		
Pollen is dark orange brown.		
A useful source of pollen throughout the season in Great Britain and important in Ireland.	All the year.	
Gourds. Pumpkin, Squash, Water Melon, Cucumber, etc.		
Cucurbita pepo, etc. (Annual)		
Pollen golden yellow.	July to	
Important in California.	September.	
Groundsel. No alternative name.		
Senecio vulgaris, jacobaea. (Annual)	March	
Pollen bright golden yellow.	onwards.	
Hellebore. No alternative name.		
Helleborus niger, hiemalis, foetidus, viridis, etc. (Perennial)	December to	
Pollen, pale yellow.	May.	
Hemp. No alternative name.		
Cannabis sativa. (Shrub)		
Useful in Eastern States.		
Hickory. No alternative name.		
Carya, species. (Tree)		
Hop. (See List III)		
Hornbeam. No alternative name.		
Carpinus caroliniana. (Tree)	February.	August.
Horse Chestnut. No alternative name.		
Aesculus hippocastanum. (Tree)		
Pollen crimson.	May.	
Ivy. No alternative name.		
Hedera helix. (Climber)		
Important late source of pollen and honey.	October to December.	April to July.
Larch. No alternative name.		
Larix, species. (Tree)		
Honey granulates in a few days.		
Lilac. No alternative name.		
Syringa vulgaris. (Bush)		
Pollen yellow.	June.	
Mallow. Marsh Mallow.		
Malva sylvestris, rotundifolia. (Perennial)	March to August.	
Maple. No alternative name.		
Acer, species. (Tree)		
Gives also a light greenish yellow honey.		
Important especially in Canada and parts of U.S.A.	April to May.	
Mulberry. No alternative name.		
Morus nigra, alba. (Bush)		
Nut. Hazel Nut, Cob Nut.		
Corylus avellana, etc. (Bush)		
Pollen pale yellow.		

	Northern Hemisphere.	Southern Hemisphere.

Useful early source of pollen and some honey. — February to March.

Oak. No alternative name.
Quercus, species. (Tree)

Pine. Various. (Tree) Early.
Some give honey of good flavour, but nitrogenous.

Plantain. Ribwort.
Plantago lanceolata, etc. (Perennial)
Useful source of pollen in Great Britain. — May to July.

Poplar. White Poplar, Black Poplar.
Populus tremuloides, niger. (Tree)
Pollen greenish. — March to May.
Important source.

Poppy. No alternative name.
Papaver, species. (Annual) March to September.
Black pollen.

Privet. No alternative name.
Ligustrum vulgare, etc. (Bush)
Useful late source of pollen and of honey of unpleasant flavour.
Pollen pale yellowish green. — June to July.

Ragweed. No alternative name.
Ambrosia elatior. (Perennial) May to August. November to February.
Pollen deep golden, honey rank.

Sage Brush. No alternative name.
Artemesia californica. (Bush)
Imported in California, Southern.

Scilla. No alternative name.
Scilla, species. (Perennial)
Pollen bluish colour. — April.

Walnut. Black Walnut, English Walnut.
Juglans nigra, regia. (Tree)
Pollen greenish yellow. — April and May.
Source of honey also.

Willow. Sallow.
Salix, species. (Tree)
Pollen dull greenish yellow. — March to May. September to November.
Of universal importance wherever grown.

Yew. No alternative name.
Taxus baccata, etc. (Tree) March and April.
Useful early source.

LIST V—SOURCES OF OBJECTIONABLE HONEY

(For Explanation, see **455 - 6***)*

Thorn Apple.
Datura stramonium. (A Herb)
Honey poisonous.

Rhododendrum ponticum. (Bush)
Honey poisonous.
Asia Minor.

Due to production difficulties this list was taken from the 1st 1932 edition of the Manual

SECTION V

HONEY-PART II

Taking and Grading of Honey

Taking Comb Honey

473. Comb honey should be removed as soon as possible after it has been sealed, as, if left, the beautiful white capping will be stained by busy feet passing over it, and by additions of propolis. After a rapid flow ceases, through change of weather, or through the fading of the flowers from which it came, unfinished racks should be removed and all finished sections taken out. They should not be left on in the hope of a change occurring. Unfinished sections may then be collected together in one rack, the least finished being put in the middle and the most nearly finished at the outside, and returned to a strong stock. If a super of partly finished sections be so given and another one of unsealed sections be placed above a board with a bee hole in it (a super clearer with the "return" open), preferably raised off the board by an empty rack, the bees will quickly empty the top lot and store the honey in the lower rack, thus securing for the bee-keeper quickly cleared combs free from stain for use as baits another year, and quickly finished sections for immediate use (see also **1285**). On removal of sections, they should be cleaned (see **1387** for use of paraffin wax) and may then be graded and stored conveniently in empty racks in a dry warm place (see **533**).

Extracted Honey

474. Combs for extracting may be removed at any time when sealed, but are generally left to the end of a flow, when they may be handled systematically in bulk. This saves labour and ensures full ripening. Combs two-thirds sealed may be taken with the sealed combs, but combs unsealed or less than two-thirds sealed should be returned to the bees to finish.

475. Partly sealed combs may be put through the extractor first, and the unripe honey gently extracted and used for feeding back. This procedure is useful at the end of the season, as it enables all the extracting combs to be disposed of at one time, and the honey returned will be stored where wanted in the brood chamber (see, however, Feeding for Winter, **959**). It is undesirable to extract unsealed and sealed honey together, especially just after a damp spell or in a damp situation, unless the honey is for immediate use and the unsealed honey is found to be of good density.

476. If several supers are removed at one time from a crowded hive, additional room may be required for the bees displaced. This may be given in the form of fresh supers if the flow be on, or of extracted combs to be cleaned up, or a box of empty combs below the brood nest, or, if late in the season, an empty box used as a lift below the brood chamber, which box may well remain through the winter.

Use of Super Clearers and the Like

477. The super clearer is inserted at least 12 to 14 hours before the honey is to be taken, being placed below the supers to be cleared of bees. See that the escape is in working order. The inexperienced can test the escape by placing a few bees in a vessel placed against one side of the escape and turning the other side to the window. Some escapes are defective as sold and others become clogged. A defective escape is worse than none at all. See that the "return" slide, if any, is closed before inserting the clearer.

478. It is best to insert the clearer when bees are flying strongly, using a minimum of smoke. In fact, if section racks can be removed during a flow, so much the better, the use of smoke being confined to a little inserted where the bodies are separated for introducing the clearer.

479. When returning combs for the bees to clear up, they are placed over the super clearer, still in position, and the return slide opened to give the bees access to the combs. After a few days the return slide is closed and the bees are cleared in 12 to 24 hours through the escape as before. If there is no "return," or if it is rendered inoperative through propolis, it will be necessary to use the super clearer for clearing only removing it when replacing the supers and replacing it for clearing purposes.

480. If several supers are piled together on a strong colony, for cleaning, the bees may collect the honey in a patch in the lowest. To avoid this, insert an empty super under the pile.

481. Great care must be exercised to avoid exposure of the honey, as this would start robbing. Later in the season it is especially desirable to remove honey with a minimum of disturbances and a minimum of exposure, or robbing may be set up, and the use of a super clearer is preferable to any other method.

482. Nevertheless, if the honey is removed late in the day, when the bees are quiet, in fine weather, it is practicable and convenient to remove supers without using a clearer, and to brush the bees back into the hive with a feather. Stand the super aside and an empty one on a base near by. Transfer the combs to the empty super as fast as cleared and keep both covered as far as practicable.

Supers Stuck Together

483. As explained under comb building (**336**) the bees are liable to join the super frames to those in the chamber below. The wax so used must be broken when removing the super. This sometimes presents considerable difficulty. This may be reduced in four ways. Firstly, see that there is not more than a bee-space allowed; sometimes excessive provision for shrinkage of new wood is made by the manufacturers; do not, however, reduce body boxes or supers on this account until they have been in use for a season and have become well shrunk. Secondly, it is generally found that fewer attachments are made if the frames in the super cross those in the body below, and the attachments made are more easily broken. Thirdly, the use of vaseline, thinly applied with a paint brush, on the top and bottom bars of all frames, gives useful discouragement. Fourthly, see that the frames, especially those in the brood chamber having metal spacers, or the like, are fitted closely, and any free space closed with a dummy. This will enable a twisting movement to be given to the super to break the connecting cells without displacing the frames. In a stubborn case, the use of a dinner knife may be called for, or a wire as used for cutting soap. The use of wide top-bars greatly reduces brace comb.

For supers stuck together see **768** and **779**.

Supers Containing Brood

484. Bees will not desert a super containing any brood. The combs with brood will have to be taken out singly in case of necessity, and the bees shaken or brushed off, but the brood will be of course chilled if not returned promptly to the hive.

Supers Containing Pollen

485. Unless the amount is negligible, it is very desirable to set aside, under cover, any combs so furnished so that they maybe returned to the bees. The bees can hardly have too much pollen for rapid spring development. Alternatively a few such combs extracted carefully may be placed near the brood nest, where they will be filled again while feeding for winter, or emptied, maybe, in the autumn.

Grading Honey

486. On removing sections or extracting combs, they should be examined for colour, the lighter tints being separated from the darker. Sections should be graded further according to appearance, etc., according to the practice of the local market.

487.a When extracting, the lighter grades may be extracted first and run off into a ripener before the darker grades, and if honey is to be sold according to source or kind, the kinds must be kept separate. In general it is, however, better to blend and endeavour to produce a uniform distinctive good product from year to year, the poorer, dark and coarse flavoured honeys, if any, being kept separate, and sold for industrial purposes, e.g. sweet making, fancy cakes, etc. Do not, however, miss an opportunity of offering a favoured kind of honey which, like a choice wine, will command a high price. KF

487. When extracting, the lighter grades may be extracted first and run off into a ripener before the darker grades, and if honey is to be sold according to source or kind, the kinds must be kept separate. In general it is, however, better to blend and endeavour to produce a uniform distinctive good product from year to year, the poorer, dark and coarse flavoured honeys, if any, being kept separate, and sold for industrial purposes, e.g. sweet making, fancy cakes, etc. Do not, however, miss an opportunity of offering a favoured kind of honey which, like a choice wine, will command a high price.

488. At present every country makes its own rules, if any, for grading. Some information is given below, based on rules made before the Great War, but the bee-keeper should make local inquiry to ascertain what rules apply in his country.

American Standards

489.a Standards are fixed by the U.S.A. Department of Agriculture and cover in full detail the grading, colour standards and packing requirements for honey. These rules provide an excellent example of good standardization; all the terms used are carefully defined, and sufficient detail given to enable a lone bee-keeper to grade his material without doubt as to compliance.

The British reader is warned that an American gallon measures 0.8327 of a British gallon, but the pound avoirdupois is substantially the same as the British. KF

489. Standards are fixed by the U.S.A. Department of Agriculture and cover in full detail the grading, colour standards and packing requirements for honey. These rules provide an excellent example of good standardization; all the terms used are carefully defined, and sufficient detail given to enable a lone bee-keeper to grade his material without doubt as to compliance.

The British reader is warned that an American gallon measures 0.8327 of a British gallon, but the pound avoirdupois is substantially the same as the British.

Australian, Canadian, British, Irish Free State and New Zealand Regulations

490. Application should be made to the local Department of Agriculture.

Extraction of Honey and Preparation of Chunk Honey

Extracting Honey, Temperature

491. Extraction should be done in a warm room (70 to 100° F. (20° to 40° C.), which should be made bee-proof. The honey also should be warm and should be allowed to stand in the warm room for an hour or two according to the temperature. If extracted at too low a temperature, not only is the labour increased, but air may be trapped in the honey and form a wasteful scum.

Uncapping Knives

492. The combs are uncapped with an uncapping knife. If the knife has a bevelled edge, note that the bevel should be towards the comb, so that the wax falls away from the comb and the knife does not cling to the damp surface of comb. For heavy continuous work knives heated by steam or electricity are used. On a smaller scale two knives may be used, one put to heat in hot water while the other is in use. To heat the knife, a tall metal vessel is required, preferably with a burner under. A knife with long blade and bevelled edge can be used cold if given a sawing motion. The straight part of the blade should be, say, 2 inches longer than the width of the comb. All knives should be as sharp as they can be made. Most knives have blades curved at the end, which is useful for uncapping sunk portions of the surface, but the bee-keeper should aim to get all extracting combs flat (**315**).

Uncapping

493. Some operators stand the comb vertically, facing towards them, the cut being made downwards. With practice the capping may be thrown off as removed. Others prefer an upward cut, the upper end of the comb overhanging the lower. The operator should then keep his hand behind the frame. It is clear, therefore, that a receptacle is required to catch the cappings, the drippings and to support the comb. Any large tray may be used on the top of a larger ripener fitted with a sieve, or a bar of wood may be placed across the ripener, grooved on the under side to fit on to the edges, and having on the top face an edge or nail point on which to steady the bottom of the frame and prevent slipping. It is desirable to have a tray or sieve in which the cappings can accumulate and drain, and a bar across on which the comb may be stood securely. The knife may be wiped on the bar, and if water-heated knives are used, one end of the bar should be wrapped with cotton rag on which the knife may be wiped before returning to the hot water. If a heated cappings-renderer is used the combs are uncapped above it. Sooner or later, something, say the knife-handle, becomes sticky with honey. Before commencing operations, be sure that you have handy a towel, one end of which has been well damped, so that sticky hands, handles and other parts, may be wiped at once and before the honey spreads over clothes and everything within reach.

494. When both sides of the comb are uncapped, the comb should at once be inserted in the extractor. If the combs differ much in weight, endeavour to arrange combs of even weight on opposite sides of the extractor, so as to maintain a balance. This saves wear and tear, much unpleasant vibration and gives quicker results. The extractor should be secured to the floor, or at least to a board on which the operator can place his feet.

Centrifugal Force

495. The force on the honey depends mainly upon the radius and speed, whilst that on the comb depends upon the weight also. It is well, therefore, to commence slowly, partly emptying both sides, so as to reduce the weight; then reverse the combs and completely empty both sides, finishing at a high speed. With radial pattern extractors, reversal is unnecessary, but the speed should be low until the weight is reduced.

496. The following table gives an indication of the maximum safe speed in revolutions per minute of the handle, for fully loaded flat combs not exceeding 1¼ inches thick, in a warm room about 80° F. with extractors of various gear ratios and radii. The gear ratio is found by counting the number of revolutions of the extractor for one revolution of the handle, or by dividing the number of teeth in the large wheel by the number of teeth in the small wheel. The radius is the maximum radius, from the spindle to the edges of the comb. Preferred ratios are shown in heavy type.

TABLE XIV

Care of Extractor

500.a   Honey is a food and should only be processed and stored in containers that are not affected or corroded by honey. The advice is now to either use food grade plastic or stainless steel and both are readily available from beekeeping equipment suppliers or food processing suppliers. It is now illegal to use an extractor not made of these two materials if the honey is subsequently sold or supplied to a third party. ID

500. *Within 24 hours of emptying the extractor or ripener, wash it thoroughly with warm water and dry in moving air. If honey is left in, it causes blackening and ultimately rusting, and rust discolours good honey. This is a case in which a stitch in time saves ninety-nine.*
Nevertheless, the bad parts of a neglected extractor may be restored with a good-quality aluminium paint, or beeswax may be used (900).

Ling and Heather Honey

501. Ling and manuka honey require different treatment, as, owing to its thixotropy or property of setting like a jelly, it refuses to flow. Heather blend may be extracted in a warm room using considerable patience to avoid speeding up while the combs are still heavy, and to allow for the slow flow. More frequently heather honey is extracted by means of presses sold for the purpose, in which the comb is inserted wrapped in straining cloth, and the honey squeezed out.

502. On the continent of Europe a machine has recently come to the front by means of which ling honey of the most stiff character may be extracted. It consists of a loosener and extractor combined, containing loosely mounted stout pins which enter and leave the cells, breaking down the jelly.

Chunk Honey

503. Chunk honey is chunks of comb honey usually sold immersed in extracted honey, but sometimes drained and sold packed in waxed paper and wax sealed. It is a well-recognized product in the American market.

With chunk honey, the cost and labour of handling sections is saved, and also the hindrance they present to the bees. The comb honey in chunk honey requires for its production the same conditions of weather, location and climate as section honey: that is a rapid flow. It should be as clean and good as that sold in sections.

504. The comb is usually produced in shallow frames, and indeed if deep frames are used it is usual to divide them by a horizontal bar into two equal halves. Full frames of super foundation may be employed, but it is customary to use starters only. Medium brood foundation is used and under the conditions of high temperature prevailing is thinned down while being drawn out. The use of starters, in fact, ensures a straight comb, as a full sheet of foundation which, of course, cannot be wired, is liable to buckle, and especially if the foundation is fitted, as it generally has to be, some time before it is wanted.

505. Starters should be 1½ to 2 inches deep. Unless the super is to be inserted below another in which the bees are working, the bees will probably work up from the bottom of the frame, and it is therefore desirable to furnish one or two of the frames at the centre with a ½ inch starter in the bottom as well. This will secure a good start and carry the bees to the top where they will start adjacent frames from the top.

506. On removing a super of combs and cutting out the chunks of honey, a starter should be left. If the frames are to be returned at once, leave up to ½ inch of comb as a starter. At the end of the season, however, cut close to the wood. The frames should be returned to the bees for cleaning as in the case of extracting combs (**698**), then fresh starters are put in next year, so as to avoid propolis and staining which is inevitable if comb is left and given for cleaning up.

TABLE XIV
SAFE SPEEDS FOR EXTRACTORS

Radius.	Gear Ratio.				
	1 : 1	1 : 2	1 : 2½	1: 3	1: 4.
5 inches	164	82	65	54	**41**
6 inches	150	75	6o	**50**	37
9 inches	I22	61	**49**	**41**	30
I2 inches	105	**53**	42	35	zG
15 inches	**94**	47	37	31	23
18 inches	**86**	43	34	29	21
–	Revolutions per minute of handle				

It is assumed that the wire screens are properly supported so that they do not bulge outwards at high speed. Some extractors are defective in respect to this, and some do not allow the comb surface to lie against the screen. With such defective apparatus lower speeds must be employed.

Cappings

497. Cappings while draining should be turned about and broken up by stirring from time to time. After extracting is over they may be put into the extractor in wire baskets, if there is sufficient quantity to be worth treating in this way. A considerable amount of wax particles will come away with the honey, and in a large commercial establishment the honey is strained in a centrifugal machine.

Use of Ripener

498. On a smaller scale the honey is run from the extractor, as it accumulates, into a ripener or other container through a strainer, separate ripeners being kept for light and dark honey. If in doubt as to the density, it should be checked as run in. The least dense honey will rise to the top in two or three days, and may be thickened by evaporation in a warm dry room, the ripener being covered with muslin or a strainer. Honey does not ripen in a "ripener" unless the air is unusually dry or is dried by heating (**437**)., The process may be hastened by hanging a small incandescent electric lamp in the ripener 6 inches above the honey. It causes circulation of air and reduces its humidity. Improvement is slow, the honey may be darkened and must not be overheated.

For honey heating, see **524**.

For treatment of cappings after draining the honey, see **400** and **497**.

Bottling

499. The small bee-keeper extracting only ripe honey in small quantities can dispense with a ripener, using a fine strainer (muslin or cheese cloth) and bottling direct from the extractor.

In bottling from the extractor or ripener, use a fine strainer, and to avoid the formatiorn of air bubbles, tilt the bottle so that the honey runs down the inside, and allow only a small fall.

For fixing starters use a wooden guide and wax them as in **811**.

Granulation and Fermentation, Heating and Sterilizing

Granulation of Honey

507. In course of time in most honeys the dextrose sugar first separates out in the form of crystals, releasing 10 per cent. of its weight as water. This puts up the water content of the remainder by about 7 per cent., so that if the honey has a density of 1.42 at 20° C. the liquid portion after complete granulation will have a density of only about 1.37, thus increasing the risk of fermentation (see **517**). The levulose separates out later if granulation is forced, and any sucrose present will also separate out in time.

Production of Large Crystals

508. For the production of large crystals the granulation should be slow and should take place in a good light. The temperature should be even and evaporation hindered.

Production of Fine Crystals

509. For the production of the much preferred fine crystals granulation should be quick (see **510**) and should take place in darkness and a variable temperature. The process is hastened by stirring and by quick evaporation. A low temperature is sometimes recommended. If the temperature is high, it is difficult to avoid the formation of larger crystals. The honey must not be exposed or it will absorb moisture (**437**). Fine granulation is best obtained by seeding the honey by adding a little finely granulated honey and stirring well.

Rapidity of Granulation

510. The rapidity of granulation depends mainly on the number of primary crystals present.

Tendency to Granulate

511. Honey from some sources granulates much more quickly than others. A very fine clear honey from Robinia, Tupelo and Milkweed seldom granulates and may keep clear for years without special treatment. Honey from Dandelion and Charlock, on the other hand, frequently granulates in 2 weeks or less. The late-season honeys are inclined to granulate quickly; Golden Rod and Willow Herb, for example, and Mustard taken late in season. In Australia honeys from Clover, Thistle and Cape Weed all granulate quickly and that from Needle wood and Hop in about a fortnight. Clover and Ling honey mixed together crystallize with a fine texture, but a mixture of clover and bell heather is apt to produce very large crystals.

Prevention of Granulation

512. The primary crystals are dissolved at temperatures from 95° to 115° F. (say, 35° to 45° C.), according to density, etc., so that heating to 120° F. (49° C.) for some hours will completely remove them. Sections containing granulated honey can be renewed by gentle heating for a long period at 105°-110° F. (41° to 43° C.), but an even temperature carefully controlled or automatically maintained is essential, or the comb will sag.

513. The crystals are melted more quickly the higher the temperature, but above about 130° F. (55° C.) the honey is changed somewhat, indeed it is undesirable to heat light-coloured honeys above 120 and for any length of time as they become much darker. In fact, light honeys stored at temperatures above 60° F. (15° C.) darken somewhat, and if stored at 70° F. (21 C.) darken badly in time. A temperature of 130° F. is not enough, however, to prevent fermentation by destroying the fermentation yeasts, so although it clears the honey the use of a somewhat higher temperature for a short time is necessary to prevent granulation and fermentation. Granulation is checked at 103° F. (40° C.), but not cured, as the crystals formed are not redissolved. A temperature of 140° to 150° F. (60° and 65° C.) is used. See heating honey, **524**. Some use 160° F. (70° C.) for ¼ hour, and then cool quickly.

Treating Honey already Granulated

514. In melting granulated honey to prevent further granulation, it is particularly important to avoid overheating. The honey cannot circulate in the vessel until fluid and is therefore readily overheated locally. Large tins, 60-lb. size, require heating for, say, 8 hours in a water bath at about 130° F., the temperature being then raised steadily to 150° F., and then lowered quickly.

Whipped Honey

515. When stirring honey to obtain fine granulation, it is sometimes whipped. This introduces air and gives a fine whitish colour, but unless the density is good the risk of fermentation is increased. Unless it is intended to introduce air in quantity it is better to stir round and round, so as to disturb the surface as little as may be.

Fermentation

516.a Usually it is not necessary to heat honey to destroy any contamination with yeast. The enzyme glucose oxidase added to nectar by the bees during processing will normally prevent yeast form multiply during the early stages of processing. Heating honey can reduce the activity of the enzymes in honey. For example the activity of distase will be reduced by half if the honey remains at 60°C for 24 hours. ID

516. The common cause of fermentation of honey is the presence of certain sugar-tolerant yeasts (Zygosaccharomyces and Torula) of which several kinds have been isolated. They are normally collected by the bees with the nectar and pollen and, of course, more readily from any exposed fermented honey. They cannot multiply at temperatures below 50° to 52° F. (10° to 11° C.), but do so with increasing rapidity as the temperature rises above that limit; but when subjected for a sufficient time to a temperature in the neighbourhood of 130° to 140° F. (54° to 60° C.) the yeasts are destroyed and fermentation ceases. A temperature of 145° F. (63° C.) will cause destruction in ½ hour. Honey containing these yeasts granulates the more readily, possibly due to the yeast particles acting as dust particles do in a crystallizing liquid.

Prevention of Fermentation

517.a Honey with a water content below 20% is unlikely to ferment and can be stored safely in jars. However, if the honey sets and leaves a wet surface, the water content in the layer may now exceed 20% and the honey may start to ferment. It is advisable to ensure that set honey has a dry surface and that the honey used to produce set honey starts with a water content below 20%. ID

517. *The yeast cannot propagate in a dense honey, as the dense honey attracts moisture strongly and robs the yeast of necessary moisture. The danger-line is not accurately known, but if the density of honey before crystallization is below 1.43 at 20° C., or if liquid honey free from crystals is below 1.40, there is danger of fermentation at temperatures above 52° F.*

518.a It is probable that nectar in flowers does not ferment because it will take time for yeast particles to fall into the nectar and more time to start to ferment. This time will probably exceed the time that nectar is present in the flower.   ID

518. *Fermentation by the yeasts mentioned is prevented also if the honey is sufficiently diluted: a honey solution containing about 80 per cent. or more of water being safe from fermentation by them at atmospheric temperature. The lower the temperature, the more the water required. The figures are but slightly different with cane sugar, so it is not clear why nectar does not ferment in the flowers. It may be prevented by some effect of light or extremes of temperature, changes of concentration, or more probably lack of time.*

519. A dense honey not sterilized, kept in a cool place and hermetically sealed, will keep good for years, but if granulated honey is to keep well, say until the next season, the density must be good and the storage dry. Hermetic sealing will not prevent fermentation save to the extent that it prevents the entry of additional moisture. Waxed paper vessels must not be used for long storage. Fermentation is aided by the presence of nitrogenous matter (**445** and **449**).

520. Honey to be sold in the liquid form will, however, keep indefinitely in closed sterile vessels if it has been heat treated as above described, so as to destroy the ferments.

521. It has been said that honey sealed in the comb is sterile; but this is not so. Its keeping properties are mainly dependent upon its density. The rate of fermentation depends mainly upon the density and temperature, but also upon the quantity of yeast particles present, and no doubt to some extent upon the kind.

522.   The greatest danger to the bee is disease; to wax, the wax moth; to honey, unquestionably fermentation. There is much yet to be learned about the causes and prevention of fermentation.

Use of Fermented Honey

523.a When yeast acts on sugar to produce alcohol the by-products are water and CO_2. It is not advisable to feed honey back to bees unless you can be certain that the honey came from the particular colony and that it has not deteriorated since being removed. Any alcohol in the honey will give bees dysentery and honey can contain pathogens that could infect the colony (thus spreading any disease). Fermented honey can be heated to kill the yeast and remove the alcohol. It can then be used in cooking recipes. However, it should not be labelled and sold as honey. ID

523. *Fermented honey may be used for slow feeding in the spring, diluted, say, 1: 1 and heated to destroy the ferments and drive off any alcohol. The acid products of fermentation can be removed in a bad case by inserting in a muslin bag an ounce of slacked lime for every 10 lb. of honey, when heating it.*

Heating and Sterilizing Honey

524. Honey is heated to prevent or greatly retard granulation and to prevent fermentation, also to clarify, this latter feature being more important with honey containing appreciable albumen, as do some of the Australian honeys, for example, the Eucalyptus honeys. Granulated honey sometimes has to be reduced to the liquid condition by heating. Honey is also heated for the purpose of sterilization. Except when diluted with water as for feeding, the honey should always be heated in a water bath, i.e. the honey container should be immersed in water, which should separate the honey from the source of heat both at the sides and bottom. This secures against excessive local heating, due to the honey not circulating freely. The temperature of the water must be controlled and should not appreciably exceed the maximum temperature to which the honey is to be raised. The process is accelerated by stirring the honey, and if this is done continuously the water may then be several degrees above the desired temperature.

Time and Temperature

525. The reduction of granulation or destruction of ferments by heating depends upon time as well as temperature, the time being greatly reduced by raising the temperature, or by reducing the quantity to be heated. Some granulated honey in sections can be restored to the liquid form by heating for some weeks at a temperature of about 105° F. (41° C.), but for commercial work in reducing granulated honey much quicker reduction is required and a much higher temperature is necessary, which would destroy sections. Temperatures up to 160° F. are usual. Greater or prolonged heating is destructive of the aroma and commercial value of the honey, and wherever possible heating should be kept within the limit of 140° to 145 F. This, incidentally, corresponds to the melting-point of commercial beeswax (**380**), thus a small piece of wax, floating on the honey, may be used as a check. For further details, see **514-20**.

Destruction of Bacteria

526.a Honey that has been treated through heating should not be fed to bees under any circumstances. Even if you can be sure that the honey does not contain any spores of American Foulbrood (paenebaccillus *larvae*) the process of heating honey produces complex sugars that are not digested by bees and in extreme circumstances are poisonous to bees. ID

526. Honey is sterilized by heating only in case there is risk of its containing foul brood germs. Honey so heated is no longer suitable for sale for table use, but may be used for feeding bees. To destroy all germs which may be present, including spores of American foul brood, it is necessary to raise the temperature of the honey to the boiling-point for *at least* ½ hour. As it is to be used for feeding purposes it is convenient to dilute the honey first by mixing thoroughly with an equal quantity of boiling water, when it may be boiled over a slow burner without danger of overheating. Such honey as a bee food is but little better than sugar syrup, and for winter consumption is not so good. It is better to use such honey for home consumption, without heating, taking great care to avoid exposing any of it to access of foraging bees at any stage, or the water in which the containers may be washed. There is no risk whatever to human beings in consuming honey containing the spores of foul brood and other bee brood diseases.

Honey destroys all bacteria in the vegetative stage by robbing them of water (plasmolysis), but does not destroy spores.

Putting up for Sale

527. When honey is to be put up at once in vessels for sale, i.e. without blending, it is convenient to heat it at the time of extracting. This is also advantageous in the case of honeys liable to granulate quickly. Where the honey is to be blended in quantity, it is more convenient to collect it from the extractors in temporary 60-lb. cans, each labelled at once to show content. Later these are heated, the contents blended in a larger vessel, and finally put up in tins or bottles ready for sale. The cans may be heated in a larger water bath for about 2 hours at 150° F. (say, 65° C.), but if granulated it will take 10 hours to reduce a 6o-lb. can. The initial temperature should be 130° F. for 7 or 8 hours, the temperature being then raised slowly to 150°.

The small bee-keeper can blend his honeys by mixing the combs containing honey from different sources, when extracting, but he should avoid blending the good with the bad.

528. Special plant is made for honey heating in larger quantities, which saves time, the honey being caused to pass over a large surface of metal, heated by water or steam, on its way to the tank, the whole being regulated so that the honey leaves the heater continuously at a temperature of, say, 150°.

529. The honey may be passed through a clarifier, a long vessel divided lengthwise by partitions reaching alternatively nearly to the top and bottom so that the honey flows over and under, the scum collecting at the top.

530. At the same time, it is a question whether it is possible to subject honey to any treatment involving heating without some loss of flavour. It is certainly easy to spoil it.

Destruction of Diastase

531. Heating honey for long periods at 140° F. (60° C.) has no effect on the diastase content, but a temperature of 150° F. (60° C.) will produce a detectable effect in a few hours, but probably no serious reduction, even in half a day. At 165° F. (74° C.) some effect is felt at once and complete destruction is likely in half a day.

532. It is said in some quarters that any heating of honey above 120° F. (say, 50° C.) diminishes the food value by the destruction of minute quantities of certain digestive ferments. In Germany before the Great War tests were applied to ascertain the condition of honey in this respect. Unheated honeys, however, differ widely, and it is impracticable to prove by test that honey has been heated above 120° F. The true food value of the digestives in question is problematical, and it is likely that, for commercial purposes, the benefit of destroying harmful ferments by heating may more than compensate for the loss.

Storing, Packing, Selling and Showing Honey

Storing Honey

533. Honey must on no account be exposed to damp. It absorbs moisture readily from damp air. See **437**. Honey in the comb requires a higher temperature or drier air than bottled honey, as the latter is better sealed.

534. For bottled honey a dry dark cupboard with an even temperature of about 54° F. (12° C.) is excellent, unless it is to be granulated, in which case, see **514**. A temperature a few degrees lower is preferable if the honey has not been heated to prevent fermentation, but is difficult to maintain. For comb honey a better temperature would be about 60° F. (15½° C.). The air is kept relatively dry if the store-room temperature is above that of the outside temperature.

All honey should be graded before storing.

535. If a really dry, cool store with even temperature is available, section honey may be glazed before storing, or protected with greaseproof paper, transparent or otherwise, each section being wrapped separately and the whole packed closely in wooden or metal boxes. It is customary, however, to store the sections so that they are freely ventilated until ready for packing for early sale. There will be some loss of aroma, and it is essential to keep out insects and dirt. Sulphur fumigation is sometimes necessary to destroy moths and also flies, but this will not destroy moths' eggs (see **361**).

536. Extracted and chunk honey (pieces of cut comb with honey to fill up) is frequently sold in tins containing, say, 5 to 60 lb., or in waxed parchment receptacles with press-in lid. The latter require special care in storing (**437**), as the lid is far from air-tight. They are better adapted for use for early sale than for storage.

Packing Section, Comb and Chunk Honey

537.a Sections are frequently inadequately packed, to the great annoyance and loss of both buyer and seller. Probably the best mode of packing is by the use of corrugated fibre containers having separate compartments for each section and arranged to take, say, twenty-four sections. If each section is first packed in its own carton, or glazed and backed, the fibre case affords sufficient protection for short journeys. But for long journeys the cases should themselves be floated in wood shavings or other springy packing and marked for careful handling.

In the USA and elsewhere there are special regulations for the packing and marketing of honey, which must be obtained and observed. KF

537. Sections are frequently inadequately packed, to the great annoyance and loss of both buyer and seller. Probably the best mode of packing is by the use of corrugated fibre containers having separate compartments for each section and arranged to take, say, twenty-four sections. If each section is first packed in its own carton, or glazed and backed, the fibre case affords sufficient protection for short journeys. But for long journeys the cases should themselves be floated in wood shavings or other springy packing and marked for careful handling.

In the U.S.A. and elsewhere there are special regulations for the packing and marking of honey, which must be obtained and observed.

538.a Comb honey cut from shallow frames may be packed for transit by a method introduced by the Kalona Honey Co. of Iowa. The comb is cut and drained, then wrapped in cellophane, dipped in melted paraffin wax and placed on a cardboard to which it adheres and which serves to support and handle the comb. The pieces are finally packed in a corrugated cardboard sectional container. KF

538. Comb honey cut from shallow frames may be packed for transit by a method introduced by the Kalona Honey Co. of Iowa. The comb is cut and drained, then wrapped in cellophane, dipped in melted paraffin wax and placed on a cardboard to which it adheres and which serves to support and handle the comb. The pieces are finally packed in a corrugated cardboard sectional container.

Packing Extracted Honey

539.a Bottled honey can be transported through the post so long as the packaging is sufficiently robust to prevent the jar being broken. ID

539. Where extracted honey is sold in small quantities in glass bottles, the bottles may be packed in fibre containers sold for the purpose in a similar manner to sections.

540.a In the UK honey is stored in bulk in food quality plastic containers of about 5 litre capacity or greater. ID

540. Honey in quantity is usually put up in lacquered tins of various sizes, which in turn are packed in crates, the weight of the whole being adjusted with an eye to convenient handling and to the schedule of shipping weights used by the railway companies. In Great Britain 28-lb. tins are frequently used, two in a crate representing ½ cwt., but tins are used also of 4 lb., 7 lb. and 14 lb. capacity. In the U.S.A., 60-lb. tins are used, but may be packed to carry 2 cwt. each and crated in pairs to weigh not more than 120 lb.

541.a Petrol tins are no longer used for storing honey. ID

541. In Australia petrol tins are frequently used and should first be cleaned by washing with hot soda water. A rough form of churn is handy for washing large numbers.

542.a Honey is transported in containers and not sold in block form in the UK. ID

542. Some dense honeys, well granulated, are so hard that they may be sold in block form, wrapped in grease-proof paper and in cool weather may be sent by post without any stiff case.

Labels for Tins and Bottles

543.a Lables on bottles and bulk containers are generally self adhesive and the adhesive is not water soluble. These can be bought from beekeeping suppliers or standard labels can be supplied by stationery suppliers. ID

543. Labels should not peel off and should withstand fungi. The paste used should keep. The paste may be made of rice flour or common flour beaten into a smooth batter, afterwards stirring in boiling water to the desired consistence. As a preservative add about half a teaspoonful of carbolic acid to the quart of paste, or add enough sulphate of copper to give a blue tinge.

544.a see 543.a   ID

544. To prevent stripping and to improve the adhesive properties some additions may be made to attract moisture. For this purpose about 1 oz. of honey or brown sugar may be used; or 2 oz. of glycerine to the quart; or about 15 per cent. soda glass (water glass).

545.a see 543.a   ID

545. Thin labels of relatively soft paper adhere better than stouter labels or better paper.

546.a see 543.a   ID

546. Gum arabic is sometimes used and, with the addition of glycerine as above and an equal quantity of washing soda, is very effective as an adhesive to glass.

Selling Honey

547. For both the small and the large seller of honey it is considered the co-operative method of selling is the best. The local man with a good local product can sometimes get a better price in the local market, but such men are beginning to feel the competition of the large co-operative organizations. The co-operative organization assists the individual by extensive advertising, reaches a far wider and, indeed often a world-wide, market, and satisfies the buyer by maintaining a better guarantee of quality and quantity than is generally possible for the individual.

548. Whatever the selling organization or methods, the seller must in any case observe the usual practices of good business if he is to succeed. His goods must be presented in an attractive way, adapted to the demand of the buyer, and where sold wholesale must be properly invoiced. The seller should be prepared to furnish samples, and the bulk must be up to sample. While such low-priced commodities as potatoes and bread are sold in an attractive manner, the bee-keeper must not expect to get good results unless he also will devote thought and care to offering his goods in an attractive form.

549. For the retail trade the seller should experiment a little. For example, in a market where the buyers are convinced honey users, but sharp buyers, large vessels of granulated honey can be sold; but where the buyers are unconvinced, are not well-to-do and regard honey as a luxury, a good market may be found for small vessels, even ¼-lb. pots.

550. Use a distinctive label, so that the buyer can ask for more of the same honey. Have that label represent a certain standard of quality which you can maintain. Never let your customer down. If you have sold even only a small quantity wholesale to a retailer, do not bid against him in his own retail market.

Blending for Sale

551. Bee-keepers should give greater attention to the blending of honeys. Some honeys from distinctive and popular sources are best sold for what they are: for example, heather honey, and so marked; but in handling large quantities it is important to maintain a certain standard and one that the public wants. The owner's own preference may be for a very delicately flavoured or again a rather coarse-flavoured honey, according to his palate and habit, but the public wants what it expects or something rather better. If you sell honeys differing much in quality, give them distinctive names or marks and watch their movements. Take the buyer into your confidence in this question of flavour and obtain his agreement, but observe his criticism. You have to compete with persons who are giving attention to flavour, aroma and blending.

552. Large quantities of sweet clover honey, for example, are bought every year on account of its mild flavour, for blending with good honeys otherwise too strong in flavour. Bass or lime honey generally runs on the thin side. Here an addition of an extra dense clover honey will rectify the density and improve the flavour.

553. A light honey added to a dark honey will show a good average colour. A really dense honey added to one too thin will reduce the risk of fermentation.

554. The article fetching the highest price is not necessarily that most costly to produce. It is always that for which there is a strong demand, and generally something rather in the, fancy line.

555. If, however, some of your honey is really poor and does not show up well in any convenient blend, then keep it for spring feeding (not winter feeding) or sell it for some other commercial use than direct consumption (see **571**).

556. For section honey the demand is for grades light in colour. The darker honeys should be extracted. Usually, the late-season honeys are dark, and there is at that time a difficulty in any case in securing good sections.

557. Finally, if you do not get a satisfactory price, make inquiries and take advice. Do not undersell the market. Look at home for the cause of trouble and remember that from the selling standpoint it is important to blend clear thought with your honey.

Pure Food Laws

558.a Most countries have pure food, and weights and measures laws, with which the seller of honey should make himself acquainted, also laws or regulations relative to marking food and especially imported food and to importation. It is suggested that the latest regulations should be accessed from the FSA site. MG

558. *Most countries have pure food, and weights and measures laws, with which the seller of honey should make himself acquainted, also laws or regulations relative to marking food and especially imported food and to importation. The following is a recent enactment in Great Britain:*

EXTRACT FROM STATUTORY RULES AND ORDERS, 1928, No. 571

1. It shall not be lawful to sell or expose for sale in the United Kingdom any imported honey, or any blend or mixture of honeys of which imported honey forms part, unless it bears an indication of origin.

2. The indication of origin shall be printed, stencilled, stamped or branded on the container, or on a label securely attached thereto, indelibly and in a conspicuous manner, in plain block letters not less than one-twelfth of an inch in height when the greatest dimension of the package does not exceed six inches, and not less than one-eighth of an inch in height when the greatest dimension of the package exceeds six inches. For the purpose of this Part of this Order the expression "greatest dimension" shall mean the height, length or breadth, whichever is the greatest, of a rectangular or approximately rectangular package, and the height or maximum diameter, whichever is the greater, of a cylindrical, oval or conical package.

3. The form of the indication of origin in the case of blends or mixtures containing imported honey shall be, at the option of the person applying the indication, either:

(a) in the case of honey derived entirely from countries within the Empire, the word "Empire"; and, in the case of honey derived entirely from foreign countries, the word "Foreign"; or

(b) a definite indication of all the countries of origin of the honeys forming the blend or

mixture; or

(c) *the words "Blended imported"; provided that the indication "Blended imported" shall be applicable to any blend or mixture of honey, even though it contain honey, produced in the United Kingdom.*

4. *This Part of this Order shall not apply to exposure for sale wholesale if the person exposing the goods is a wholesale dealer.*

5. *The provisions of this Part of this Order shall come into force at the expiration of six months from the date hereof.*

War-time and Post-war Regulations

559. War-time regulations have been concerned mainly with the control of prices and of sugar for feeding. Price control is likely to be continued for a time after the war. The bee-keeper should make himself familiar with any local or national regulations he may be called upon to observe.

Honey for Show-Extracted Honey

560. Honey for show will be judged by flavour, density, colour and all that the judge can see. It must, therefore, be absolutely clean and clear and well put up in bright, clear, flawless bottles with clean polished caps and clean wads, got up according to the practice and rules of the show. The honey must be well ripened before taken, and should be transferred from the extractor to a ripener standing in a warm, dry room. This will assist the separation of impurities and allow the more dense honey to settle, but do not place the ripener where uneven heating will cause circulation. See, however, **437.**

561. The honey should be run off slowly through a well-dried, well-warmed strainer, more than once if necessary, avoiding trapping air bubbles. Old flannel is good for straining. Prepare more bottles than are wanted and select the best after the most minute inspection. Warm the bottles before filling, and fill above the shoulder to allow of skimming off any scum. Store in a cool place when bottled.

562. The glass bottles may be given a final polish with soft paper moistened with methylated spirit. The wads, if of cork, should be waxed or covered with waxed paper to prevent cork dust contaminating the honey. Waxed cardboard wads are superior to cork.

563. Granulated honey must be fine in grain. It is liable to contract away from the glass leaving frost-like markings. It is best to fill the jars from a larger store vessel, the honey being gently warmed through, using a water bath, until it will flow. Pour it or ladle it in and leave it for a few weeks where it will not be subjected to sudden changes of temperature.

Section Honey

564. In Great Britain the best cappings come from sainfoin or heather and excellent honey from clover. Lime is to be avoided, as the density of the honey is lower. Generally flora on heavy soils give good density.

565. Section honey must be the very best fancy grade, the wood-work, made absolutely clean by use of the scraper and fine sandpaper, and the whole got up according to the practice of the district and rules of the show.

566. It is a good plan to use paraffin wax on the outside of sections (**1387**) and to paint the insides with melted beeswax which encourages the bees to build out comb to all four edges and makes it easier to remove any stains on the projecting wood.

567. Any honey in unsealed cells should be carefully removed with a camel's-hair brush, as it may weep. Sections should be carefully examined by transmitted light for lack of uniformity and other defects.

568. The ambitious exhibitor must visit shows, examine exhibits and ask questions.

Points for Judging

569a. The many different honey shows throughout the UK each have a schedule and set of rules which must be adhered to. A judge will then check for cleanliness, viscosity (in the case of clear honey), fineness of grain (for set honey), aroma and flavour. A points system is seldom used in the UK, other than for display and composite type classes, a comparative method being used in which only the best will win, and allows for entries to be downgraded if the required standard is not reached. MD

569. *The practice of different shows and different judges must be studied, but the honey cannot be too good in any particular. Flavour, density and appearance are the three most important features in order of merit from the buyers' and from the sellers' standpoint. There are, however, difficulties in fixing a .standard of flavour.*

570a. TableXV. An historic document from the 1930s listing the desirable qualities of honey in certain countries. There is no present day equivalent.

570. *Table XV gives the principal qualities judged and some indications of the pre-war practice of certain countries in assessing these.*

TABLE XV
POINTS USED.IN JUDGING HONEY

	Victoria.	W, Australia.	British Columbia.	Irish Free State.
Section Honey:	–	–	15	20
Aroma and Flavour	–	–	30	–
Cleanliness	–	–	25	20
Absence of Pop Holes	25	25	15	–
Uniformity of Capping	–	–	15	–
Thinness of Capping	–	–	–	10
Weight and Filling	25	25	–	25
General Apperance	25	25	–	25
Neatness	25	25	–	–
Extracted Liquid Honey:				
Aroma	10	10	–	–
Flavour	40	35	15	25
Density	30	20	35	30
Colour	10	20	25	10
Brightness	5	5	25	–
Condition	–	–	–	20
Appearance	–	–	–	15
Clearness	5	10	–	–
Extracted Granulated Honey:				
Aroma and Flavour	50	35	–	35
Density (Firmness)	–	30	–	–
Fineness of Grain	–	–	–	35
Appearence	–	–	–	15
Colour	30	10	–	10
Regularity of Grain	20	25	–	–

Uses of Honey

Use in Cookery and Confectionery

571. Before the introduction of sugar, honey was the sole sweetening agent used in cookery and confectionery. Owing to its superior flavour, moisture-retaining properties and .dietetic value, it is again coming into favour for such uses. Honey which is too strong in flavour for ordinary use may be used to advantage in this way. In cake-making particularly, honey is superior to invert sugars, molasses, glycerine and common sugar, giving a superior flavour and appearance and proving most valuable as a moisture-retaining agent.

Recipes for Cooking, Using Honey

572. Many hundreds of recipes have been published in which honey is used. The reader who masters the contents of the paragraphs following may make his own recipes, as almost any good recipe in which sugar is employed can be converted into one employing honey; moreover, the reader may make the conversion correctly, which is not always done.

Sugar Equivalent

573. It should be noted that honey contains about 20 per cent. of water and this should be allowed for when substituting honey for sugar, by reducing the water or other liquid used. For every cupful of honey used, use one-fifth cup less of water or other liquid. A cupful of honey goes further in cooking, however, than a cupful of sugar. It is economical and effective to use honey in substitution for half the sugar, and the total may be reduced, say one-eighth to allow for the honey going further, measure for measure. With these allowances any recipe for cakes and bread, and many for sweets of various kinds, may be adjusted for the use of honey and improved by its use. Honey goes well with spices, dried fruits, nuts, delicate flavourings, such as rose and orange-flower water, milk and cream, but not with vanilla.

Neutralizing Acid

574. Advice is sometimes given to use bicarbonate of soda instead of baking powder, so as to neutralize the acid. This is unnecessary; honey goes well in recipes with either soda or baking powder, the amount of acid being negligible from the standpoint of cookery.

Relative Sweetness

575.a The sweetness of honey varies considerably according to proportions of fructose, glucose and sucrose it contains. Taking cane sugar (sucrose) as 100, fructose is 173 and glucose 74. This accounts for the sweetness of the honeys from fruit trees, which contain a relatively large amount of fructose.   ID

575. *The sweetness of honey varies considerably according to proportions of levulose, dextrose and sucrose it contains. Taking cane sugar (sucrose) as 100, levulose is 173 and dextrose 74. This accounts for the sweetness of the honeys from fruit trees, which contain a relatively large amount of levulose.*

Icing

576. Excellent icing may be made with honey or part honey and part sugar, making the necessary adjustment as in **573** for the water content.

Preserves, Sauces, Drinks

577. Honey may be used to advantage in preserving as a substitute for all or part of the sugar (**573**), and in sauces, jellies and fruit drinks.

Ice Cream

578. Honey may be used to great advantage in making ice creams, contributing to the flavour as well as to the sweetness. For this purpose it is best to replace the sugar entirely with honey, a blend of honey and sugar being less successful. By weight use about 25 per cent. more honey than you would sugar, say 16 to 18 per cent. of the total weight of ice cream The mixture is somewhat harder to freeze and melts rather more readily than if sugar is used, but the flavour may be greatly improved. The result depends upon the honey. Clover, alfalfa and lime (bass) are all good, the last-mentioned having a slight minty flavour. The stronger flavoured honeys are not so popular for ice-cream making. Avoid vanilla flavouring, also chocolate, and of the fruits, strawberry is among the less successful, though other fruits are good and especially pineapple.

Toilet Preparations

579. Besides its direct use as a food and indirect use as a sweetening and moistening agent in cooking, honey is much used in toilet preparations for the skin and hair, acting as a skin food and being useful in place of glycerine as a moisture absorbent.

Anti-freezing Mixture

580.a   It is advisable to use commercially available anti freeze solutions for motor cars etc. These are specially formulated to prevent damage and using honey for this purpose is a waste of a high value commodity.   ID

580. Honey also has been used in anti-freezing mixtures in the radiators of motor-cars, where it has the advantage that it does not evaporate or attack metalwork. Subject to there being no leakage, it is only necessary to make up the level with soft water from time to time. A mixture of half honey, half water may be employed, but where very low temperatures are met with, it is advisable to add alcohol. For temperatures of -15to -20°, stir in two parts honey to one part boiling water, boil and skim if necessary, then add 10 per cent. of alcohol wood spirit.

SECTION VI
THE APIARY-MOVING BEES

The Apiary

Location near Honey Sources

581. Bees will fly several miles for honey (**411**), but it is not satisfactory to keep bees commercially except in places where there is an abundant source of supplies within a radius of, say, two miles, and preferably less. This radius should include sources such as are in List 1.

582. On account of the radius limitation, apiaries are generally limited to about 100 stocks, and when this limit is exceeded out apiaries are started. In really good districts 200 stocks can be kept at one spot, and in the exceptional area of the Australian eucalyptus forests several times this number may be kept without disadvantage.

Pollen and Water Supplies

583. While access to good honey plants is essential, it is important also to have ample supplies of pollen, especially in the spring, and to remember that the economical distance for flight for pollen-gathering in the early days of brood rearing is only about 100 yards. Bees will fly much farther if they have to, but not without loss. Honey they can take from store, but fresh pollen and water they must get, even in poor weather, when breeding fast; as the season advances pollen may be gathered within say half a mile radius (see also **411**).

Prevailing Wind

584. It is advantageous if the apiary can be so located in respect to the principal source of harvest that the prevailing wind is towards the apiary so that flight unladen will be against the wind and laden with it. Similarly, it is advantageous to have the apiary at a lower level than the crops. Nevertheless, bees have been known to rise 5,000 feet after nectar. It is also advantageous to have a site with a slope towards the morning sun.

Floods, Fires and Vibration

585. Avoid land which may be flooded, and in certain districts forest fires also are to be dreaded and avoided. Bees have been kept on railway embankments, but they object to vibration and are apt to build much brace comb (**316**), and are more apt to swarm.

Fencing and Shelter

586. The site should be fenced so as to secure against intrusion of animals, and where the bees fly over neighbouring land it is well to have a high fence to direct them upwards. This need not be a close fence, as the bees prefer to avoid obstacles by rising above them if the way is clear. Wire netting of not more than 1½ -inch mesh is effective.

It is a great advantage to have a wind-break against cold winds blowing towards the hive entrance, and tall trees at one side giving shade during the middle of the day.

587. When placing a few hives in a garden, select a site or sites giving protection and shade of this kind. If a hive is to stand by a path, have the back of the hive facing the path, the front some wind-break, and trees or a wall giving some mid-day shade.

In an exposed position a wind-break for winter can be made with rot-proof coarse canvas or sacking mounted near the hive and partly surrounding the entrance.

Liability to Neighbours and Others

588 A   A beekeeper may be held liable in Nuisance or in Negligence.   To keep an unreasonable number of colonies or to keep them in an unreasonable place, for example too close to the boundary of another's land, could be found to be a Nuisance if the bees caused damage to the neighbour. Similarly damage caused by negligent manipulation when examining a hive could give rise to a claim in Negligence, particularly if the bees were known to be ill-tempered.   D.S

588. *An apiary of any size should not be established within 2 miles of another. It is not fair to the man on the spot or good for the newcomer.*

The owner is liable for damage done by his bees to persons or animals. He should not keep ill-tempered bees or make good bees ill-tempered. It is wise, however, to insure against accident. Particulars of insurance schemes should be obtained from the Secretary of the nearest Bee-keepers' Association.

589.   Inquiry should be made for particulars of any legislation relating to apiaries, through the nearest Bee-keepers' Association, or this failing, from the appropriate Government Department. For example, in Ontario, apiaries must be registered.

590. In Australia, in the forest area, application may be made to the Conservator of Forests, Melbourne, for a site of about 10 acres in which to establish a bee farm. For a further payment of about four guineas, a sole licence may be obtained for a "bee range" of one-mile radius securing the occupier of the farm from competition in his area.

Out Apiaries

591. Many of the above observations relative to location in general apply with increased force to out apiaries, as inspection and oversight cannot be so good as in the home apiary. When out apiaries are infrequently visited, bees tend to acquire a sense of safety by isolation, and resent the intrusion of animals, including man. Such bees are more apt to sting than are those in the home apiary, or one regularly visited. Pettigrew advocated installing a scare-crow, moved occasionally.

Apiaries in Orchards

592.   Where fruit-growing is extensive it is very desirable to introduce bee-hives into orchards. In favourable cases a gain of as much as £50 ($250) per acre has been recorded by the fruit-grower, due to better fertilization brought about by the bees when seeking pollen. Complete pollination is necessary to secure that apples and pears shall be fully developed on all sides. Hives should be distributed at the rate of one per acre.

593.   For orchard work skill is required on the part of the bee-keeper to produce strong breeding colonies early enough in the spring and in extensive fruit-growing areas, to maintain them successfully as the fruit-blossom fails. It is frequently necessary to move the hives to other districts at this time.

Payments by Fruit-growers

594. Fruit-growers are willing to pay competent bee-keepers for maintaining bees in their orchards. In British Columbia a payment of $5.00 (£1) per hive is usual, and where the conditions are difficult more is paid. In Great Britain, payments up to 25*s.* per hive have been obtained.

The grower of apricots and peaches should not pay for bees, as he has to thin the fruit in the ordinary course, but he may keep bees to benefit by the crop if the district is favourable also for a later flow of nectar.

Package Bees for Orchard Work

595. Owners of orchards will sometimes pay up to 20*s* to 40*s.* ($5 to $ 10) for a package of bees designed so that the bees can take care of themselves for a few weeks. On receipt the package is set up with the entrance open and after pollination is over the bees are destroyed. Such owners generally learn later to save the bees and become bee-keepers (**1332-48**).

Spraying in Orchards

596. If poison sprays are used during the nectar flow many bees may be poisoned. Spraying should be done before the blossoms open and repeated just after they drop. This method is effective against the Codlin moth. The bee-keeper should come to an understanding with the owner of the orchard re spraying. For further details see **1549-51**.

Town and Garden Sites

597. Bee-keeping in gardens and towns is carried on as a pleasant and profitable hobby, but hardly as a means of livelihood. Many of the observations above as to location apply with equal force to such apiaries, and especially the necessity of considering one's neighbours. Most of what follows as to arrangement also applies, but the bee-keeper will be able more readily to select suitable sites for a few hives in a garden than he can make suitable sites in a large apiary.

Preparation of Stands

598. Permanent sites for hives should be prepared, the ground being cleared and weed-killer used if necessary; also a good layer of ashes applied, or better still, cement or cement slabs, set in the ground, so that the grass or weeds between may be mown. Sheep with a good coat of wool may be safely and profitably employed to keep down the grass in a large apiary, if they are well dipped so that they will not have to rub against the hives. In small bee gardens, the ground may be sown over with arabis, aubretia or other hardy, vigorous and short-growing spring plants, and will provide some pollen handy for early use. Gooseberry-bushes or other crops wanting but little attention in the season may be grown, as well as apples, etc.

When setting out the hive sites, and later the hives, a level should be used. A useful rough level may be formed by placing a saucer of water on a short board.

Arrangement of Hives

599. At times when temperatures are low, the bees will not fly unless the sun shining on the walls of the hive has caused the bees to stir. They will then send out a few bees who will report faithfully. The bees may be trusted to utilize the conditions to best advantage. Thus it is not important that the hives face the sun, but helpful if the early sunshine falls on one of the thin walls. The considerations in **586** are the important ones. In locating hives by trees, evergreen should be avoided, but an apple-tree giving shade from noon onwards is excellent. Sunflowers, or Jerusalem artichokes, planted on the right spot, will give shade at the right time in the hot weather. It is better to avoid having many high branches right over the hives causing unnecessary disturbance in the winter from dropping water.

600. Notwithstanding any general protection of the apiary as a whole from wind, there should be some wind-break to individual hives or rows of hives, so that the prevailing cold winds of winter and spring do not blow in freely at the entrance. Free ventilation is important, but cold winds cause far more loss of bees and stores than many bee-keepers realize.

601. Hive stands should not be arranged in serried ranks, but care taken to secure that individual hives are readily distinguishable from their neighbours, or excessive drifting will occur and queens will be lost on returning from their mating flight.

602. Bees locate their hives much as human beings do their houses. It is what is seen during the approach that is most helpful to the bees in locating their homes. The hives may be arranged in lots of 1, 2 and 3, with odd spaces between, and in particular, distinguishing marks should be provided by bushes or other noticeable objects towards the front. A few bricks or large white stones may be used to advantage to help location. Coloured objects are also helpful, or large discs of colour on the bodies near the entrances. It is not much use painting supers different colours, as it is not convenient to maintain a colour scheme when supering. Bodies also should not be coloured differently, as this hinders certain manipulations.

603. The hives in rows or groups should be arranged so that flight from those in one row is not materially impeded by an operator working on hives in the row in front. It is undesirable to work near the line of flight when the bees are troublesome.

604. When hives are located in a small garden or plot, well sheltered by hedges or fences, it is convenient to arrange them around the sides so that access is obtained from outside the group, the hives all facing inwards, using the best aspect for a majority of them.

Honey House or Shed

605. The bee-keeper even with the fewest hives requires a place in which to keep his appliances and in which to handle honey, and is not always welcome with his belongings in the rooms of the home. The largest bee-keeper will have permanent or temporary buildings for extracting, packing and storing, for appliances, a workshop, offices and other conveniences. For an apiary of 40 to 50 hives a honey house 20 feet by 15 feet is none too large. It must be bee-tight but well ventilated, any ventilating openings being covered with wire cloth preserved with, say, aluminium paint; or perforated zinc may be employed.

Windows

606. It is convenient to have the windows to open inwards, the openings being covered outside with gauze carried about 6 inches above the openings and supported 3/8-inch away from the frames at the top so that bees working upwards within may escape. Some use windows pivotally mounted so that they may be rotated through 120° or more to let imprisoned bees escape, but they cannot then be left open.

Walls and Floor

607. A closed wooden shed with single roof is apt to become exceedingly hot. It is better to use a wooden frame with double walls and roof. The inside may be match-boarding, convenient for the reception of shelves and fittings and the outside asbestos boards. A light brick or breeze block building with boarded and tiled roof is better still, but considerably more costly. The floor must be flat and solid to carry heavy tanks and an extractor. Creosoted wood is used or cement or wood blocks on cement. The principal windows should not be exposed to strong sunshine.

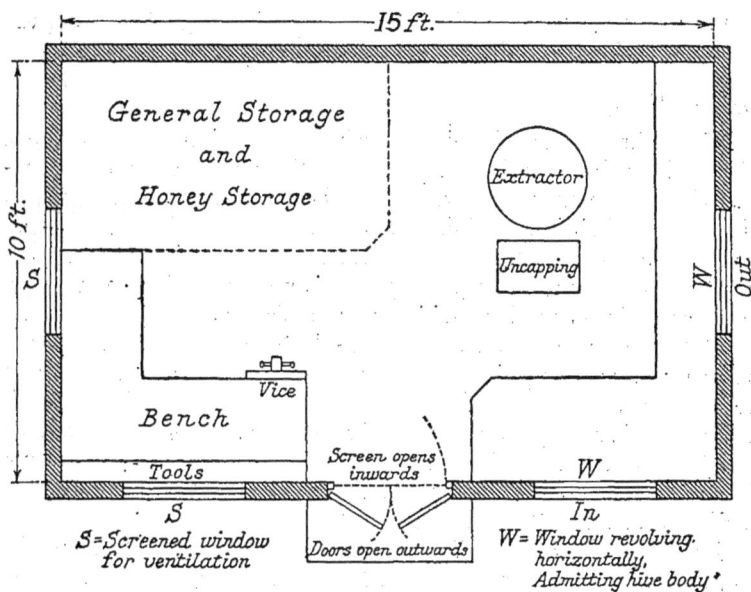

General Storage
and
Honey Storage

Extractor

Uncapping

Vice

Bench

Screen opens
inwards

Tools

S

S = Screened window
for ventilation

Doors open outwards

W

In

W = Window revolving
horizontally,
Admitting hive body *

FIG. 4.—HONEY SHED.

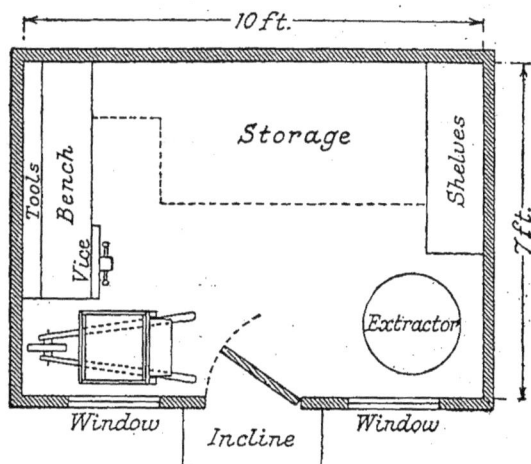

Storage

Shelves

Tools

Bench

Vice

Extractor

Window

Incline

Window

FIG. 5.—HONEY SHED.

On a sloping site in a region where cellar wintering is necessary, the cellar may be built into the ground and the honey house over.

General Arrangement

608. Arrange to minimize the labour of handling heavy weights. The ideal is to bring in the heavy supers at the upper level, uncap somewhat above the extractor and run off the honey to tanks at a lower level with access again below for filling from the tanks, thus utilizing gravity to the full throughout. The labour cost, however, of lifting a few tons of honey a few feet in lots of 30 to 50 lb. is not great, thus most honey houses are built on one level with perhaps a platform at one side or end.

609. More and better work can be done and more material stored in a well-planned honey shed than in one twice the size occupied without prearrangement and allowed to get into the state of disorder which is inevitable. Have a place for everything and everything in its place.

610. The outline plans (Figs. 4 and 5) are suggestive, but do not show the full use of wall space by shelves and cupboards, or storage above roof beams or under bench and under platform, if any.

Empty combs are generally kept in hive bodies, section racks and the like, tiered one above another from floor to roof with a cover on top. The pile should be moth-proof (see also **357-69**). Moths do not attack foundation, but it is undesirable to store foundation fitted in frames longer than is necessary before use because it will warp more or less. For Apiary Appliances and Clothing, see Section IX and **846**.

Winter Cellars

611. These may or may not be required. Their construction and use is dealt with in another section (see **989-1002**).

Starting an Apiary

Starting in a Small Way

612. Bee-keeping, like other industries, should be started in a small way. Do not expand until you have experience. Before commencing a beeyard as a commercial venture, it is well to gain experience as a pupil and hired assistant, preferably in one in a similar district to that you will occupy.

The amateur should commence with one or two hives, but after a few years and according to aptitude and opportunity, 15 to 50 colonies may be kept as a profitable spare-time hobby. The bee-keeper must study the written word, subscribe to a good local journal. He should become acquainted with neighbouring bee-keepers and join any local association.

Having satisfied himself that a certain type and size of hive suits his methods and district, he should keep to the one pattern, thus ensuring interchangeability of all parts. There is no bee-keeper with a mixed lot of hives but will readily admit that he often wishes they were all alike.

One-Man Apiary

613. Experience in U.S.A. shows that the most profitable type of apiary is a one-man apiary, the one man managing 350 to 400 colonies. One man with a junior assistant can manage up to 500. It may be advantageous to work bee-keeping in conjunction with some other type of agricultural activity.

Essentials of an Apiary

614. The essentials of an apiary besides colonies of bees, are spare hives, bodies, frames and foundation, and for personal equipment, a veil, smoker, hive-tool and notebook. Next in importance come receptacles for honey, an extractor, uncapping knife, super clearers, ripener and wax renderer. The bee-keeper should avoid multiplication of gadgets and devices requiring special manipulation unless he intends to make their use part of his standard practice. The aim should be to settle down to a standardized and simple method of procedure which suits his circumstances. Complications may be avoided by studying the craft and taking thought in advance. Everything should be planned in detail before it is done, even to the planning of alternatives in case of likely miscarriage.

Influence of Good and Bad Years

615. In many lands and in particular areas, the honey harvest varies greatly from year to year according to the weather. In many parts of Great Britain and of Australia, for example, good years and bad years fall in groups. Fortunate is the bee-keeper who can rely upon a good harvest from year to year. It is a form of fortune that may well be sought after by the intending commercial bee-keeper. A large commercial venture should not be made in a variable district just after a period of good years, as bad years are likely to follow.

Obtaining Bees

616. Beginners frequently start with a swarm. This is an old fashion and there used to be a superstition against paying for the swarm. The beginner may follow the superstition of not buying a swarm by purchasing from a reliable source package bees (**1332-48**) or a nucleus (**618**) instead. A swarm is liable to contain an old queen, may carry disease, and may come of a strain liable to excessive swarming. Package bees may be bought early in the season from a warmer region. Even if hived on foundation they will build up quickly, if fed, and will give a good account of themselves quickly. A nucleus may be bought about as early as package bees and can be built up quickly by feeding, to give a return later in the season. A full colony costs more but requires perhaps less skill in handling to advantage. It may be divided (**1446**), thus providing later two or even more (**1455-8**) stocks in condition to go through the winter without further outlay. Driven bees and queen can be bought cheaply late in the season in some districts and if hived on combs can be built up to winter safely (**1349**).

The established bee-keeper may make his increase in medium and poor seasons, or by purchase of package bees, according to circumstances.

Colonies and Nuclei

617. Bees with a queen and brood established in a hive are described as a colony. When found in nature the word "nest" is generally employed. A nucleus is the beginning of a colony. The word "stock" is also used for an established colony.

618.a These are the current guidelines for the supply of nuclei as agreed between FERA and breeders in 2011.

A nucleus is a well balanced colony between 3 and 6 standard brood combs. It should have bees, stores, brood and a fertile, young, laying queen. The frames should be clean and complete with no wax foundation and as little brace comb as possible. At least 50% of the combs should have brood present in all stages and it should occupy at least half the total comb area, with no brood cycle break. No more than the total brood area should be drone brood. There should not be any active queen cells.

Food requirements for nuclei will vary. As a guide a 6 frame nuclei should have two frames of stores and a full frame of pollen. There should be a good balance of adult bees of different ages and in a 6 frame nuclei 3-4 frames should be well covered. The bees should be good tempered when handled by a competent beekeeper. The brood should be healthy and show no signs of disease at any stage. A small number of cells showing chalkbrood is acceptable in UK nuclei. There should be no signs of wax moth on the combs.

Adequate instructions/ guidance on hiving the nuclei should be readily available to the customer. PS

618. *Nuclei and stocks offered for sale should always contain a fertile queen. All combs*

should be well covered with bees and those containing brood should be well filled. With colonies of six or more combs, all but two should contain brood in all stages, and these two should contain honey and some pollen. With 3- or 4-comb nuclei one comb only should contain stores and the remainder brood. This is important, because the price varies with the number of combs, whereas the value lies mainly in the bees and brood. The prices will vary also with the quality of the queen.

The above figures were embodied in standard practice in Great Britain for a number of years, but the rule has recently been altered to the less definite one that two-thirds of the combs shall contain brood. The fact that 4, 5, 7, 8, 10 and 11 will not divide by 3 suggests that what is intended is "the equivalent of two-thirds" of good combs of brood and the remainder stores. The term "brood" as here used includes "eggs," but there should not be an excess of young brood.

Partnership and Renting

619. Partners may, of course, go shares in everything. Generally, however, a sleeping partner (S) is sought by a working bee-keeper (W) who lacks capital. If W finds the site and all labour, then S should pay for the bees, hives, all material and appliances, and food where necessary in time of shortage, i.e. all outgoings. It is considered that S and W are then on a footing of equality as to returns, and an agreement based on the above should provide that each takes half the crop and half the value of the increase, leaving stocks equal in value to the original lot bought or carried over. W has to pay for hives and other materials for the increase. In case of sales of bees, each takes half the proceeds, but it is convenient for S to pay for W's share of increase kept, so that the apiary remains his property, he having paid for all hives, materials, etc. W will not work for increase unless he can handle a larger number of stocks or sell bees to advantage.

Moving Bees

General

620. It is frequently necessary to move stocks of bees, either in re-arranging an apiary, as part of a system of management, or for transport to a new locality. Bees have a strong homing instinct, some strains possessing it in a more marked degree than others. Nevertheless, although under circumstances detailed below bees will generally take note of a change of position of their hive, the fact remains that after removal considerable distances, or after removal from a cellar where they have long been confined, many bees fail to note their new location and are lost, or if not actually lost, drift to another hive, thus upsetting calculations and introducing risk of spreading disease.

Aids to Noting Location

621. It is well, therefore, in all cases where bees are moved, to take special steps to assist them in observing their new location.

If the new and old sites are within flying radius (**411**), some bees who have noted their new location may, nevertheless, forget it and use an old line of flight to the old site. Bees that have been well shaken during transit and imprisoned for some hours will take note of their surroundings when released. Those moved quietly and released at once are likely not to. Queenless bees, requeened at the time of removal, readily note their new site. The bee most likely to go wrong is one flying out fast and freely; so put a hindrance in front of the entrance to make her stop and take notice; a sheet of glass is sometimes used, leaning against the front and giving access at the sides. Stuffing the entrance with grass, which the bees have to remove, is effective, but see that ventilation is not stopped, especially in hot weather. A more certain way is to imprison the bees for 24 hours in a cellar or other cool place, but with plenty of ventilation, as there will be considerable disturbance and activity within the hive when they discover their imprisonment.

Distinguishing features in the foreground of the hive are as important as aids to location as distinguishing features in the appearance of the hive itself.

Influence of the Weather

622. In the coldest weather, hives may be moved in the apiary if they be lifted and put down very gently so as not to disturb the bees. If a longer move is necessary, it should be made when the weather is suitable for flight, but preferably not too hot, as in hot weather it is difficult to provide adequate ventilation for confined and disturbed bees.

623. When the weather is such that daily flights occur, hives to be moved in the apiary must be moved very short distances per day so that the bees may observe the change of location. The entrance should not be moved more than about 12 to 18 inches sideways in one day, or more than 3 feet backwards. If another hive is so located that it may readily be mistaken for the hive moved, it will be necessary to make smaller moves, as bees seeking the new position will drift to the other hive. If a hive is to be rotated it should not be turned through more than half a right angle between flight days.

624. If, therefore, it is desired to move a hive to a different part of the apiary during flying weather, it is better, unless there is a good honey flow, to close the hive one evening and ventilate as for a long journey and move it the next day, shading meanwhile if exposed to sunshine. The internal disturbance and the use of a glass screen (**621**) will cause the bees to take note of their new position.

Long-distance Moves

625. When a stock is to be moved any considerable distance steps must be taken to secure the bees, to secure adequate ventilation, and to secure the hive. The hive should be prepared after flights have ceased for the day, or in the early morning before flights have commenced, so that all flying bees are at home. It is necessary to secure ample ventilation, as the bees become excited by the vibration of removal and there is much activity in a confined space. A ventilating screen should be furnished both for the top and bottom. It is best to close the mouth with some solid block which can be removed to give flight on arrival at the new permanent location. If attempt is made to ventilate through a screened entrance alone, there is danger of suffocation, as the bees make for the entrance position in their efforts to escape.

626. If a screened floor-board ventilator is furnished it should be opened. Failing this a temporary floor board must be furnished with ventilating screen. The top of the body should be covered with a ventilating screen in place of the cover. A super clearer of the ventilated pattern will serve. If the cover is used to keep off sun or rain, spacing pieces must be inserted beneath it so that the top ventilation is not stopped. However screened, the bees travel more quietly if the screen or screens are shaded as just indicated.

Critical Distances

627. The most critical move is one within the original flying range or to a distance such that the new and old flying ranges will overlap. This generally leads to a number of, shall we say absent-minded, bees finding their way to the old site. If hives have to be moved in warm flying weather when the stocks are strong, the queen and part of the brood, with enough bees, may be moved one day and the remainder 2 or 3 days later, re-uniting by the newspaper method. The additional disturbance makes the bees alert to observe the new location.

Preparation of One-piece Bodies

628. Where one-piece bodies are used it is convenient to have handy some boards of the type shown in Fig. 25, the battens being on the outside when the boards are in place, and so spaced that the bottom of one hive will stand safely on the top of another. (See also **888**.) Where two bodies stand one on another, the joint may be secured with sticky tape, or a 2-in. strip of gummed paper, sold in rolls.

Preparation of Hives with Outer Covers

629. If the bodies are in two parts with loose outer covers the bees may be secured in the inner bodies and the outer covers used for shade only. If the outer cover is relied on for security the inner parts must be so blocked or wedged that even if the blocks or wedges shake loose, the bodies cannot come apart.

Securing Frames

630. Finally, while packing down, the frames must be securely fixed against swinging. However fixed, the bees will probably build a lot of brace comb when the journey is ended, but if not well fixed the bees may be crushed, the queen injured and great excitement and loss entailed. With self-spacing frames or well-fitted frames with metal ends, it is sufficient if the top screen fits against the frames or ends holding them down, and each body holds the frames in the one below. Failing this, strips of wood must be inserted to do so and these may have to be screwed in place. They should lie along the rows of lugs.

631. Travelling-boxes for delivering stocks and nuclei by train are generally furnished with racks to secure the bottoms of the frames, making a better job.

Heavy Hives

632. Heavy hives may be lifted by means of rope slings and a pole across the top, the pole preferably running the same way as the combs. The pole may be shouldered by two men.

Cart, Motor and Railway Transit

633. When moving bees in a cart it is usual to place the hives so that the combs lie across the cart, as the principal jolts are in this direction. When using a motor or train, the severest jolts are liable to occur length-wise of the vehicle. A light van will travel with a light load such as bee-hives much more smoothly than a heavy lorry. Corners should be taken slowly, and care used when starting and stopping.

In moving whole hives it is necessary to carry them the right way up, but if the frames are properly secured, the combs travel more safely upside-down, and this is readily managed with self-contained bodies furnished with top and bottom covers screwed on.

A skep may be covered with hessian, preferably having a wire screen securely sewn in, and they may be inverted into a box or crate for transit.

Moving to the Heather

634. In taking bees to the heather it is important to observe all the above recommendations re securing the bees, the ventilation and the hives. In addition, the hive should be so constructed as to admit of all these things being done and with a minimum of disturbance on arrival at the new location.

A Swarm

635. A newly-hived swarm will always note its new location, but if moved after some flying has occurred, it must be treated as an established stock. It is desirable, therefore, to hive a swarm in the position it is to occupy.

SECTION VII
HIVES AND THEIR ACCESSORIES
Hives and their Parts

Use of Skeps and Early Types

636. Bees are still kept in skeps made of straw, or reeds, or of osiers plastered with mud and cow-dung, in hollow logs, in earthenware pipes, and in wooden boxes, but the modern movable frame hive has practically superseded all the earlier types. Straw skeps are convenient for taking swarms and as a temporary housing for them. Their use occasionally enters into modern systems, but owing to the difficulty of controlling disease with hives with fixed combs, the use of skeps, boxes and the like as permanent hives is poor practice, and indeed in many places is prohibited by law.

Size of Hive in General

637. To obtain the best results the size of hive used must be suitably related to the prolificacy of the strain of bees, to the abundance of honey and to the method of management.

Even the skeppist altered the size of his hive by the use of an eke of straw to eke out the size, and frequently some form of super for surplus. The skep proper ranged in capacity from about ½ to 1 cubic foot, which may be contrasted with the single body of an M.D. hive, which has a capacity of approximately 2 cubic feet. An exceptionally prolific queen may fill 3 cubic feet of brood chamber. The relation between laying capacity of the queen and number and size of brood combs required for her is dealt with in **50**. The question of frame size is dealt with in **335-341** and in Section VIII.

Hives too Small and too Large

638. If the brood chamber be too small, either the queen will be crowded into the supers or the bees will swarm. If, however, the brood chamber is too large, there is a danger of loss of surplus. It might be supposed that excess stores might be left in the brood chamber from year to year and that the bees would then adjust the size of the brood nest in the brood chamber to suit the laying capacity of the queen. This may be the case in some districts and with some strains, notably where there is a sustained flow throughout the season, but in general it is not so. The bees having access to excess stores are liable to turn the stores into bees at a time which suits their ideas but not the idea of the bee-keeper. Thus a brood chamber can be too large.

Size in Relation to Management

639. The question of size in relation to honey flow and management is mainly, but not wholly, one of adequate provision of supers. On the one hand, the bees cannot fill space that has not been provided, and, on the other hand, if the brood chamber becomes crowded with stores swarming will certainly follow. The use of a brood chamber of ample dimensions is favourable to the omission of queen excluders. Strong stocks of bees with ample brood-chamber capacity will work in sections without any necessity for a queen excluder and will overflow into them from the brood chamber without the excessive crowding that is necessary with small stocks to persuade them to take to the supers.

640. Where the bees are managed by force of circumstances or by disinclination on a "let alone" plan, a brood chamber erring considerably on the large size is advantageous, but an alert bee-keeper will snatch a harvest where the "let alone" man lets it pass by him, and for this purpose it is essential that the brood chamber should not be too large. The hive body should be of a size to suit the strain of bees and district and preferably somewhat on the large side, but unless bees of fairly uniform prolificacy are used it will be desirable to help out the weaker lots from the stronger to equalize them, or to provide dummies on occasion (**686**) to occupy part of the brood chamber.

641. A brood chamber consisting of a single body taking, say, 10-11 British Standard or 8-10 Langstroth frames, is quite inadequate for the modern profitable bee; its use must inevitably lead to swarming. Two, and on occasion, three such bodies should be provided. Even those having bees of the hardy but non-prolific type should tempt them with an additional brood chamber (**1394**), or at least a shallow chamber.

The double brood chamber lends itself to several plans of management for minimizing swarming, widely and profitably employed. On the other hand, with suitable bees, the single chamber of adequate dimensions lends itself to let-alone methods which economize labour; moreover, the bees themselves undoubtedly prefer large one-piece brood combs.

Types of Movable Frame Hive

642. Hive bodies are generally made rectangular to receive rectangular frames, with a bee space at the ends, also between the frames in tiered bodies, and a larger clearance between the bottom of the frames and the floor board.

643. The brood body is frequently made square inside or outside to suit the ideas of those who like to employ combs arranged either parallel to or at right angles to the entrance. A form approximately square affords the least cooling surface for a given external surface, but sides in ratio 2 to 3 furnish only

about 4 per cent. less cubic capacity for the same side wall area. The depth of the frame should preferably, however, be less than the length, i.e. the frame should preferably have its larger dimension horizontal. If the frame is shallow, the bees are forced to place surplus above the brood chamber in their attempt to keep the brood nest of normal shape. The M.D. body having a ratio of approximately 2 : 3 in height to length gives about the ideal proportion, although good results are, of course, got with Langstroth hives having a ratio nearer 1 : 2 generally used double, and even with hives with square frames 1:1. With deeper bodies the uncontrolled surplus in the upper part of the body tends to act against the bee-keeper, as does a body too large (**638**).

644. The type known in France as the "automatic" is not used by English-speaking bee-keepers although made in England many years ago. The brood frames used therein are trapezoidal and one side of the body is extended to form the flight board so that all dirt and rubbish falls away automatically. In other continental hives hinged bottom boards are used for ready cleaning, also floor boards consisting of a rack of wood strips of triangular section between which the dirt falls out. The simpler and cheaper plain bottom board used by English-speaking bee-keepers is, however, more readily cleaned by the owner, and bees with Italian blood, in general use, are better house cleaners than the brown bee of the continent of Europe.

645. Double hives have been advocated able to carry two stocks, with or without some communication between them. The advantage is claimed that the two clusters support each other in winter, but in general the labour and skill required to manage double hives, bees being what they are, is not offset by the advantage gained when all goes according to plan.

646. Important differences are found in hives in the arrangement of combs either parallel with or perpendicular to the entrance, in the arrangements for expansion either horizontally or vertically, in the use of single or double walls and of external cases. These features are given detailed consideration below.

Perpendicular v. Parallel Way (Cold v. Warm Way)

647. Some hives are built so that the combs are mounted at right angles to the entrance, i.e. lying parallel with the sides of the hive and running from front to back. This is the universal practice with standard American hives and is called also "cold way." Others are built so that the combs lie parallel with the front or entrance, described as "warm way." Yet others have square bodies able to be placed either way on their floor boards, and some use these "cold-way" in summer and "warm-way" in winter. Practical experience indicates that in general there is very little to choose between the two plans as measured by returns obtained on comparative tests. There are, however, certain differences worth noting.

648. With the parallel arrangement the bees start at the front and extend the nest backwards, storing honey at the back of the brood chamber. With the perpendicular arrangement the bees store on either side of the brood. There is rather more risk of bees with inadequate stores becoming separated from their stores in very cold weather with the latter arrangement. With combs not too short and a bee passage over the top there is nothing to choose in this respect. The cluster is better protected from cold wind warm-way, but ventilation in hot weather is easier cold-way, and if the hive be screened from cold winds this gives the balance of advantage to cold way (see also **708**).

649. With the parallel arrangement access by the bee-keeper to the combs from the back is somewhat easier. Only one division board is used and this is at the back. With the perpendicular arrangement, access from the side is easier. It is said that with cold way the bees are more liable to rob wax from the corners of the combs for use elsewhere, making the damaged part good later with drone comb.

Horizontal v. Vertical Extension

650. With the more usual vertical arrangement of hive bodies and supers both the brood chamber and the super space can be extended. The parts used for expansion are interchangeable as between one hive and another, and the supers can be kept clear of brood. The ready transfer and rearrangement of supers and of brood chambers gives great flexibility to methods of management. Again, there is no limit to vertical extension, which on occasion can be made to provide accommodation for several hundred pounds of honey, while for wintering the capacity can be reduced, leaving a brood and store chamber of convenient form and size and no excessive space to be kept warm.

651. These are some of the advantages of the modern hive invented by man. The arrangements differ from any found in nature in that the combs are divided vertically, each box of combs being separated by a small space from the one below or above. This separation is unnatural. In nature, comb is built in one piece vertically, and the bees signify their disapproval of man's division by carrying on a constant warfare with the bee-keeper, he endeavouring to keep these horizontal tiers of comb separate and the bees endeavouring to join them into one continuous whole (see **336** and **778**). A bee nest in a hollow tree is expanded vertically, whereas in a roof, horizontal expansion is frequently to be found. There is not, however, much to be learned of practical import from nature in this matter.

The "Long Idea" Hive

652. It is, nevertheless, worth while considering what is to be said for the horizontal arrangement, or "long idea," as the Americans love to call it. The "long idea" had its greatest exponent in de Layens, and its use reached the greatest perfection in his school. Large numbers of de Layens hives are still to be found in France. The frames are square or somewhat deeper than they are wide and 20 to 30 can be placed in the hive, side by side, and perpendicular to the entrance. Two entrances are furnished in the front, but one is closed save in hot weather. The bees tend to produce brood near the open entrance. Others have overlooked the value of this feature in keeping a relatively compact brood nest and have tried a single central entrance without much success. There is no lifting or carrying save the carrying of combs. Every comb is accessible at any time on removal of the quilt and without disturbing any other. The combs being long, and guided at the bottom by pegs or staples between them, can be tilted one at a time for examination without removal from the hive after removing the two end ones. The size of the brood nest is quite unrestricted and swarming is infrequent. The labour of management is reduced to a minimum. The hive is good where let-alone methods are practised and the crop not too irregular. Simplicity and ease of management is an argument in its favour.

653. De Layens used no division boards or excluders, but it is quite practicable to use, in this type of hive, in the form of special division boards, most of the excluders, screens, clearer boards and the like employed in modern methods of management with the more usual vertically extended hive. While securing the advantages of such methods one retains the unique advantages that every comb and every item within the hive is individually directly accessible at all times, from the top, and that every comb is a large comb.

654. In some modern forms of body taking only 20 frames, one or more supers of 20 frames of shallow comb maybe used. Division boards are also used to conserve the heat. Doolittle claimed that the return in relation to volume of brood was less than with the vertical type, but in suitable districts the return per hive is at least as good and the labour less. The M.D. and British Deep frames are suitable for this system, but the system as practised in France has not been taken up by English-speaking bee-keepers, the hive suffering the disadvantage of one with too large a brood chamber (**638**). This objection can be overcome by the use of an excluder division board.

Single v. Double Walls

655. The problem of keeping the bees warm, or cool, as the case may be, is dealt with at length under packing, ventilation and temperature control, in **708-11**. In the present section all that is done is to distinguish the several types.

656. The terms single-walled and double-walled should refer to the individual bodies. A double-walled body or chamber must not be confused with a body protected by a separate outer case. A body should be described as having a double wall if the wall is constructed of two layers of wood with an enclosed air-space between.

657. Many hives have bodies single-walled at the sides parallel with the frames, and double-walls at the ends. The British National Hive (Fig. 15) is of this type. The modern forms of American hives are single-walled throughout, and there are British bodies made which are interchangeable with the National and having single-walled bodies. They are sometimes referred to as "single-walled National" bodies.

Outer Cases

658. The use of outer cases also involves a question of packing. It is touched upon here mainly from the standpoint of types.

An outer case of some sort is necessary if the hive is to have additional packing round the sides in winter. Outer cases have been used to provide shade in summer. During the active season such outer cases hinder and delay operations. In cold weather an outer case without packing material is of no material assistance in keeping the bees warm and may lead to the loss of a cleansing flight in a prolonged cold spell by hindering the warming up of the inner body during a short warm spell (see **709**).

659. The supreme example of the outer case is the bee-house, a weather-proof structure carrying the hives proper and with walls perforated for entrance to the individual hives. All manipulations are carried on within the house.

660. The use of the bee-house is much favoured in Switzerland, where the winters are severe and the takes of honey per hive are not large, but the Dadant Blatt hive, a form of the M.D. made to metric dimensions, and other forms of independent hives, are in increasing use throughout Europe. It is interesting to note some interest being taken in bee-houses by small bee-keepers in Great Britain, more especially in portable bee-houses which can be moved bodily from one source to another during the season. The owner of a bee-house is substantially independent of the weather; he examines his bees with a minimum of disturbance, and the bees are amenable to examination in a half-light.

Fillets or Plinths

661. These are strips of wood used to cover the joints in tiered outer cases. The fillet or plinth is placed so as to project below the lower edge and overlap the box below, thus keeping rain away from the joint. The upper edge of the fillet should be bevelled the whole width so as to throw off rain and not let it accumulate at the joint between fillet and body. The projecting part of the fillet is cut away inside to allow a clearance all round, as no two bodies are of just the same size (see Fig. 6).

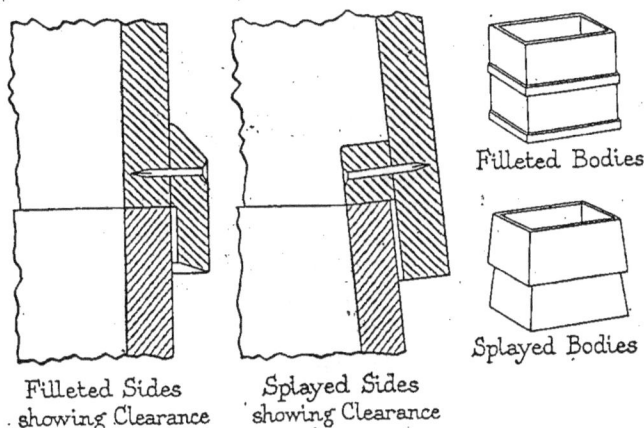

FIG. 6.-HIVE FILLET AND ALTERNATIVE CONSTRUCTION.

662. Fillets should not be put on the hive proper or any kind of break joint which prevents the free entry of the hive-tool to separate the bodies and prevents rotation of the body before separation, a motion frequently necessary to break attachments made between the combs of one body and another.

663. Fillets are costly to make well, and are mechanically the weakest part of the structure. Their use is generally avoided in Great Britain by the use of sloping walls as illustrated in Fig. 6. These splayed bodies, or more properly outer-cases, are sometimes described as telescopic, but they are no more telescopic than those with plinths.

Parts and Fittings

Floor Boards

664. The functions of the floor board are to hinder the bees from extending the combs below the frames in an irregular manner, to keep out the wind and other enemies, and to give the bees ready access from the entrance to all parts. Bees live comfortably in all weathers in a hive with no bottom, provided the walls extend well below the cluster, but will then build irregularly. The lower parts of the combs are normally used for breeding, and it is safe and advantageous to leave from ¾ to 1½ inches below the frames. Skill in management is required, in fact, to get combs built out *to* the bottom bar throughout (**315** and **351**). It is a good plan to insert an empty body between the floor board and brood chamber while the hive is wintered out of doors, thus improving ventilation and reducing draughts.

665. The margin in space below admits of giving the floor board proper a slope towards the entrance, starting ¼ to ½ inch clear of the frames at the back and finishing 1 to 1½ inches clear at the front edges. Such a floor board is so mounted in a frame that the frame sides may be set level to receive the brood chamber while the board itself slopes from front to back. They are also made reversible and sometimes so made as to give more clearance when used one way up than the other. Where the floor board does not slope in its own frame it may be given a tilt towards the front when levelling, so as to throw off water and assist the bees in house cleaning, but this plan is not good except where the frames are perpendicular to the entrance.

Legs and Stands

666. The floor board is raised from the ground to bring the brood chamber to a more convenient level for handling, and also to keep the entrance clear; sometimes also, to assist in keeping out ants. It should not be raised more than a few inches as the wind velocity is much lower near the ground than it is a little way above.

667. Floor boards or frames for them used to be provided with legs.

Legs are relatively costly and apt to rot, and hamper storage and transit. It is more convenient to use a simple framed floor with a separate stand. The stand may consist of bricks or tiles or two octagonal drain-pipes, or sometimes two parallel beams of wood or iron are set up to receive two or more hives. Again, pairs of cast-concrete blocks have been used, solid or hollow, of inverted V-section with flattened tops. Such supports may be 4 to 7 inches high. If ants are troublesome, it may be necessary to use legs standing in small bowls of heavy oil with projecting collar to keep rain from the oil, but kept clear of weeds. Alternatively posts may be driven into the ground to support the floor and an old tin inverted over each to hold a layer of axle grease. The ants cannot cross the grease. Posts made of Sweet Chestnut or Robinia last in the ground for many years; those of deal or other soft woods require impregnation with creosote.

Alighting Boards

668. The floor board extends, say, 2 to 6 inches beyond the brood chamber to provide an alighting board. Unless a substantial projection is provided, it is good to add an extension piece, which may slope to the ground and will save bees in bad weather.

669. With modern hives, having no porch or other shelter to the alighting board, rain falling on the front of the hive accumulates on the floor and tends to enter. It also increases the humidity of in-flowing air. Such hives are drier if the projecting portion of the floor is cut off. In a wild state bees do not find ready-made alighting boards or porches and do quite well without. On the other hand, an alighting board is convenient also as a promenade for guard bees and its provision has the advantage that it is easier for the bee-keeper to study their coming and going and accompanying revealing actions.

670. In the British National Hive (**746**) the floor board projects as in the American pattern. In addition, a stand is sometimes used, as also with the Langstroth, which provides a sloping extension. Some economy may be effected, and the objection to the projecting floor be removed, while retaining the advantage of a landing stage under observation, by combining the floor board and stand as shown in the author's design in Fig. 7.

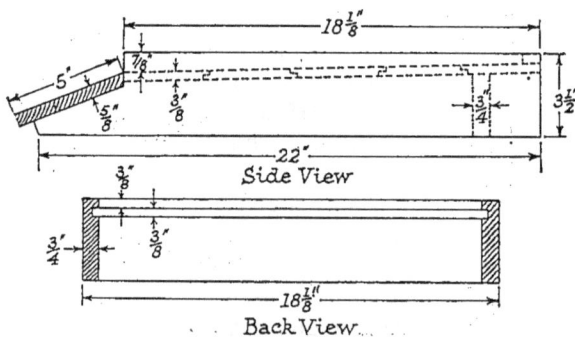

FIG. 7.—COMBINED FLOOR BOARD AND STAND.

Special Floor Boards

671. A special floor board is made to receive a brood chamber placed on top of a stock from which it receives heat and sometimes bees as well. This board is used to keep a small stock warm, for queen raising out of season, and other special- manipulations. It is constructed with a flat bottom, acting as cover for the hive below and shaped at top to receive a brood chamber, the entrance being cut in the thickness of the board in a part projecting to act as flight board. The centre is cut out and covered top and bottom with wire gauze. The two layers are used so that the queens above and below cannot reach each other even tongue to tongue. Sometimes pieces of queen excluder are used so as to admit the passage of bees, these pieces not being opposite each other, but there is some risk to the queen with this arrangement, and it is better to adjust the division of bees between the bodies by other means when there are queens above and below. The screens should be not less than ½ inch apart.

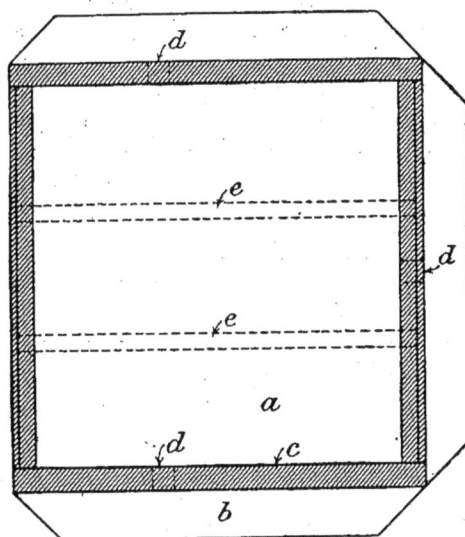

a. Base board.
b. Extensions of base for alighting.
c. Walls of brood chamber.
d. Entrances at base of walls.
e. Division boards.

FIG. 8.—SPECIAL FLOOR BOARD FOR NUCLEI.

672. A special floor board is also made for use in queen raising for placing over a strong stock as above or placing on its own stand. It has entrances in two, three or all four sides and is arranged to receive a body which is divided into two, three or four compartments to take two or three combs each, for use in fertilization of queens and to carry spare queens in nuclei (see Fig. 8). If the board has a level surface, it is more readily cleaned. The division boards should then reach to the board and don't forget packing under their lugs if there is a bee-way there.

A somewhat similar floor board, but with entrances on both faces, is made for use with the Snelgrove system of swarm control (**1171**).

Entrances

673. The simplest form of entrance fitting for a movable frame hive is that employed in the American pattern of hive and consists of a bar of wood fitting the gap between the front of the brood chamber and the floor board. It is removed to give a full-width entrance and is cut away on one of its faces, say, 5 inches x ⅜ to give a reduced entrance. If made square in section an alternative narrower entrance, say ¾ x ⅜, can be cut away on a face at right angles to that with the wider gap.

674. Where an outer case is used the entrance proper is in the outer case and should be connected to one in the brood chamber in such a way that the bees are prevented from obtaining access to the general space between the hive and the outer case. The inner entrance should be of full width and the connecting passage also. The outer case is frequently furnished with slides by which the entrance may be controlled. The "Swiss" entrance consists of metal slides so arranged that when the entrance is reduced to prevent robbing, ventilation is provided through holes or slots perforated in the slides and too small for a bee to pass through.

675. To keep out mice the entrance proper should not be more than 3/8 inch in height, but this limitation is only necessary when the bees are wintered out of doors. When outer cases are used for wintering only, they have a fixed entrance of appropriate dimensions, but this can be reduced if desired by the use of a heavy block placed on the alighting board (for Top and Middle Entrances, see **758-61**).

676. The use during winter of the full entrance in the British National and American hives is finding increasing favour. To keep out mice the full entrance is covered with coarse perforated zinc having an entrance way cut, say, 5 inches by 3/8. When removing the square bar to throw open the full entrance the author substitutes one made in accordance with Fig. 9, consisting of a strip of oak 7/8 x 1/4 inch in section built up to 7/8 inch square by adding small spacers on the top and bottom, thus giving two practically full-width entrances each limited to 3/8-inch. When there is danger of robbing, the entrance is temporarily reduced by the addition of one or more wood blocks placed in front of it.

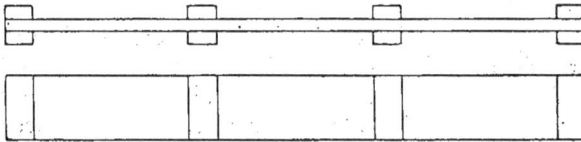

FIG. 9.—MOUSE-PROOF ENTRANCE BLOCK.

Brood Chambers

677. The brood chamber is designed to take eight to eleven or more frames plus a division board or dummy which may be prised out first, enabling the frames to be separated before removal. A 10- or 11-frame body is most commonly used. See also **652**.

As the bees do not raise brood in the outer sides of the outermost combs, or indeed in these combs at all, unless overcrowded, there is no real occasion to provide space for a division board at the side of the frames. It is more convenient to loosen the frames for examination by lifting out an outer one and leaning it against the hive than by removing a propolized division board. If the body provides, say, ½-inch space for a division board, substitute two strips of hard wood, 3/8 x ¼, using one on either side of the batch of frames.

678. The body is made one bee space deeper than the frames so as to retain a bee space between frames when bodies are placed one above another. This bee space should be below the frames, the top surface of body and frames being flush.

In American hives the bee space is above the frames. This has the advantage that it extends over the lugs, which cannot then be propolized to the body above; also, the bees have access to the joint between bodies, which they can propolize to keep out water. It is true that they propolize the frame ends to the side walls, but this has the advantage of holding them down when the upper body is lifted, and assists in holding all firm and steady when the upper body has to be moved to break comb connecting the frames in the two chambers.

The same advantages are got in the British National hive by leaving a bee space over the lugs, in the bottom of the body above, which space also serves for the release of any bees (and perhaps queen), which may have collected around or under the lugs. See Fig. 15.

Shrinkage of Wood

679. The makers of new hive bodies sometimes allow too much margin for shrinkage of the wood, with a result that the space between bodies remains excessive. It is difficult at all times to prevent combbuilding between bodies (**651**), but especially so if the clearance between frames is less than 3/16 or more than 5/16 inch. Bodies and supers should be checked after the first season's use and corrected in height if necessary.

Supers

680. Where the frames used are not more than about 9 inches deep, the same frames are frequently used, both for brood and for supering for extracting, so that one size of body alone is needed. Many bee-keepers avoid extracting from combs in which breeding has occurred and keep brood combs filled with honey as winter stores. Where one size of comb alone is used, however, and with some systems of management, honey is frequently extracted from old brood combs and the honey suffers no loss of colour or flavour thereby. Where shallow frames are used for extracted honey, the queen sometimes lays in them, especially if supers are given much in advance of requirements and a large entrance is used. This may be prevented by the use of a queen excluder beneath the supers (see, however, **702**). Shallow combs are more readily uncapped than deep ones, and a super of shallow combs is lighter to handle. Where heavy crops are not expected the use of shallow combs gives adequate but more gradual increase of space.

Frame Runners

681. In the simplest hives for use with frames with short lugs of the American pattern, the side walls are just rebated to receive the lugs. The lugs bearing on the surface of the rebate are liable to be securely propolized in place. For this reason, the rebate is generally cut deeper and a metal runner mounted on the edge, projecting upwards to support the lug and leaving a bee space between the under side of the lug and the rebate. The area of contact is then so small that the frames cannot be glued down in a manner to give trouble.

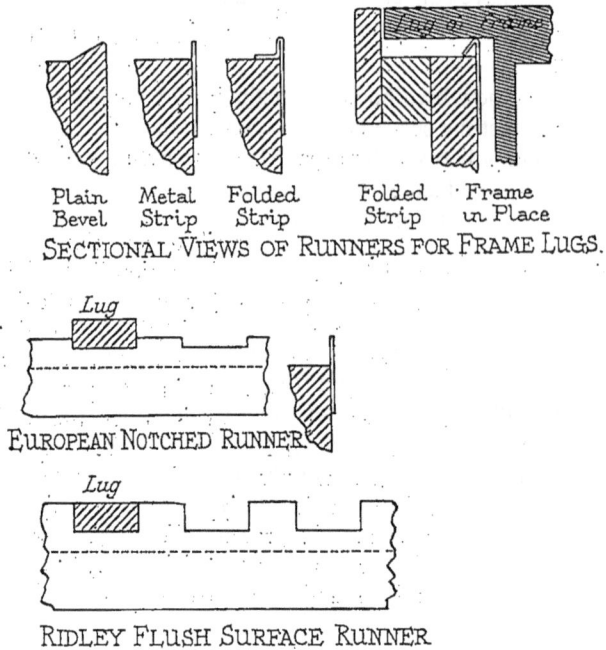

Plain Bevel — Metal Strip — Folded Strip — Folded Strip — Frame in Place

SECTIONAL VIEWS OF RUNNERS FOR FRAME LUGS.

EUROPEAN NOTCHED RUNNER

RIDLEY FLUSH SURFACE RUNNER

FIG. 10.—FRAME SUPPORTS.

682. Where longer lugs are used, as in the British frame, necessitating a built-up rebate, the board forming the wall next the frame end frequently projects to form the runner, its top edge being bevelled to present an edge 1/16 inch or more wide. Runners so formed cost less than metal runners and, if vaselined (**876**) when spring cleaning, are as convenient in use. See Fig. 10.

Propolizing of Lugs

683. It is customary to dimension the hive body so that the top bars of the frames are an easy fit between the end walls, so that the frames take their position from the end walls, thus securing a bee space at both ends of the body of the frame. If a bee space were allowed beyond the ends of the lugs, apart from other considerations, there would be a risk of frames becoming displaced endwise during manipulation.

Metal Runners

684. On the continent of Europe metal runners are frequently used, notched to receive the frames to save metal ends or other spacing means on the frames themselves. A modification by Ridley is excellent for super work for use with plain frames without any other form of spacing. He employs a notched metal runner, the notches being deeper than usual; in fact, the metal between reaches to the level of the tops of the lugs. These runners hold the frames securely while the body is being twisted to break comb built from body to body (**483**) and reduce propolization. The author has fitted such runners to the top and bottom of supers, one set giving standard spacing and the other wide spacing.

Section Racks

685. Supers for comb honey are described as section racks, being fitted with runners on which the rows of "sections" stand in rows. Between the faces of the sections "separators" are used to guide the bees to finish the surfaces of the combs flat and parallel with the edges of the section boxes. Separators may be of metal or of wood, the latter being preferred by the bees especially during an early flow. Plain flat separators are used with bee-way sections. Where no bee-way is cut in the section the separator has strips of wood on either face to give a bee-way. Similar strips are furnished at each end of the row of sections, one set on the body of the rack and one on the board which is slipped in at the end of the row and presses all together by means of a spring. Wood separators or fences with such strips make a better job than the plain ones, but a bee-way section, if clean, is generally found more attractive to the buyer. It appears larger.

Dummies

686. A dummy is a dummy frame. It may be of any thickness but hangs as a frame does, clear of the walls by a bee space. From the stand-point of conserving heat a dummy is but little better than an empty comb, whereas an empty comb can be occupied by the bees as soon as they require it. A dummy, however, if spaced a bee space from the end wall, will not be fastened down by the bees, and is preferable to a division board when the body is full.

Division Boards

687. A division board is made wider than a frame so as to touch the sides, and is as deep as the body in which it is used. It is used to reduce the capacity of the brood chamber when only a few combs are in use, and, on occasion, to divide the chamber into two.

The division board should be an easy fit; the bees will make good the joint. It is advantageous to have a baggy fold of American cloth over the ends, so that the board will enter easily, but fit snugly. When it is desired to divide a hive body completely, the space, if any; below the division board must be closed with a strip of wood and any space beneath the lugs behind the runners closed with wood or felt.

688. Division boards are sometimes made of queen excluder, or of wood with a window covered with queen excluder, and are useful in certain manipulations (see **242**).

Quilts and Inner Covers

689. The top of the frames are closed with a fabric quilt which fits all over and may be drawn back for inspection, or by a wooden board with a bee space beneath and which must be removed bodily for inspection. A quilt makes a good joint with the division board when contracting a chamber, for example, to receive a nucleus, and provides some upward ventilation. Except for weak lots, however, there is a lot to be said for the inner board; it is obtaining increasing favour; it is a permanent fitting, whereas flexible quilts require renewal as they get damaged or overloaded with propolis (see also Ventilation, **728** , and Packing, **714-16**).

Material for Quilts

690. Quilts of light texture, of cheap material, or laid on an uneven surface, are soon perforated by the bees. Strong unbleached calico and ticking are suitable, but sail-cloth and deck-chair canvas last longer. Of the non-ventilating type a cheap American cloth is excellent, but should be well packed above. Transparent sheet celluloid is nice when new, but after cleaning it soon becomes opaque. The tops of the frames should be scraped if necessary before laying on a new quilt. For packing above the quilt or inner board, see **714-16**.

On running out of quilting the writer has substituted a smooth surface grease-proof paper with success, but it will hardly bear usage again on removal, although usable through a long winter.

Glazed Inner Cover

691. Inner covers or crown boards are sometimes made of framed glass, with an opening for feeding, or of wood having one or two glass panels. When part is of wood a feed-hole may be made in the wood, adapted also to receive a bee escape.

A good deal can be seen through a glass panel without disturbing the bees. Where a wooden crown board is used one may still use glass or celluloid to cover the feed-hole, thus giving some view of the interior. In winter it is very necessary to have ample packing in contact with and above any glass panel.

If a glazed inner cover be used, it is well to vaseline the edges of its frame, so that the glass will not be broken when prising up the frame. Further, the clearance between glass and the frames below should be 3/16 to 7/32 inch. If made ¼ inch or more, much wax is built between glass and frames.

Hive Roofs

692. The simplest form is a flat cover with narrow rim fitting over the top of the hive but preferably spaced from it (**715**). If made with a deep rim, say 8 or 9 inches, it provides some additional protection to a hive having no outer case and serves better as a temporary stand for hive bodies during manipulation.

The deep roof cannot be blown off. The shallow roof should be secured with rope in winter, or a couple of bricks should be put on top.

The flat-topped roof is sometimes made with a sloping top to throw off rain. Such a roof is not so useful when off the hive, and if metal covered, there is no appreciable gain obtained in weather-resisting properties by making the top slope. A favourite pattern in Europe has sloping sides like the roof of a house, giving a free air-space in the top of the hive, but of no service when off the hive. The appearance is attractive, but such a roof is not so easily kept weatherproof as the rectangular type. It provides some room for a bottle feeder, but so does the deep rectangular pattern, or an empty super. When used on an outer case (**656**) the roof is generally fitted with a bee escape.

The Hive as an Ornament

693. From the purely ornamental standpoint, the old-fashioned hive with its fancy roof, porch and, maybe, coloured plinths, was more attractive in the garden than the strictly utilitarian American features adopted also in the British National hive; indeed, one has heard that certain large stores will not stock the more practical and economical type for this reason. Hive-makers might consider catering for this market by offering and illustrating such hives, with a W. B. C. type of roof at a higher price, also a porch as an extra, to be fastened to any body with two screws and located above the full-width entrance. Again, a stand such as that in Fig. 7 could be offered, with fashionable legs. If the purchaser expands his apiary he will soon learn what to discard.

Keeping out Water

694. T'he principal function of the roof is to keep out the rain and to provide shade and additional warmth. It is best covered with sheet zinc overlapping the sides by, say, 2 inches. The zinc should be secured with flat-headed galvanized wire nails. The corners, if cut for fitting, should be lapped and soldered, but they may readily be folded without cutting. Other thin sheet metals are used, including galvanized iron. The cut edges of tinned or galvanized iron should be tinned with a soldering iron, as it is here that rust starts.

695. The gable-pattern roof is generally covered with calico put on to a well-painted surface and itself well painted, but a good quality of bitumenized roofing felt is more readily applied and does not require painting. Do not use tin-tacks; they work out in time. Flat-headed wire nails are better.

Keeping out Bees

696. The roof should be bee-proof in all parts save for the bee escape, if any, and that lets bees pass only outwards. Deep roofs especially should be examined to ensure that they do not allow bees to pass upwards, when they may find a weak spot in the quilting and commence robbing.

Escape Boards and Super Clearers

697. There are many pleasures in bee-keeping, but the climax for most bee-keepers comes with the removal of a bumper harvest of honey. The removal of honey during a honey flow presents but little difficulty, but its removal after a honey flow when breeding has been checked and the bees are lying quiet to economize food, but ready instantly to deal with robbers, is a different story. The use of the escape board or super clearer comes in here. This is a board fitting the top of a hive body and provided in its thickness with a trap through which bees can pass one way only. This board is inserted below the supers to be removed and the bees pass down during the night, leaving the super substantially clear of bees for removal the next day. Sometimes 2 days are required for a real clearance. The board generally has a rim providing a bee-way over the under surface of one or both surfaces when the board is in place (see also **473** *et seq.*, also **691**).

Return Hole

698. Super clearers are generally provided also with a clear hole through which the bees can pass freely but closed by a slide actuated from without. This hole is normally closed, but is used to allow access by the bees to combs which have been extracted and require cleaning up. The slide is prone to be stuck down by the bees, so the means for withdrawal must be substantial. The writer prefers a plain hole near the edge normally covered with a square of metal which the bees secure, but which is readily prised off with a hive-tool. To close, the super is just lifted with the hive-tool and the square slipped in place, the position of the hole being indicated by a mark on the outer edge of the clearer.

Faulty Bee Escapes

699. Super clearers should be tested before use. Some are sold which will not allow a bee to pass through, particularly those with celluloid tongues, which tongues are frequently too stiff. Trap a few bees in a glass or basin and place against the clearer with the exits side towards the light.

Ventilated Escape Boards

700. Super clearers are sometimes made with openings covered with wire gauze on both sides allowing the passage of heat upwards. Such a board has other uses (see **102** and **1472**). The bees, however, do not pass down so readily as with a plain board, and there is more risk of the bees starting to pass down the honey unless the gauzes are well separated.

Canadian and Improved Pattern

701. A simple and efficient design, with no moving parts to get out of adjustment, has been introduced in Ontario and is shown in Fig. 11 A crown board is used having an entrance hole near one edge, covered with coarse wire gauze on the lower side, the gauze having a bee-space provided by the use of supporting strips of wood. The strips nearly meet but leave a bee entrance. On attempting the return journey the bees make direct for the hole and are held up by the gauze. Possibly a future generation of bees will discover the open return path, but the author's present experience with the device is satisfactory save that it necessitates too large a bee space beneath the cover. He has overcome this difficulty by cutting an opening in the crown board, shaped as in Fig. 12, which is made 5/16 inch thick. The top side is completely covered with thin sheet metal, having a 1-inch hole over the middle of the large part of the opening in the wood. The wire gauze, of stout wire, 0.024-28 inch diameter and 6 per inch, covers all the opening on the lower side save for patches 5/16 inch diameter at the ends to serve as exit. Recent experiments by the author suggest that a satisfactory clearer on this principle can be made to dimensions giving interchangeability with the Porter bee escape.

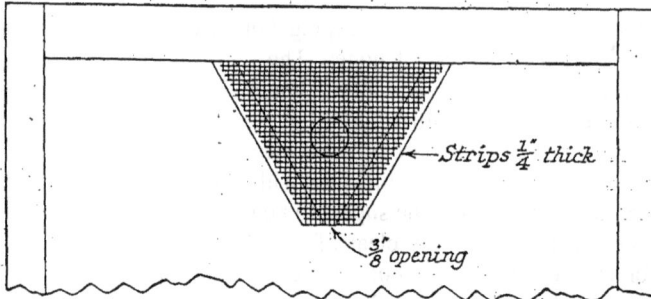

FIG. 11-CANADIAN BEE ESCAPE BOARD (UNDER SIDE).

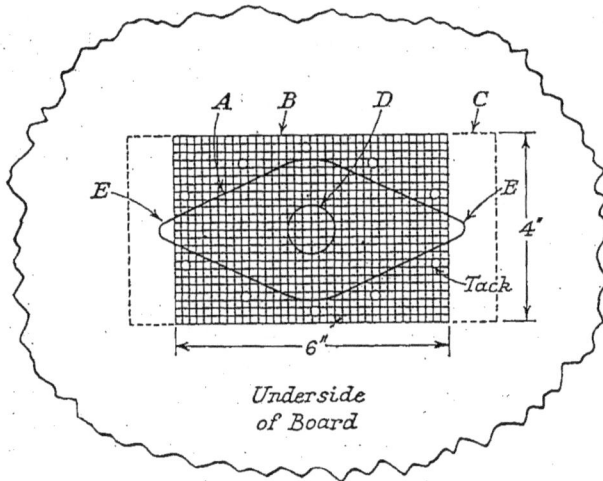

FIG. 12-IMPROVED CLEARER BOARD.

A. Opening in framed board $^5/_{16}$" thick.

B. Wire gauze, 6 mesh, over opening on under side.

C. Thin ply-wood, covering opening on top side.

D. Entrance hole in C.

E.E. Exit holes $^5/_{16}$" X $^5/_{16}$", not covered by B.

Frame projects $^5/_{16}$" on under side and ¼" on top side.

Return hole 1" diameter (not shown) is normally closed by placing a square of glass or thin over it.

wood

Queen Excluding

702. The bees like to store some honey in brood combs at least in the upper corners and in the outermost combs of the brood nest. The queen does not have occasion to travel over these parts in expanding the brood nest, and provided the brood nest is not overcrowded and the bees are not driven by cold and the depletion of stores in the supers to move the brood nest upwards, and provided there is no drone comb above, the queen may be effectively excluded by a barrier placed over the brood nest giving clear passage to the bees only at the corners and edges. Such a barrier is sometimes made of wood, but one may use one fitting snugly on the frames. A piece of American cloth (enamel cloth) is very suitable, placed smooth side downwards on the frames after the tops have been scraped. It may have the corners cut away 2½ inches X 2½ inches and should be about an inch smaller each way than the inside dimensions of the brood chamber, so as to allow a free passage-way all round. The bees will not build comb on to this as they will on perforated excluder. Drones have free passage past such an excluder.

Provided there is no drone comb above, 2 inches of honey over the brood is alone adequate as an excluder, so long as the queen has room to lay below.

Queen Excluders

703. Those who have not learned the art of doing without excluders (see e.g. **1393**) may prefer something more positive in action and will use an excluder which will admit the passage of workers but will prohibit the passage of the queen and unfortunately of drones also.

704. The earliest form was made of perforated sheet zinc with slots 5/32 inch (4 mm.) wide. The width 5/32 is 0.157 inch. This should be about 0.162. The queen is hindered by the size of her thorax, so it should be noted that a virgin queen cannot pass an excluder.

705. A form constructed of parallel wires securely soldered in or on to cross-bars finds increasing favour. The wires should be set with a normal clearance not exceeding 0.165 inch, which figure is in general use in U.S.A. where stiff wires are used, and in case they become slightly out of parallel, the user should correct any noticeable error. In the well-known Root pattern there are alternate strips of wood and of stout parallel wires with bee space on both sides of the wires.

Use of Excluders

706. The sheet-metal- or celluloid-pattern excluder should be laid across the frames so that the slots cross the frames. It is less liable to distortion on removal if then stripped off across the frames.

The wire excluder requires framing. It may be used either way across, but is used generally with the wires across the frames. In framing the wire excluder a bee space is allowed, but the sheet-metal pattern lies direct on the top bars of the frames and the top bars should be scraped when inserting the excluder.

A new excluder of sheet zinc showing the slightest burr from the punch should be rubbed with coarse emery-paper.

707. If drone brood is placed above a queen excluder, provision should be made for the drones to escape. They will fly back to some hive, no matter which. A hole may be provided to be afterwards filled with a cork, or the quilt may be lifted occasionally in sunshine until a fortnight after the last drone is due to hatch.

Packing, Ventilation and Temperature Control

Packing in General

708. It will have been seen from the particulars given above that different constructions of hive provide in different measure for keeping out the cold and keeping out excessive heat. The provision is frequently supplemented by some form of packing, above, below, and around. To avoid repetition in dealing with roofs, floor boards, and hive walls this question was deferred and is now discussed as a single issue but with several aspects.

Too Much and too Little Packing

709. One meets the bald argument that heavy packing keeps in heat in cold weather and keeps out heat in hot weather and must, therefore, be good at all seasons save for inconvenience in manipulation. This, however, is not the whole story. It leaves out the important consideration of the stimulating effect of sunshine penetrating the hive walls. This action of the sun stimulates breeding in the spring to an extent that frequently more than compensates for the loss of additional stores consumed to maintain the temperature during the cooler periods.

Bees can be kept in unpacked hives out of doors in severe sustained winter weather if of a hardy disposition and if kept dry. Similarly, bees can be kept heavily packed in mild climates or put in cellars. Thus, it is not a question of what is possible so much as of what is most profitable.

During prolonged cold spells the use of packing can be made to save food but tends to loss of bees, sometimes the whole stock.

Degrees of Protection

710. If a prolonged period of cold weather has to be faced, unbroken by short warm spells and lasting for several months, then cellar wintering is generally practised. The bees may thus be kept at a temperature at which not only is the loss of stores a minimum, but the excessive accumulation of waste matter in the bodies of the bees is limited. Bees cannot consume food indefinitely without a cleansing flight. If this is prevented many bees are forced to die and dysentery is set up. In the spring, when the frosts are breaking, the hives are brought out of the cellars and the bees brought into activity during warm spells. Free exposure of hive walls to sunshine is then advantageous. Later, during very hot weather external shade is desirable (for cellar wintering, see **989** *et seq.*).

711. When severe frosts are not as prolonged, or prolonged frosts are not so severe, and especially if mild spells intervene, the hives may remain out of doors protected by temporary external packing. The bees can then work later in the season and start earlier than if wintered in cellars.

This practice has been widely followed in Canada and parts of the U.S.A. A mistaken effort was made some years ago to introduce it into Great Britain, where it is quite unnecessary. In recent years, even Canadian bee-keepers have been experimenting with unpacked hives. The consumption of stores during winter is increased somewhat, but simplicity is achieved. Damp due to condensation is defeated by the use of a top entrance.

712. Where the winter is relatively mild and frequently broken by a warm day, hives may be left out of doors without packing, if well protected from cold winds but freely exposed to the sun.

713. If, however, there is a lack of early pollen and no early harvest to be expected, the bees may be retarded with advantage by the use of heavy packing and a darkened entrance, or even by the use of cellars, but there are other methods of management applicable to such conditions (see **1036** in conjunction with **1277** *et seq.*).

Packing at Top

714. From the standpoint of packing, the top of the hive is the most vulnerable point: vulnerable to high temperature by exposure to the sun which endangers the attachment of the combs at the weakest point; vulnerable to cold because the condensation of moisture most readily occurs at the top and is most detrimental to the bees at that point. If the top is ventilated, then heavy packing serves to conserve the heat, which would otherwise be carried away by the escaping air. If the top is not ventilated it is the more important to prevent condensation there. In cold weather, heavy packing at the top is most important. In hot weather, protection at the top is most valuable, and the information following applies.

715. Where a quilt is used, it is customary to employ several layers of carpet felt, old carpet, or even newspapers. With an air-space between roof and packing, this is sufficient for the warm season, but for the winter and spring it is good to add, or substitute, two or three inches of porous packing in the form of a cushion of pine sawdust or cork chips, chaff or the like, which will allow through ventilation and stop condensation. During a cold spell, the upper side of the packing will be found damp, but the lower side remains dry. On a warm day the upper side dries out. With the British National hive one can use a wooden rim of standard outside dimensions to carry the packing between a canvas top and bottom. This can be removed in one piece. A feed-hole is readily contrived by framing a central opening.

716. If a crown board is used, packing is still desirable even in summer, and with a glazed crown board it is still more important. In winter absence of top ventilation is compensated for, at least in part, by the use of a full entrance, but top ventilation is the reliable way of removing excess moisture. The feed-hole may be covered with porous material and soft packing used above as in the preceding paragraph.

Side Packing

717. The avowed object of side packing is to maintain a higher temperature within the hive, and has been the subject of much misconception, due in part to certain unfortunate experiments on hives having closed entrances and failing in other ways to reproduce working conditions.

Now in cold weather when the winter cluster is formed, experiments have amply demonstrated that in a hive with an open entrance freely exposed to the outer air, the temperature of the inner walls and floor approximates to that of the outside air. It may be one or two degrees higher but not more, but this is a negligible improvement. The facts are that it is the bees forming the outside of the cluster that constitute the packing that maintains the temperature within the cluster at this time; the hive walls serve primarily to protect the duster from being cooled by wind or excessive draught.

Now when breeding commences on any considerable scale a different set of conditions prevails. Larger amounts of food are consumed and of heat are produced, and economy of stores may be obtained by the use of side packing. To be really serviceable, however, it should be 6 or 8 inches thick. The use of substantial packing of this order in really cold situations may have up to 15 lb. of honey, but the use of a light outer cover, such as that of the W.B.C. hive, in Great Britain effects no measurable economy. The moral is - if packing is good and necessary, use ample; in the alternative, use none.

718. Where heavy wall packing is desired, it is preferably of a temporary character as it should be in place only during the cold months. Straw matting is cheap and readily stored, but is not often made in these days. Wood outer cases are good and, if made in sections, are readily stored, but tarred paper can also be used more than once as a support for leaves, bracken and the like. Whatever construction be used, the packing material must be protected from damp.

Wooden Winter Cases

719. These are large packing-cases made to take either two or four hives. The case consists of bottom, sides and top, made separately and bolted together so that when out of use the parts can be stacked compactly. They should be uniform in dimensions, including position of bolt holes, so that any side will fit any base and so forth. The top should have a rim fitting over the sides and should itself be covered over top and rim with sheet zinc, galvanized iron or bitumenized roofing felt so as to keep out all wet. Wooden battens on the floor raise the hives so that some packing may be placed underneath. The hives are placed close together, their ordinary covers, which project all round, being omitted. Space is left round the sides for a heavy packing of dry leaves or other handy, dry, close packing, 5 to 8 inches thick according to the district, and finally the top is packed 6 to 10 inches deep to the top of the case.

720. Openings are made in the sides opposite the hive entrances with some sort of slide to adjust the size of the entrance, and wooden tunnels provided through the packing. There should be no ledges on the outside on which snow and ice may build up so as to close the entrance. The tunnel should preferably be inclined upwards towards the hive entrance. Alighting boards sloping about 45° are furnished reaching from the ground to the bottom of the case, and therefore well below the entrance.

721. A four-hive case requires much less material per hive than does a one-hive case, but the latter can be made of lighter stock. The entrances in a four-hive case face in opposite directions, whereas with one or two-hive cases the entrances in a row of hives can be made to face all one way. A two- or four-hive case offers less cooling surface per hive than does a one-hive case. For equal protection a single hive requires packing more heavily all over than four hives in a single case, in the ratio of 8 inches to 5 inches.

722. It is convenient to feed, where necessary (**967-71**), after inserting the hives in the cases, so that the bees have the benefit of the extra protection before the cold weather sets in. It is convenient to put the top packing into a large cover of sacking, making a one-piece cushion to fit the whole top.

Tarred Paper Covering

723. Tarred paper may be used in place of wooden cases and in a similar manner to hold the packing. A stout quality is required and the top at least should have no open joints. Pieces may be stuck together with hot asphalt put on with a brush. If the whole is well corded and neatly made, the side joints can be made with 4- to 6-inch overlaps and the edges of the top piece folded over without cutting so as to shed the rain. The cord used should be tarred or creosoted.

724. It is convenient to use a rectangular wooden support at the bottom projecting 5 inches all round the hives. The tunnels are put in place (**720**) and tacked to the floor boards. A wooden frame of four sides hinged at two opposite corners and hooked at the other two may be made large enough to surround the packing and to reach level with the top of the hives. This is put in place temporarily to support the paper while packing. The wood bottom and bottom tarred paper are put in place, then the sheets for the sides with any joints placed at the flattest parts, then, after putting the hives and tunnels in place, the wooden frame is built round. Cords for tying over may be inserted with the paper. If the front is packed first the frame will be pushed away from the tunnel entrances. Pack the back first, therefore, then sides and front. The whole is finally securely corded front to back, side to side, and round the sides. Openings for the tunnels should be made and may be improved by tacking thin strips of wood over the outside.

Packing at Bottom

725. Owing to certain experiments with totally closed hives, misleading ideas have got about as to the importance of packing the floor board. Most of the heat lost by the winter cluster is, or should be, lost by convection and ventilation.

726. The temperature of the hive floor has a negligible effect on both convection and ventilation. Even if a foot thick, its temperature is still controlled mainly by that of the cold air entering on the same level by the entrance. After consideration of general experience and experimental evidence, the writer is not convinced that there is any measurable benefit to be got under working conditions from a packed floor board. A windbreak is much more important than this packing. A free space between floor and combs undoubtedly suits the bees and reduces draughts. If it is decided, nevertheless, to pack, then the bottom of the floor board may first be tarred and a second layer of tarred wood used with air-space between. If a winter outer case be used (**658**), protected floors are unnecessary save in the winter packing.

727. Bees will winter successfully in a hive with no bottom at all if protected from draughts, i.e. if the hive walls extend a long distance below the cluster.

Ventilation in Cold Weather

728. Ventilation is controlled by the bees except during the period of quiescence, when it depends upon natural air currents controlled by temperature, moisture, chemical content and especially by wind coming through the entrance. Foul air charged with carbonic acid is heavier than pure air, but moist air is lighter than dry air and warm air than cold air.

729. The bees produce heat by combustion of food aided by bodily exercise, bees within the duster going through certain exercises for the purpose. The gaseous products of combustion are mainly carbonic-acid gas and water vapour. The hot moist air leaving the cluster, rises; spreads over the cover, reaches the side walls, cooling as it goes and falling down the walls, finally crossing the floor and rising again. At the same time, some leaves the upper part of the entrance and is replaced by cold air entering and passing over the floor. Some air also escapes at the top unless the top is hermetically sealed, as it may be substantially if of impervious material. If wind is passing the entrance a more rapid change of the air within the hive takes place.

730. If the top is sealed, then a relatively large entrance is necessary, even a full width with a strong stock, but if a porous quilt and packing is used the entrance can be reduced to 1 inch. It is very desirable to use a substantial entrance. With considerable top ventilation a 4-inch entrance can be used in the summer, but if only a small hole is used at the top a full-size entrance may be used. With a substantial top entrance the bottom entrance should be closed, save in hot weather.

A tunnel entrance (**720**) requires less protection than a short entrance passage. A double entrance with chamber between acts like a silencer to puffs of wind.

731. If some British bee-keepers suffering from winter losses in the presence of plenty could be persuaded to try and kill their bees one winter by using excessive ventilation they would be surprised at the number of colonies they would save thereby. Stocks cannot be killed by cold air alone so long as it is only in slow motion in the hive.

Ventilation in Warm Weather

732. When the bees are active the consumption of stores is increased and the air supply for breathing must be much greater. The bees then take control of ventilation, some acting as ventilating fans by the use of their wings. If too large an entrance is provided no ventilators will be seen at work, and brood rearing near the entrance may cease. If the entrance is adequate for the bees to leave and enter freely in the height of the day, the bees can manage the ventilation, but in hot weather, especially if there is but little shade, the bees will welcome a full-width entrance.

During a honey flow many bees will be seen assisting the natural ventilation to remove the excess moisture from the honey even with a full-width entrance. The presence of bees fanning at the entrance at any other time suggests that the entrance is inadequate.

Floor Ventilators

733. Sometimes a ventilating hole is cut in the floor board of the hive, covered with perforated metal and fitted with a slide below to close it. The slide is not of much importance. Bottom ventilation is useful if the entrance has to be reduced to prevent robbing, but the perforated metal is not easily cleaned and hinders cleaning of the floor board. A ventilated entrance slide is better adapted for the purpose. For travelling it is, however, advisable to have ventilation in floor and roof or sides and roof (see **888**).

Temperature Control

734. While the bees can raise the temperature in the hive by consuming food and by exercise, they can also lower the temperature by taking in water and evaporating it by ventilation. As much heat is absorbed in evaporating a pint of water as would have to be supplied to raise it from freezing- to boiling-point six times over. The evaporation of nectar in the hive, and the removal of the vapour by the fanning bees, absorbs large quantities of heat at a time when the outside temperature is high. If occasion arises to confine bees, they should always be given water so that they may not become overheated. Water may be syringed into a comb or sprinkled heavily on the ventilating gauze used when confining.

735. In very hot weather bees having no intention or provision for swarming may be seen to leave the hive in large numbers and circle in the air as though intending to swarm. After a short while they will return, but many-will remain outside; young bees may be seen crowding out on the alighting board. Such activity is a sign of overheating and a clear indication that shade is required and should be improvised somehow.

Hive Patterns and Materials

Particular Makes of Hive

736. Size of hive has been discussed above in **637-8**, both in relation to the hive as a whole and to the portion containing the brood. The present section is mainly concerned with dimensions and peculiarities of particular makes of hives. The larger chambers are intended to be used singly; the smaller should be doubled. In relation to size of frame this difference is pursued again in **789**. In relation to management it is pursued in **906-10**.

Standardization of Dimensions

737. One of the main objects of standardization of hive parts and fittings is to secure interchangeability. As hitherto carried out this result is not in fact attained, so that to secure interchangeability it becomes necessary to purchase from one maker only, or to purchase a sample on approval before buying from a different maker. This is not only because the authorities frequently omit to specify certain of the essential dimensions, but mainly because they do not consider and provide for the effect of unavoidable errors in workmanship. No dimension can be worked to exactly, and some are subject to change with time, for example, by shrinkage. The maximum error will depend upon the materials employed and method of manufacture and should be covered by "tolerances" acceptable to the user and placing no undue burden on the manufacturer. This is universal practice in the world of engineering.

Present-day Practice

738. Not only are there no agreed limits, but frequently the wrong dimension is the one standardized. For example, on the continent of Europe it has been the practice to specify the inside dimensions of frames, no limits being given for outside dimensions or supporting lugs. The result is complete chaos so far as interchangeability is concerned. Again, in converting inch measure to centimetres, the figures have been rounded to even dimensions so that the Dadant Blatt parts, nominally M.D., are far from interchangeable with M.D. parts.

The Langstroth frame has been nominally standard for years but is found with two or three lengths and widths of top bar, and widths of side bars.

739. Certain dimensions for American hive bodies have recently been agreed amongst the leading American makers, also dimensions of top bars of frames. In Great Britain frames have been standardized for many years, and certain dimensions have been agreed for hive bodies, but not enough to secure interchangeability.

The latest dimensions and some earlier dimensions are given under the names of the various makers below.

740. In most hives of modern design the tops of the frames are flush with the tops of the side walls of the chamber, super clearers and queen excluders being framed to cover the whole surface of the frames and walls, but in American hives a bee space is allowed above the frames in the chamber, the frames being flush at the bottom, or nominally so (see **678**).

The Langstroth Hive

741. This hive is given first place in honour of the introducer of the design with bee space all round the frames (see Inventions and Discoveries, **1570-1** and **1583-4**), and because it is the most widely used by English-speaking bee-keepers the world over.

Important dimensions of the bodies have recently been agreed in the U.S.A. It would be useless to detail dimensions used in the past, but certain dimensions used in Australia and British Columbia are given in the table below.

The new U.S.A. figures allow for shrinkage after manufacture and use.

A shallow body is also made to take the shallow extracting frames which are 6¼ inches deep.

The B. C. body is used with an outer cover not standardized.

The floor boards of these hives extend 2 to 4 inches beyond the body.

742.

TABLE XVI
DIMENSIONS OF LANGSTROTH HIVES (see Fig. 13)

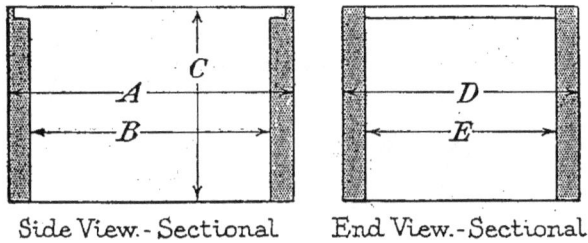

Side View.- Sectional   End View.-Sectional

FIG. I3.-DIMENSIONS OF AMERICAN HIVES.

Dadant Hives

743. The Modified Dadant (or M.D.) hive is the form of Dadant hive now generally used. It admits of supering with Langstroth bodies on occasion and of the use of Langstroth covers. The Dadant Blatt of the continent of Europe is not interchangeable (see **738**) but has otherwise the same good features. The difference between the M.D. and the Langstroth is that the former takes brood frames 2½ inches deeper and is generally designed to take eleven frames 1½ inches wide, or ten with a substantial dummy.

744. The M.D. body as built to match the new American Langstroth series has the following dimensions:

TABLE XVII
DIMENSIONS OF M.D. HIVES (see Fig. 13)

British Standard Hives

745. These are made to take the British Standard frame which has longer lugs than the American pattern. These lugs are very handy for handling, but involve some complication of the hive bodies to accommodate them. This complication practically disappears where double-walled ends are used (see Fig. 14).

Metal end on lug

Distance from
top of body to
top of runner is
made $\frac{7}{16}"$ to allow
for metal end

Section of Single-walled Body for
use in Hive with Outer Case

Width here made $17\frac{1}{16}"$ to allow
clearance over lugs

Air
space
closed
at
bottom

This piece let into grooves
in side walls

Section of End of Double-walled Body

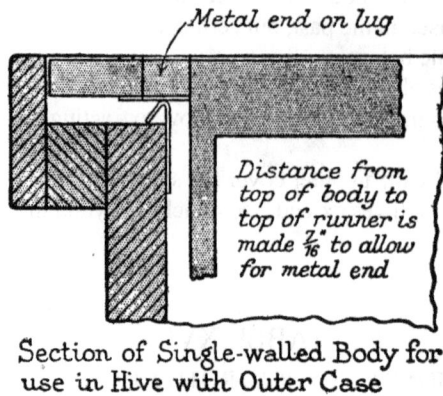

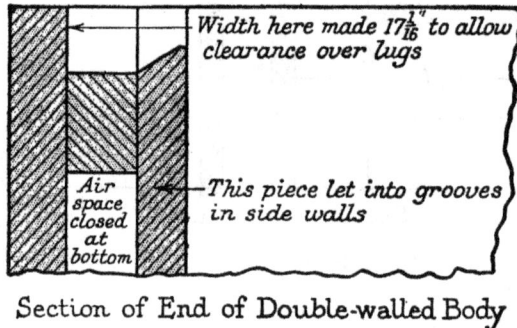

FIG. 14.-DESIGN OF BRITISH BROOD CHAMBERS.

British National Hive

746. This is illustrated in Fig. 15. It is generally similar to the American hives but, of course, made to take British frames. All the bodies are made 18 1/8 inches square, the extra room required for the long lugs of the British frame being utilized to provide double walls against the ends of the frames. The side walls are single and 3/4 inch thick. The clear width inside accommodates eleven frames with 1.45 inch metal ends plus a dummy or division board, but will equally well accommodate 12 frames spaced 1 3/8 inch bare measure.

The bee space is allowed below the frames according to British practice. Provision is made for a bee space over the lugs, but for those who think they must, some ¼ inch strips of wood are furnished, which may be used to close it.

The floor, which is reversible, slopes from back to front and accommodates a full width bar of wood, 7/8 inch square, as entrance block, which can be used to reduce the entrance either to about 4 inches or 1 inch by 3/8.

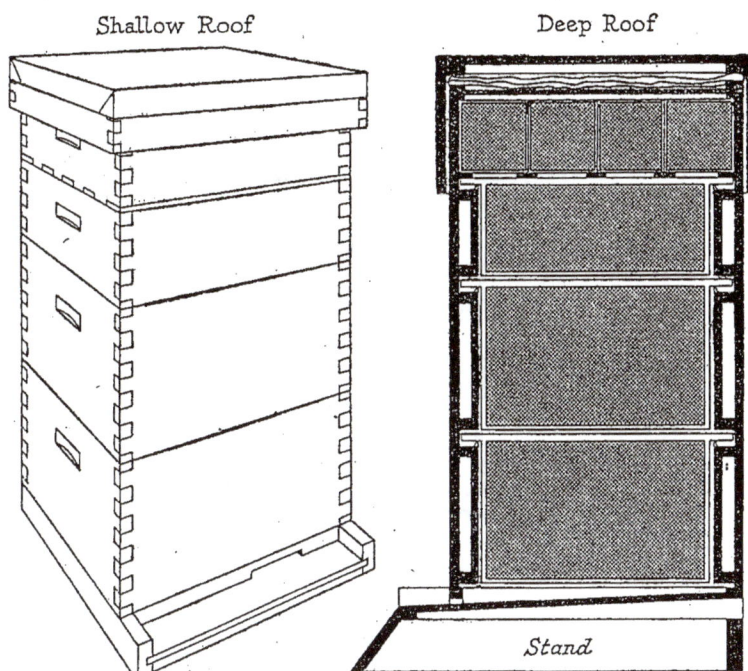

Shallow Roof   Deep Roof

Stand

Fig. 15 BRITISH NATIONAL HIVE

747. The roof or cover is flat topped, metal covered, and by means of fillets inserted round the top is prevented from sitting closely on the crown board or quilt. The air-space so provided assists in hindering the inflow of heat in summer or loss of heat in winter.

Two patterns of cover are offered, one shallow, the other deep. The deep cover affords additional protection to supers, feeders, packing trays and the like at the top, and when inverted on the ground, serves as a temporary support for removed bodies, at a more convenient height than is given by the shallow pattern. These covers should be ventilated to let out moist air. This can be done by boring a quarter inch hole through metal and wood sloping upwards from outside to within the top of the cover in the air-space. The deep cover does not require tying down for winter.

748. A stand is offered, which receives the floor board and provides an extension of the alighting board.

749. For discussion of this hive and modifications, see **657**, **669-70**, **676-8**, and **693**.

Hives to Old B.B.K.A. Standards

750.a The only hive that is really standardized in the UK is the BS National. These are still constructed to BS 1300 of 1960. Although this standard is no longer active it is still recognised as the cornerstone of National Hive manufacture.

The WBC was never standardized, lifts, floors, roofs etc are rarely interchangeable from one supplier to another. Similarly Smith, Commercial, Dadant and Langstroth hives can vary slightly from source to source. Although floors, roof, excluders and crown-boards of Commercial Hives are interchangeable with BS National Hives. PS

750. *For some years hives have been manufactured to early designs, a measure of*

interchangeability in which was secured by standardization of certain dimensions by the British Bee-keepers' Association. The above heading is used because it is likely that about the time this edition is issued the standardization of hives, frames and the like will be under consideration by the British Standards Institution.

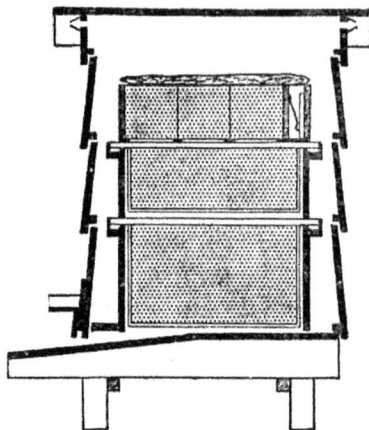

Fig. 16 DESIGN OF W.B.C. HIVE, WITH OUTER CASE

The W.B.C. Hive

751. The following dimensions were standardized by the British Beekeepers' Association for W.B.C. and "Single-walled" Hives, it being recommended that floor boards be provided with legs secured to the inside of the floor joists and supported by crosspieces, and that the roofs be covered with painted calico (see Fig. 16), (see also **667** and **692** however).

752. The outer case to be 20 inches square, outside measurement, and lifts for same 8 inches deep. These may be parallel walled with fillets or taper pattern, and 5/8 inch thickness finished.

Brood chambers, 16 inches 14½ inches, inside measurement.

Entrance full width and to go back at least 6 inches inside the hive.

Shallow frame supers, 15 inches X 14½ 1/2 inches X 6 inches deep, inside measurements.

Section rack, 15½ inches long X 15 inches wide, inside measurements.

Half-inch space between bottom of the brood frames and the floor board.

Brood frames to be B.B.K.A. standard (**791**).

The" Single-walled" Hive

753. This refers to a hive of the general type illustrated in Fig. 17 on the right hand. There are only two single walls. Outside dimensions, 18¼ inches by 17¼ inches with telescopic lift 10 inches deep. This lift is reversible to fit over the body in winter, but in summer serves as an outer case for an inner super.

Entrance sunken in floor board and made full width of hive.

Half-inch space between bottom of the standard frame and the floor board.

All material for outside cases to be not less than 5/8 inch thick, finished.

Other British Hives

754. A number of single- and double-walled hives are offered (see Makers' Catalogues) to take the standard frames, of dimensions differing from the above standards, but these are likely to be superseded in time by those with standard dimensions.

755. I have added illustrations in Fig. 17 of cheap and effective forms largely used for rural districts in Scotland and Ireland respectively.

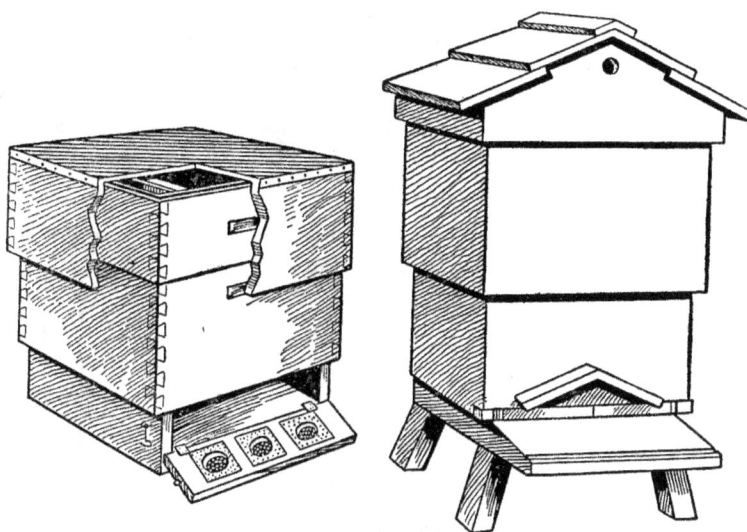

(By courtesy of Messrs. R. Steele & Brodie, Wormit, Scotland)

Scottish "Heather" Hive.   Irish C.D.B. Hive.

Fig. 17. IRISH AND SCOTTISH HIVES.

Observatory Hives

756. These are sold by appliance makers, especially arranged and glazed so that the work of the bees may be watched. They are used mainly for exhibition and sometimes for scientific observation. Rectangular hives may be purchased with one or more sides glazed and covered with removable shutters. They are interesting to show to visitors and of interest to beginners.

Use of Glass Walls in Hives

757. In the first edition of this work, particulars were given of the use of glass-walled hives based on certain work then recently done in Russia. Experimenters in Great Britain and elsewhere have been unable to obtain comparable results, and it would appear that a case has not been made out to justify further development of such hives.

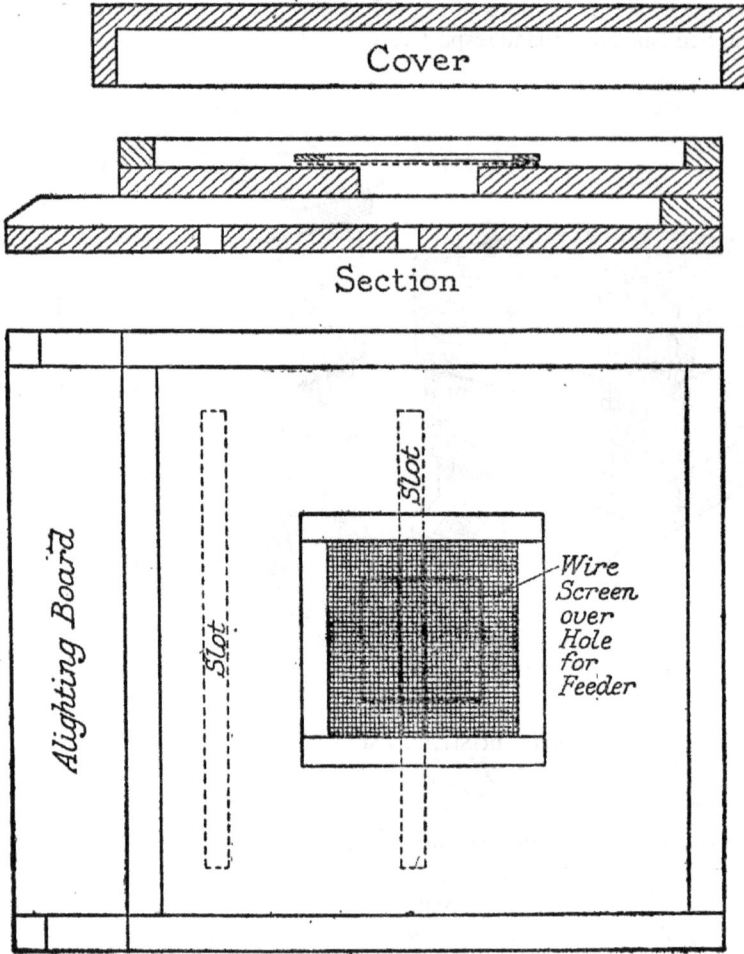

Cover

Section

Plan.- Cover Removed

FIG.18.-TOP-ENTRANCE HIVE.

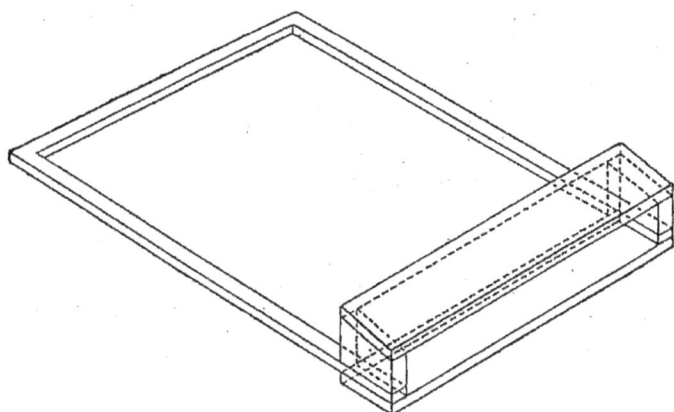

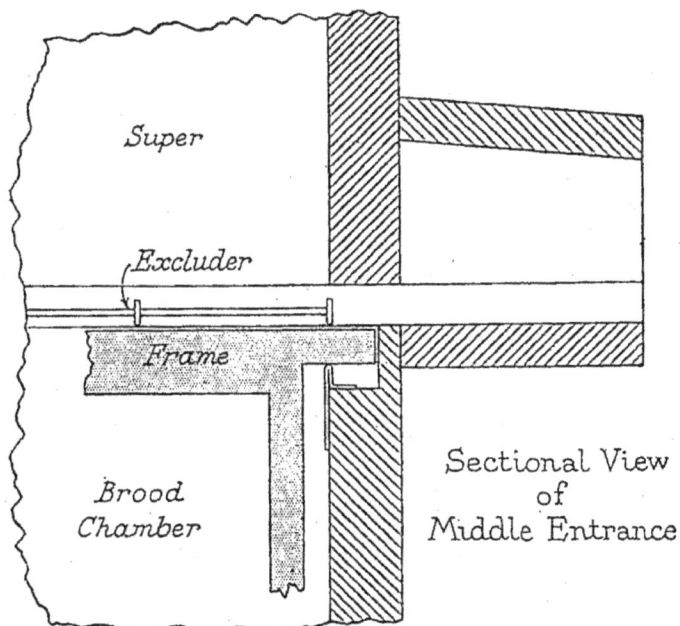

A three-sided frame ¾″ × ¼″ rests on the body walls and projects at the front, forming part of the side walls of the entrance porch.

Narrow projecting ledges of tin, not shown, carry the excluder, which reaches within ⅛″ of front. The bees enter between the frame lugs, see sectional view.

Entrance blocks are used as required.

FIG. 19.-MIDDLE-ENTRANCE HIVE.

A three-sided frame 1" x t" rests on the body walls and projects at the front, forming part of the side walls of the entrance porch.

Narrow projecting ledges of tin, not shown, carry the excluder, which reaches within t" of front. The

bees enter between the frame lugs, see sectional view.

Entrance blocks are used as required.

Top- and Middle-Entrance Hives

758. In nature, bees generally close a top entrance in favour of a lower one. Middle entrances have been known for generations, being used in Holland and North Germany.

The hives now to be dealt with are movable frame hives, modified by the provision of a floor which closes the bottom of the hive and with special entrance boards for use at the top, or at the middle, as the case may be. The middle entrance leads to a difference as compared with top-entrance hives in that by the use of a queen excluder, the bees are forced to breed below the entrance and the queen is excluded from the supers.

Details of the special constructions will be found in Figs. 18 and 19.

759. The middle-entrance design shown is such as to expose a large surface of the brood chamber. In recent years entrances have been employed similar in plan to the top entrance shown in Fig. 18, but with the upper part open to receive supers. Such a design becomes a middle entrance hive when the supers are in place, although still described as top-entrance.

760. The recent interest in constructions of this type has arisen more especially where hives are wintered out of doors under severe conditions of frost and where damp is also a serious source of trouble. Many stocks are lost in winter through choking up of the tunnel entrances in the heavily packed hives, and dysentery is assisted by damp and fungoid growths. Top and middle entrances cannot be choked, and as damp air rises more readily than dry air, the contents as a whole are kept much drier and also more fully ventilated (see also **730-1**).

761. If one is to use top and middle entrances throughout the season one requires reasonably good-tempered bees, for on removing the entrance to obtain access to the brood chambers some confusion of the bees is introduced and necessarily some disturbance of the guard bees. This is minimized where the modified top entrance is used. This complication does not occur where the top entrance takes merely the form of a small flight-hole serving mainly for ventilation, the principal entrance being retained below, or where the bottom entrance is gradually opened up in spring and the top one reduced to a mere hole.

By these latter methods of working one avoids trouble such as robbing, which may be attempted if the hive is too freely opened up. If a bottom entrance is used part of the time the bees have an easier job house cleaning, but it is astonishing how clean good house-cleaners will keep a bottom even when there is no bottom entrance.

The technique of the use of top and middle entrances can be developed to great advantage. It is claimed with good probability that swarming is reduced. Swarming through overheating of the brood chamber should be minimized, and through overcrowding of it, as excess bees collect in the chamber above the entrance.

Materials Used in Hive Construction

762. The use of straw, osiers and the like will not be dealt with. We are concerned with movable frame hives made generally of wood. Light-weight woods give better insulation than the same thickness of a heavier wood, the air cells in the former being better insulators than the wood or resin-filled cells in the latter. Some of the heavier woods last longer, but if protected from decay or other covering the light-weight firs in general use will give ample life and are cheap. In general also the light-weight woods are more porous than the heavier ones, causing less condensation.

763.a Most hives are made using cedar wood, which does not need treating for many years, or from pine of varying quality which may well need treating or painting.

Creosote has been found to have adverse properties. JC

Creosote.

H315 Causes skin irritation.
H317 May cause allergic skin reaction.
H319 Causes serious eye irritation.
H350 May cause cancer.
H361 May cause damage to fertility or the unborn child.
H411 Toxic to aquatic life with long term effects.

763. *Lightness of weight gives advantage in manipulation and also a saving in cost of transit. Freedom from knots is important, not only in small parts such as frames where knots may lead to fracture, but in side walls; large knots are apt to fall out and may become the cause of unequal shrinkage. Some of the relatively light, knot-free woods, free from excessive shrinkage, such as cedar and Oregon pine, are in favour for hive bodies, and especially where creosote and the like is used in preference to paint. Californian red wood is similar and proof against white ants. In Australia there are excellent woods available. Cypress is one of the best woods for bottom boards and their frames.*

Frames are generally made from a light-weight pine, selected to avoid knots. Sections are always made of bass wood (lime).

*Materials for covering hive roofs are dealt with in 692-5 and for outer coverings in **658-60 and 719-24.***

Objects of Hive Painting

764a, 765a, 766a, 767a and **768.a** Using any wood preservative always check very carefully with the manufacturers if it is suitable for painting hive parts.

Creosote.

H315 Causes skin irritation.
H317 May cause allergic skin reaction.
H319 Causes serious eye irritation.
H350 May cause cancer.
H361 May cause damage to fertility or the unborn child.
H411 Toxic to aquatic life with long term effects.

764. The principal object of hive painting is the preservation of the timber, but other wood-preservatives, such as creosote or cuprinol, last longer, are cheaper to apply and do not close the pores of the wood.

Secondary objects of painting are decoration, also distinctive colouring. There is no objection to the preservation and decoration of outer cases with paint, save the higher cost of labour and materials. In hot situations, especially if there is lack of shade, it is advantageous to paint the outer cases white or other light colour. The roofs of exposed hives may be painted white with advantage.

Keeping out Damp

765. Protection from damp has been touched upon under packing (**708-24**), but must here be considered in relation to painting. Paint owes its efficiency as a preservative to its action in preventing growth of destructive fungi, but it serves also to keep out damp and wet. A properly prepared and painted hive is damp-proof. Now keeping out wet is essential, but keeping out moisture also involves keeping in moisture, and unless there is a prolonged rainy season, the drying due to wind and sun more than compensates for wetting from rain; the hive contents are kept dryer than if the pores are sealed. If moisture tends to enter through small cracks or excessive porosity the bees will correct this by the use of propolis as a filler and varnish, where required.

766. Creosote and proprietary wood-preserving solutions are much more quickly applied than paint, they penetrate the joints and end grain better than paint. They last longer, one good treatment lasting a number of years, whereas paint requires renewal every three years. On renewing the coat there is nothing to strip. Wood-stains added to colourless preservatives give a more uniform and decorative finish.

767. If cedar wood is used, little or no preservative is required, save on the base and alighting board which may collect sufficient dirt to nourish fungi and the like.

768. Painted parts may be used as soon as dry. A little talc powder or powdered chalk applied to faces coming into contact will prevent sticking. In fine weather, new bodies may be used without painting, being removed for treatment after the active season. Where creosote or the like is used, the parts should be freely exposed to the air for about three weeks before use. It is advantageous to assemble and treat in the winter the bodies required for the following season.

Painting Hives

769.a ,770.a, 771.a and 772.a.
Lead based paints are not allowed.
Zinc based paints are not available.
Most modern paints use safe pigments (except marine anti-fouling.) JC

White lead (Lead carbonate).

H 302 Harmful if swallowed.
H332 Harmful if inhaled.
H360 Causes damage to organs (CNS and Reproductive).
H373 STOT RE2 Prolonged exposure may cause damage to body organs.
H400 Acute toxicity to aquatic life.
H410 Very toxic to aquatic life with long term effects.

Red lead (Lead oxide).

H 302 Harmful if swallowed.
H332 Harmful if inhaled.
H360 Causes damage to organs (CNS and Reproductive).
H373 STOT RE2 Prolonged exposure may cause damage to body organs.
H400 Acute toxicity to aquatic life.
H410 Very toxic to aquatic life with long term effects.

769. *Where parts are to be painted, all joints should be painted during assembly. Soak patent stopping into the end grain that will be in the joints, and before or after assembly, coat any knots there may be with patent stopping, or weak glue with white or red lead mixed in. After assembly, drive in all nails with a nail punch so that the heads are below the surface. See that the hive is dry. Smooth off all rough parts and round off sharp edges with sandpaper. Give the first, or primary coat of paint, then stop all nail holes and other holes or cracks, if any, with a stopping of equal parts of putty and white lead, or one of the stoppings sold in patent-top tins for use by coach builders, finally smoothing off with fine sandpaper. Plastic wood is a good stopping.*

770. *For a primary coat use red lead and boiled oil in the ratio of 2 lb. red lead to not less than 3 pints oil. Do not add turpentine as the risk of blistering is increased, and boiled oil is better than raw oil and driers. The paint recommended can be purchased from a paint-shop though not sold ready-made in tins. The primary coat should be a thin coat and allowed 48 hours to dry. For a second coat a ready-made paint can be used and for a third coat (omitted from edges which tend to stick together) the same or a ready-made enamel paint may be used, but these should be paints sold for outside work and guaranteed made of linseed oil, turpentine, and mainly white lead.*

771.a JC

White lead (Lead carbonate).

H 302 Harmful if swallowed.
H332 Harmful if inhaled.
H360 Causes damage to organs (CNS and Reproductive).
H373 STOT RE2 Prolonged exposure may cause damage to body organs.
H400 Acute toxicity to aquatic life.
H410 Very toxic to aquatic life with long term effects.

Turpentine.

H226 Flammable liquid and vapour.
H 302 Harmful if swallowed.
H304 May be fatal if swallowed and enters airways.
H312 Harmful in contact with the skin.
H315 Causes skin irritation.
H317 May cause allergic skin reaction.
H319 Causes serious eye irritation.
H332 Harmful if inhaled.
H411 Toxic to aquatic life with long term effects.

Zinc oxide.

H400 Acute toxicity to aquatic life.
H410 Very toxic to aquatic life with long term effects.

---

771. For a home-made paint the following proportions are good: White lead, 10 lb.; zinc oxide, 5 lb.; raw linseed oil, 1 gallon; turpentine, about 3/4 pint; and 5 lumps of patent drier about the size of a walnut, or about 8 oz. japan drier. Add raw sienna until the white has a cream shade.

772. Enamel and hard-gloss paints should not be used on the walls of brood chambers even in hot dry climates. Such paints are apt to blister in that position. They last well, however, on outer cases. Aluminium paints, made with a varnish body, have exceptional water-proofing properties and last well for the same reason. Use two coats, after stopping.

Re-painting a Hive

773. When re-painting hive parts, in good condition, rub down with a stiff bristle brush, washing if necessary, and when quite dry give two thin or one normal coat. If in bad condition, give one thin coat and then stop cracks or holes, rub down with sandpaper and give a final coat on exposed faces. If old and rough the paint must be stripped with a blow lamp and scraper and the whole treated as a new hive. Putty may be used for stopping, but a filler is sold in the paint-shops.

774. Remember always that there are certain weak spots requiring special attention. If there are wooden legs, rot commences at the bottom and they may well be tarred. The next weakest spot is at the junction between the floor board and first body or outer cover. The floor board is best tarred underneath and paint should be carried ½ inch over the joint and say 1 inch within the entrance. It is good to sand the flight board while the paint is wet. The edges of bodies should be painted, the first two coats extending over, say, 1 inch of the inside, but avoid heavy paint on edges apt to stick.

775. When re-painting, the bodies may conveniently be piled one on another, turned alternately, so that edges cross, but for new work with edges to be painted the work must be done in stages. It is convenient to support the bodies on a projecting bar or baulk, or they may be strung on a horizontal bar supported at both ends, long enough to carry several bodies.

Colour of Hives

   776.

White lead (Lead carbonate).
H 302 Harmful if swallowed.
H332 Harmful if inhaled.
H360 Causes damage to organs (CNS and Reproductive).
H373 STOT RE2 Prolonged exposure may cause damage to body organs.
H400 Acute toxicity to aquatic life.
H410 Very toxic to aquatic life with long term effects.
Zinc oxide.
H400 Acute toxicity to aquatic life.
H410 Very toxic to aquatic life with long term effects.

776. Bees have preference for certain colours, as shown by observations of drifting. The colours appear to bees differently from what they do to us, as bees are blind to rays at the red end of the spectrum and can perceive short waves extending well beyond the violet. Lead white appears white both to bees and humans. Zinc white, however, fails to reflect the ultra violet and therefore appears strongly coloured to bees. Experiments on hives show that bees have a distinct preference for dark as against light colours. Dark coloured blue is a favourite, then comes black and brown. Lead white is not disliked but zinc white, light green and pink are not favoured. It is preferable therefore to use form rather than colour to assist the bees to find their homes (277).

SECTION VIII
FRAMES, SECTIONS AND FOUNDATION

Frames in General

777. The size of the hive has been discussed under "Hives" in **637-41**, to which reference should be made. We now turn to the general consideration of the size and proportions of individual frames.

Preference for One-piece Combs

778. Except in the little-used "long idea" hive (**652**) in which frames are used side by side in one body, both for breeding and storing, hives are divided vertically by horizontal divisions into two or more sections, and the combs have to be similarly divided. This is unfortunate, as it is contrary to the instinct of the bee. To the discomfort of the bee-keeper, the bees are constantly trying to have matters the way they want, by re-uniting the combs in frames which are above one another. A "bee space" is left between, but this does not prevent the union. If the bee space is much under or over size the trouble is increased, as in the former case the bees will seal it up and in the latter case will utilize it for storage or brood comb (**679**).

779. The trouble is lessened if a thick top bar is used, say ¾ to 1 inch, instead of the 3/8 bar sometimes found, and especially if the bee-keeper will give the top and bottom bars a brush over with vaseline (petroleum jelly). This not only hinders comb building over the bars, but makes it easier to remove any propolis or wax the bees may deposit. A pot of vaseline with a stiff paint-brush in it should be kept handy where frames are assembled, and frames and other hive parts cleaned. The edges of bodies and parts of clearer boards and the like may be lightly vaselined, but it is well to keep vaseline off painted bodies, against the time of re-painting.

780. The instinct of the bee is so strong that, if frames are not in line, the bees frequently build trunk-like extensions from the top of one frame extending to and spreading over and into the frame above, wax being robbed from other places for the purpose.

781. When the brood nest is not divided the nest can be expanded by the queen without hindrance. The food is stored next the brood, mainly around the upper part of it. When the food supply is right, there are generally two or three rows of cells between brood and stores. If there are more it indicates shortage of stores; if less, it indicates that the incoming stores are pressing on the brood and more space will be required.

Combs in a State of Nature

782. In nature, the combs reach the hive walls, being suspended from the top and secured at the sides to within a few inches of the bottom of the comb. Owing to the form of the hexagonal cells, the weight of the comb tends to draw it out, distorting the cells and drawing in the sides. This distortion is prevented and the side pull resisted by fixture to the side walls. It is, however, evident that extension to the side walls serves no other important purpose as, for example, in controlling ventilation, as the bees will respect a bee space at the sides of frames right up to the top, the extension to the frame itself giving all the mechnical support needed at the sides. On the other hand, if the frames in the several bodies are not in alignment, and especially if they are staggered, the control of ventilation of individual frames must be rendered more difficult.

The Form of the Brood Nest

783. In maintaining a continuous comb surface there is, however, a limit; one does not find a fully developed bees' nest consisting only of one large comb, but of several in parallel.

Let us consider the occupation of these parallel combs by the bees. In winter the cluster shrinks in size and tends to a spherical shape offering the least external cooling surface. Actually the bees take advantage of the combs as packing, and the cluster is longer measured perpendicular to the combs than it is parallel with them. Again, the vertical dimension is somewhat smaller, giving a larger area in contact with the stores above.

784. Now, as warmer weather comes, this formation breaks up and a different one is formed. Brood is spread over the combs in ever-enlarging ovals, the bees occupying a much larger diameter parallel with

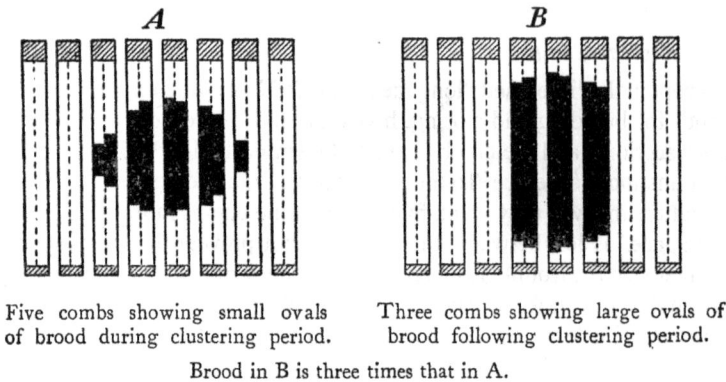

Five combs showing small ovals of brood during clustering period.

Three combs showing large ovals of brood following clustering period.

Brood in B is three times that in A.

FIG. 20.-FORMS OF BROOD NEST IN SPRINGTIME.

the combs than perpendicular to them. Indeed, sometimes, for a short time, fewer combs contain brood, although there is much more of it, as shown in Fig. 20. The author has observed a rough relationship of 7 inches diameter for three frames occupied by brood, a ratio of 8 to 5 in width and thickness. This is in marked contrast to the winter cluster formation, in which the ratio is nearer 5 to 8. The combs now form the principal thermal protection and a calculation, too mathematical for demonstration here, indicates that fewer bees per square inch are required to pack the comb faces on the outside of the brood nest than to keep up the warmth by more or less closing the spaces between the combs on the exposed margins between combs (see **798**).

785. The brood nest extends, increasing in diameter and occupying a larger number of frames, until the inner walls of the hive are approached. These walls are, however, of but little service as direct packing for the brood as the bees keep the space outside the frames and combs for traffic. In this way, the brood reaches its maximum width across the frame about the time five or six British frames or six or seven American frames are occupied. On further extension of the nest no increase of diameter is possible, but by the time this stage is reached there is some excess of bees available for night and other cold spells. The cluster, however, grows more freely if its expansion is unimpeded, and so far the arguments are in favour of large combs, larger even than M.D. combs, but also of smaller spacing than the M.D. (see also **798**).

Unimportance of Cubical Form

786. In cold weather there must be some advantage in minimizing the surfaces of the internal walls of the hive through which heat escapes. Now the cube form gives a lower ratio of surface to volume for a given capacity than any other rectangular form, but the argument for body dimensions approaching the proportions of a cube has been pushed too far.

In the first place, some heat is actually conserved by raising the body on a lift inserted in winter between the floor and brood chamber, although such an addition largely increases the surface and causes a marked departure in proportions.

In the second place, it should be noted that proportions departing considerably from the cubic give nearly the same ratio between surface and volume for a given capacity. For example, if, instead of making the length, width and height equal, one of these dimensions be made equal to half that of the other two, the capacity being kept the same, the surface is only increased by less than 6 per cent. For example, a cube 15 X 15 X 15 inches has a volume of 3,375 cubic inches and surface of 1,350 square inches. A body 18.9 X 18.9 X 9.45 has the same capacity and a surface of 1,429 square inches, less than 6 per cent. more. This is an extreme difference.

See also **717**.

Further Advantages of Large Frames

787. The larger the frame, the fewer frames there are to handle and the less is the cost of the frames necessary to carry a given amount of brood. Frames might be larger than the M.D. The Quinby was larger. Very large combs, however, require to be well supported by wiring, and, if much larger than the M.D., would be awkward to handle with safety.

Until queens are even more prolific than those this century has known, it is unlikely that anything larger than an M.D. frame will be used, but 15 or more may be used in one hive body.

A Working Compromise

788. If only a few very large frames were used, the top surface exposed for supering would be far from square and relatively small. Thus a compromise has to be made, and this points to a chamber nearly square, the smaller dimension being across the frames. The depth should be about half the major dimension for double brood nests, or three-quarters for a single nest. Small departures in either of these proportions make but little difference. The bee-keeper knowing the habits and prolificacy of the strain he expects to keep and studying the above, will have no difficulty in finding standard frames and bodies very well suited to his needs. He had better err in the direction of assuming that queens are going to be more prolific and that larger stocks are going to be the rule.

Form in Relation to Management

789. The problem of frame size and proportion has been discussed already, under hive bodies, which in turn determine frame size (see **639-43**).

Frames for Supering

790. Where large harvests are the rule and extracted honey is worked for, Langstroth or British Standard frames are frequently used for supering, especially where these frames are used in the brood nest. A super of shallow frames will weigh, however, say 35 to 60 lb. according to dimensions and number of frames, which is a more convenient weight, at least for the amateur, to handle. The comb in the shallow frame, moreover, is more suited to uncapping with a single stroke of the knife.

Where harvests are small and the flow irregular, the shallow frame is the most suitable and especially for snatching a harvest during a short flow.

TABLE XVIII

Dimensions of Frames see (Fig 21)

Make of Frame.	A	B	b	C	c	d	e	f	g	h	i	j	k	l	m
British Standard Brood* . . .	17	14	$13\frac{1}{2}$	$8\frac{1}{2}$	$7\frac{7}{8}$	$1\frac{1}{2}$	$\frac{3}{8}$	$\frac{3}{8}$*	$\frac{1}{4}$	$\frac{1}{4}$	$\frac{7}{8}$	$\frac{7}{8}$	$\frac{7}{8}$	$\frac{7}{8}$	—
British Deep Brood . . .	17	14	$13\frac{1}{2}$	12	11	$1\frac{1}{2}$	$\frac{3}{8}$	$\frac{3}{4}$	$\frac{1}{4}$	$\frac{1}{4}$	$\frac{7}{8}$	$\frac{7}{8}$	$\frac{7}{8}$	$\frac{7}{8}$	—
British Standard Shallow . .	17	14	$13\frac{1}{2}$	$5\frac{1}{2}$	$4\frac{7}{8}$	$1\frac{1}{2}$	$\frac{3}{8}$	$\frac{3}{8}$	$\frac{1}{4}$	$\frac{1}{4}$	$\frac{7}{8}$	$\frac{7}{8}$	$\frac{7}{8}$	$\frac{7}{8}$	—
British Commercial . . .	$17\frac{1}{2}$	16	$15\frac{1}{4}$	10	9	$\frac{3}{4}$	$\frac{3}{8}$	$\frac{3}{4}$	$\frac{3}{8}$	$\frac{1}{4}$	$\frac{7}{8}$	$\frac{7}{8}$	$\frac{7}{8}$–1	$\frac{7}{8}$	—
Langstroth Brood	$18\frac{3}{4}$	$17\frac{5}{8}$	$16\frac{7}{8}$	$9\frac{1}{8}$	$8\frac{1}{8}$	$\frac{9}{16}$	$\frac{3}{8}$	$\frac{3}{4}$	$\frac{3}{8}$	$\frac{1}{4}$	$\frac{7}{8}$	$1\frac{3}{8}$	$\frac{7}{8}$	$1\frac{1}{8}$	$3\frac{1}{2}$
Langstroth Shallow . . .	$18\frac{3}{4}$	$17\frac{5}{8}$	$16\frac{7}{8}$	$5\frac{1}{2}$	$5\frac{1}{4}$	$\frac{9}{16}$	$\frac{3}{8}$	$\frac{3}{4}$	$\frac{3}{8}$	$\frac{1}{4}$	$\frac{7}{8}$	$1\frac{3}{8}$	$\frac{7}{8}$	$1\frac{1}{8}$	$3\frac{1}{2}$
M.D. Brood	$18\frac{3}{4}$	$17\frac{5}{8}$	$16\frac{7}{8}$	$11\frac{1}{4}$	$10\frac{1}{8}$	$\frac{9}{16}$	$\frac{3}{8}$	$\frac{3}{4}$	$\frac{3}{8}$	$\frac{3}{8}$	$\frac{7}{8}$	$1\frac{1}{2}$	$1\frac{1}{32}$	$1\frac{1}{4}$	$3\frac{1}{2}$
M.D. Shallow	$18\frac{3}{4}$	$17\frac{5}{8}$	$16\frac{7}{8}$	$6\frac{1}{4}$	$5\frac{1}{8}$	$\frac{9}{16}$	$\frac{3}{8}$	$\frac{3}{4}$	$\frac{3}{8}$	$\frac{3}{8}$	$\frac{7}{8}$	$1\frac{1}{2}$	$1\frac{1}{32}$	$1\frac{1}{4}$	$3\frac{1}{2}$
Victorian Deep	19	$17\frac{1}{2}$	$16\frac{3}{4}$	$9\frac{1}{8}$	$8\frac{1}{8}$	$\frac{3}{4}$	$\frac{3}{8}$	$\frac{5}{8}$	$\frac{3}{8}$	$\frac{3}{8}$	$\frac{7}{8}$	$1\frac{3}{8}$	$\frac{7}{8}$	$1\frac{1}{8}$	$3\frac{1}{2}$
Victorian Shallow	19	$17\frac{1}{2}$	$16\frac{3}{4}$	$4\frac{1}{2}$	$3\frac{7}{8}$	$\frac{3}{4}$	$\frac{3}{8}$	$\frac{3}{8}$	$\frac{3}{8}$	$\frac{1}{4}$	$\frac{7}{8}$	$1\frac{3}{8}$	$\frac{7}{8}$	$1\frac{1}{8}$	$2\frac{1}{2}$

* Frames with heavier top bars are likely to supersede these.

Dimensions of Frames

791. The usual dimensions of frames are given in Table XVIII and Fig. 21.

792. The British frames were standardized with a 3/8 inch top bar, but this is on the light side and ¾ inch bars are frequently used. There is a British Deep frame, a modification of the Standard, being 3½ inches deeper. The ¼ inch side bars are rather light for this depth, although made of tougher wood than is required for, and used in, American frames. The British Commercial frame was introduced by Simmins, who urged that the Standard frame is too small, but is practically superseded by American sizes where large frames are wanted. British frames are made also to take Dadant foundation.

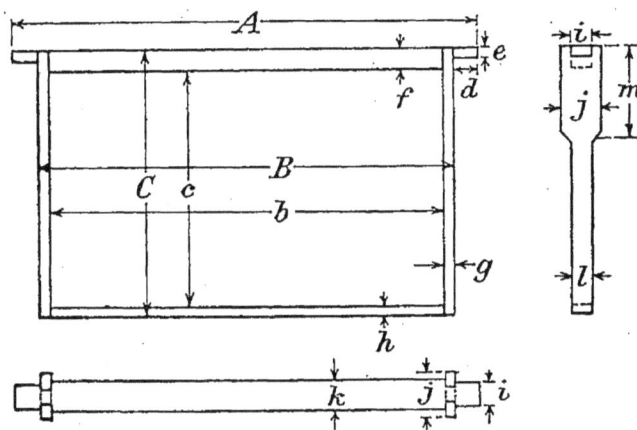

FIG. 21.—DIMENSIONS OF FRAMES.

The British Shallow frame is frequently made 1¼ inch wide, giving more support for the comb when used with the 1 7/8 inch wide spacing and giving a good surface to which to cut down, when uncapping.

793. Agreement has been reached recently among American manufacturers as to the dimensions of Langstroth frames. The M.D. Brood frames are similar, save in depth and spacing.

The frames scheduled are those with the "short" top bars. A longer "full length" top bar is also recognized and is made 19 1/16 inches long. The M.D. frames are scheduled with 3/8 inch bottom bar to take Dadant foundation. The same bar is used with Langstroth frames for the same purpose.

The Quinby frame, now practically replaced by the M.D., was 19⅛ X 11 inches and is sometimes referred to as the Jumbo.

The American top bars are sometimes made 1 inch wide in the centre portion.

794. The Australian frames are made in a variety of sizes not differing seriously from the American sizes, save in the depth of the shallow frame. The sizes given are those standardized in Victoria, but dimension B is also made 17⅝ as in the U.S.A. The top bars are sometimes 1 inch wide and sometimes ⅞ inch thick. Frames of the Hoffman pattern have no "V" edge, and the "V" edge is being abandoned in the warmer States in U.S.A.

Comb Area in Frames

795. Figures are given for the area of the comb space in frames in Table XIX, based on the dimensions "b" and "e" in Table XVIII, and, of course, requiring correction where top, bottom or side bars are used differing in dimensions from those scheduled. *To obtain the total comb surface exposed on both sides the figure given must be doubled.*

TABLE XIX
Comb Area of Frames, in Square Inches

Make of Frame.	Number of Frames.							
	1	2	3	4	6	8	10	12
British Standard Brood . . .	106	213	319	425	638	850	1063	1275
British Deep Brood (12 in.) . .	148	297	445	594	895	1188	1485	1782
British Standard Shallow . .	66	132	198	264	395	527	659	791
British Commercial	137	274	412	549	824	1098	1372	1647
Langstroth Deep	137	274	411	548	823	1097	1371	1645
Langstroth Shallow	89	177	266	344	532	709	886	1063
M.D. Deep	171	342	513	683	1025	1367	1709	2050
M.D. Shallow	86	173	259	346	519	692	865	1038
Victorian Deep	137	274	411	548	822	1095	1369	1643
Victorian Shallow	65	130	195	260	389	519	649	779

Spacing of Combs

796. The Langstroth Brood frames were spaced 1 3/8, but are frequently made 1 1/2 inches. Dadant advocated 1 1/2 inches, claiming better wintering and less swarming. The British metal end was designed for 1.45 inches, which approaches the Dadant practice; as now made, these ends are generally oversize, and, in fact, combs in British Brood chambers approximate to 1 1/2-inch spacing. (See, however, **798**.)

797. Shallow frames are frequently given the same spacing, but the American shallow frames are also made for spacing up to 2 inches and the British wide metal end gives nominally 1 7/8, but generally in excess of this. The use of a wider spacing in the super gives economy in wax building, reduces the number of combs to be handled in extracting without increasing the area of cappings to be removed, and is said to discourage breeding in supers, but water is evaporated more readily from shallower cells.

798. It is questionable whether wide spacing of brood combs, even of large size, is an advantage. For ventilation, a few fanning bees can keep all the air in the hive in motion. The 1 3/8-inch spacing used in the Langstroth hive is used successfully even where bodies are tiered to six feet or more. If wider spacing is given, the wider gap disappears in any case above the level of the brood, when the combs are drawn out for honey storage. Experiment in Great Britain has shown that stocks build up more quickly in the spring with the 1 3/8 spacing than with the wider British or M.D., care being taken to use the same number of bees in each case. This may be attributed to the smaller number of bees required to seal the gaps between combs to maintain the temperature. The British 1.45 inch was merely one-tenth of 14½, the width of the hive body along the length of the frames. The use of 1 3/8 spacing in a British National body enables 12 frames to be inserted, providing spaces for 10 per cent. more brood in one body. The metal ends should be nominally 1.365 inch to secure 1 3/8-inch centres of frames in use.

799. Attempts have been made to discourage drone comb by reducing the normal spacing, but this hinders ventilation and is not an effective remedy.

800. Spacing is generally determined by the width of the frames, the well-known Hoffman design, illustrated in Fig. 21, having the upper portion of the end bars expanded. This serves not only to keep the frames correctly spaced but to hold them parallel, an important consideration in widely spaced deep frames, for if the spacing is increased anywhere, due to lack of alignment, the bees are liable to build out burr comb from the side of the hive to utilize the space, hindering manipulation.

Where the bees use much propolis, metal spacing pieces on the frames are preferred.

801. The British metal end secures the spacing only at the top, and where deep frames are used considerable care, must be exercised to secure that they are assembled squarely and that they are hung truly from the cross-bars.

802. The space between frames and side walls is nominally ¼ but frequently 5/16 to 3/8 inch. Sometimes a minimum gap of 3/16 to ¼ is ensured by the use of metal studs projecting from the ends of the frames. This reduces the risk of injury to bees when removing frames. The studs should be fixed near the bottom of the frame, say, 1½ to 2 inches from the bottom (for spacing Between and Below Frames, see **679** and **796**)

Improved Metal Ends

803. The W.B.C. metal end is difficult to remove and gives trouble by distortion. The author is experimenting with a metal end of his own design shown in Fig. 22 (*a*). The bees lift the flaps of the W.B.C. end by inserting propolis under the edges, but they will not lift a straight strip bent over the sides as in the author's comb indicator. This led to the design shown, in which a strip 3/8 inch wide is bent to fit the top and sides of the 7/8 cross-bar, and then extended and bent over to form ¼ projections on either side to butt against those on either side. The author uses metal 1/32 inch thick. The bees fasten the metal on both edges of the top and sides, yet if desired the piece is readily lifted off with a hive tool. When separating frames the hive tool is inserted between the metal ends if the propolis is hard.

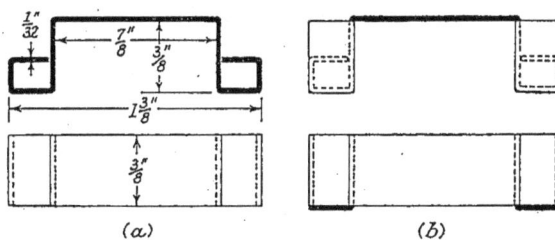

FIG. 22.-IMPROVED METAL END.

As shown at (*a*), the bees have free access to the lug, which the author prefers, and he uses a bee space above the lugs and metal ends, all as described in **678**. If there is no bee space above the lugs the metal end should be placed on the lugs from the under side so that they cannot be lifted off. They are easily loosened from the runners by inserting a hive tool to slide the frame along them.

If it be desired to exclude the bees from the lugs it would be necessary to employ a punched piece in place of the strip, giving wings to be folded over as indicated in Fig. 22 (*b*).

The ends illustrated are shown 1/32 inch thick, but those in Fig. 22 (*a*) can be made of half this thickness if spring steel strip is used, capable of being bent one way once without breaking and without retempering.

Divided Frames

804. The bee-keeper should avoid unnecessary multiplication of frame sizes or he will lose the advantages of interchangeability. Standard brood combs are on the large side for nuclei for queen mating. The difficulty has been avoided by dividing the brood frame into two pieces clipped or hinged together, but most bee-keepers prefer to use standard brood or shallow combs. The author described and illustrated a novel divided comb in the first edition of this work.

Assembling Frames

805. Frames must be assembled all square and free from warp, and securely nailed to make a permanent job. Special blocks or assembly; jigs are sold by the appliance dealers which are of assistance in quickly assembling the parts squarely before nailing. Fine wire nails are used and the cement-coated ones hold best.

806. The Hoffman frame has wide ends as illustrated in Fig. 21, and, to minimize propolizing, one edge of each frame is sometimes given a V section. When assembling, with the face of frame towards the observer and top bar at top, the left-hand end should show its V edge, and right-hand end its plain edge. By adhering to this plan one secures that any two adjacent frames even after turning either end for end, will always so come together that there is one V edge at every meeting-point.

Fixing Foundation

807. Foundation should not be inserted in frames in a cold room and preferably not more than a few weeks before it is required, as, however stored, it tends under the influence of changes of temperature to bulge between the wires used for fixing, the distortion being cumulative. The result is a permanently crooked comb, the use of which leads to an increased tendency to build drone comb. Frames with wired foundation awaiting use should be stored hanging vertically to minimize warping of the wax.

Split Top Bar

808. In Great Britain a top bar is sometimes used with a saw-cut through it, into which the sheet of foundation may be slipped. The cut may be opened by a mechanical device exerting pressure on the ends of the bar, or, with a little practice, is quickly opened by inserting a small flat key and turning it at right angles. After the foundation is inserted it may be secured by two or three wire nails, or by the used of melted wax. The top of the split bar offers a favourite location for the wax moth to lay its eggs, and for this reason, and to secure the advantages of a deeper and stronger bar, a bar with unbroken top face is to be preferred.

Top Bar with Wedge

809. A good method of fixing, widely used, is that of driving a strip of wood of wedge section into a groove adjacent to a central groove into which the foundation is tucked. For this purpose a hard wood block is useful with a rebated end, the projecting end bearing on the side of the frame and the rebate being wide enough to overlap the wedge and stand clear of the foundation. The block is moved along as it is struck. The wedge is a little apt to spring out. The remedy involving least time is to put two or three spots of liquid glue on the wedge before insertion. Some break the wedge into three pieces and find they hold better than one piece, no adhesive being used.

Special Top Bars

810. Heavy top bars are made for use with wired foundation, in which the lower half is so cut that part can be removed and later nailed to hold the wires and foundation securely. These bars have no groove to attract wax moths (see Fig. 23).

Fixing with Molten Wax

811.a 812.a and 813.a

Modern frames make this procedure unnecessary. It is not a good idea to introduce adulterants into beeswax (813) as this can interfere with comb building. JC

811. One of the best and oldest methods of fixing foundation, or starters, to the top bar is by the use of liquid wax. The frame is laid on a board, having a block on it fitting inside the frame against the top bar, and of such a thickness as to support the foundation, when laid on it, just central on the top bar. Wax heated a little above its melting-point is now poured along the joint. A spoon may be used, kept in the molten wax and having its tip squeezed up to form more of a spout. Alternatively, a conical-ended metal tube sold for the purpose may be used, having a small hole at the lip from which the wax emerges. If the temperature of the wax bath is kept just right, the work can be done very quickly with this implement. Others use a thin flat stiff paint-brush which is very handy, and some go so far as to wax the inside of the frame with it, enticing the bees to build the comb out to the frame. Such wax is not wasted, but there are other ways of securing the result (351). The wax may be melted on water and the brush dipped through to the water to remove superfluous wax.

812. A wax melter and pourer is sometimes used consisting of a vessel with narrow spout surrounded by a water bath extended nearly to the spout. Quick work may be done if the temperature is right, but if left standing the spout is apt to cool.

Useful Tips

813.a

Turpentine.

H226 Flammable liquid and vapour.
H 302 Harmful if swallowed.
H304 May be fatal if swallowed and enters airways.
H312 Harmful in contact with the skin.
H315 Causes skin irritation.
H317 May cause allergic skin reaction.
H319 Causes serious eye irritation.
H332 Harmful if inhaled.
H411 Toxic to aquatic life with long term effects.

813. Another "worth while tip" is to reduce the melting-point of the wax temporarily by the addition of a good oil varnish first thinned by the addition of an equal volume of turpentine. The diluted varnish may be stored for the purpose, and one adds to the molten wax about one-fortieth part of the varnish. This works well in a melter. The turpentine eventually evaporates and the varnish hardens so that there is a gain of strength.

814. Another tip occasionally worth while is to make up candles of beeswax or purchase those made for church purposes. Wax is poured from the candle as desired, and if the wick is kept of suitable length there need be no soiling of the wax. For rapid clean work two or three candles should be to hand for use in rotation.

Cutting Foundation

815. Foundation may be cut with a sharp knife against a straight edge, but it is useful to have a board the exact finished size required and to cut all round it.

However it may be secured, the foundation should be so cut to size as to hang a little clear of the side bars and 1/8 to ¼ inch clear of the bottom, to allow for stretching when the bees get to work on it unless a divided bottom bar is used.

Wiring Frames

816. In addition to fixing at the top, all foundation in brood frames should be wired. It is not necessary to wire shallow frames, but desirable to do so. Wiring serves to secure the building of a flat comb and helps to support the comb in warm weather and against shock or accidental mishandling. The comb should not need internal support when in a properly constructed extractor, but the wiring may help here also.

817. Except where pre-wired foundation is used it is important that it should be stretched tightly so as to give proper support. Every wire in a properly wired frame should give a twang if plucked. The ends of the frame offer the best support for wiring; moreover, vertical wires, unless crimped, do not give such a good hold in hot weather as horizontal wires, but wires sloped at different angles give good support.

The wire used is tinned iron wire, 0.0125 to 0.0150 inch in diameter (No. 28 or 30 British S.W.G.) being very suitable.

If the bars are pierced for wiring and the wood is very soft it is useful to employ metal eyelets, which may be purchased very cheaply for the purpose. Their use enables the wiring to be done more quickly and more securely.

Methods of Wiring

818. The simplest wiring consists of horizontal wires 2 to 3 inches apart, with a smaller space, say, 1½ inches, next the top bar, and a small space, say, 1 inch, from the bottom bar. The top part needs the best support and the free bottom end of the foundation is also in need of close support. The wire may be in one continuous piece, the first end being secured under the head of a small tin-tack or gimp pin. The first wire must be pulled tightest, as the tightening of later wires tends to reduce the tension on those already done. The last end is also secured by a tin-tack or gimp pin.

819. An old method of wiring British Standard frames is that shown in Fig. 23, the first fixture being made at A and the wires being inserted in the order shown, the last to pass being looped through the upper wire so that, when tightened, all is drawn taut. The bottom wire presents no difficulty in this respect and should not be necessary. Some insert staples or nails bent into hooks, but there is no saving in time and the bees dislike stout wire or nails and the like, even more than the thin stuff.

Parallel wiring is now generally employed as shown in the adjacent illustration and is suitable for electric embedding. The additional wire, shown dotted, draws taut the top wire as indicated and ensures against distortion of the comb at its weakest point, but it cannot be electrically embedded so conveniently and is not generally used.

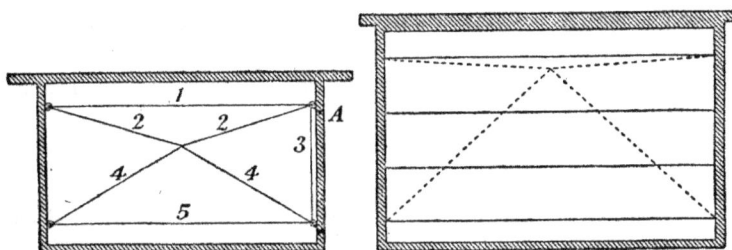

Methods of wiring frames.

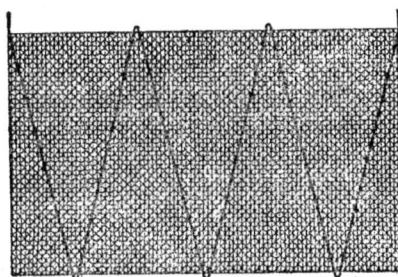

Lee's wired foundation.

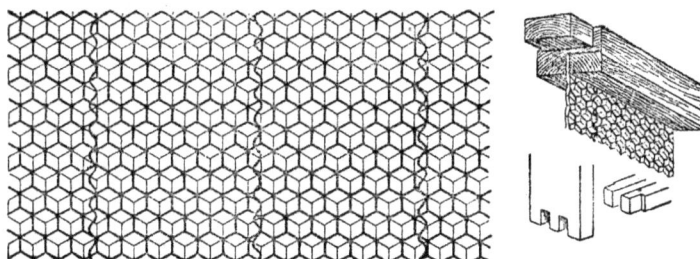

Dadant's wired foundation.

FIG. 23, WIRING FRAMES, GOES HERE. ED.

Embedding the Wire

820. The wiring should be done before the foundation is fixed to the cross-bar and the wires should afterwards be embedded in the foundation. For this purpose it is convenient to employ a board as indicated in **811** for use for waxing foundation, having a central block of suitable thickness practically filling the inside area of the frame. The foundation is most quickly embedded by means of an electric embedder sold for the purpose, by which the wires are electrically heated. Failing this, an embedding wheel sold for the purpose may be used. It is used warm, being dipped in hot water. A small bradawl with a rounded end having a groove for the wire can be used. A copper soldering iron with grooved tip holds the heat well. A plain grooved wheel, used cold, is satisfactory if used in a warm room.

Wired Foundation

821. The labour of wiring frames and embedding the wires is now generally avoided by the use of pre-wired foundation. Examples will be found in Fig. 23.

Lee's foundation can be inserted with practice about as quickly as the Dadant pattern, and the thinner wires offer less difficulty to the bees.

Dimensions of Sections

822. The wooden cases in which comb honey is made for sale are called sections and are made from a tough, knot-free wood, known as bass or lime, by special machinery designed to secure the necessary accuracy of workmanship.

The size most employed is probably the 4¼-inch square, but a no-bee-way section is made 5 X 4 inches to hold the same amount of honey, and both larger and small sections have been used.

823. The American pattern of 4¼-inch section, being the pattern in most general use, has an overall width of 1 7/8 inches with bee-ways, or 1½ inches with no bee-ways. The corresponding 4 X 5-inch section has a width of 1 3/8 inches.

824. The British-made section runs 1 15/16 to 2 inches wide, and the larger figure is required if sections of full weight of the Irish standard are to be produced, but the American sizes are also largely used in Great Britain.

Bee-way and No Bee-way

825. The pattern with bee-ways is preferred to the no-bee-way section in appearance as a finished article, but is not so easily cleaned and occupies more space in packing. The no-bee-way section must, however, be glazed, whereas glazing is not necessary with the other, and the transparent wrappings now obtainable for them serve the purpose of protecting the honey and exhibiting it. The two-bee-way section is more easily cleaned than the four-bee-way and offers better protection. The extra bee-ways in the four-bee-way section are not necessary and are of doubtful value.

Folding Sections

826. It is a mistake to fold sections dry. The corners will certainly be weakened and are liable to give way at an awkward moment. If the strips of wood are laid out in a row, the "V" grooves may be quickly moistened, preferably on the outer side, with a camel's-hair brush, or the stock may be left overnight in a closed box in the presence of water, so that the air becomes saturated; for example, by placing a damped cloth over them and a wetter one on top and replacing the cover. The appliance dealers sell a convenient wooden jig for the rapid and accurate assembly of sections.

Inserting Foundations in Sections

827. Sections are sold with plain or split top, the split serving to grip the foundation. The split top is not very satisfactory, the tendency being for the foundation to hang crooked and also to offer an irregular surface on the top. If two strips of foundation are to be used, this being the best practice, the split top does not serve (see **830** below).

828. Sections are sold also with grooved sides, the foundation being cut by gauge, so that it may be slipped into the grooves. There is some demand for these, but the design most generally employed is the plain-topped, plain-sided section in which the foundation is fixed in two pieces, top and bottom, by the use of melted wax.

829. Where there are quantities to be handled, the Woodman foundation fixer is in general use, obtainable from appliance dealers.

830. A favourite method of fitting foundation is to employ two strips 3¼ inches and 5/8 inch wide respectively, the wider strip being fixed across the top and the narrower across the bottom, leaving a gap of about 1/8 to 3/16 inch to be filled by the bees. This procedure helps to secure sections built out to the wood, the worst part generally being the bottom. For fixing with wax the foundation should rest on a block, fitting the section and of such thickness as to hold the foundation central in the section. The waxing is done as when fixing foundation in frames (see **811**). Fix to the top only except the bottom strip, which should be fixed at its ends as well. Another tip sometimes given to secure well built combs is to paint the inside of the section box with wax.

831. After assembling a rack of sections it is a. good plan to paint over the top and bottom faces quickly with hot paraffin wax applied with a wide soft brush in a few strokes. This enables the sections to be cleaned very readily later.

Choice and Care of Foundation

832. Foundation most acceptable to the bees has a good aroma and a good supply of wax at the base of the cell walls. In testing for aroma comparison should be made between specimens at the same temperature. Foundation is made in light, medium and heavy weight. The last named has a heavier base, and the bees are so apt to leave the heavy wall in the base, starting at once on the cell wall, that no benefit is got from the heavier kind to offset its increased cost per sheet. The most economical make, viz. medium weight, runs 6 to 6.1/2 square feet per pound, the heavy weight running 5 to 6, and light weight 6 1/2 to 7 1/2. In Australia the figures are perhaps ½ square foot larger. Foundation made in France generally runs heavier, the heavy and medium grades giving respectively 4 and 5 square feet per lb. Foundation for sections runs 10 and 11 square feet per lb. for thin grade, and 12·4 to 12·8 for extra thin. Wired foundation runs rather heavier than here indicated.

833. Table XX below gives the approximate number of sheets of foundation per pound in different sizes and common measurements of sheets supplied.

834. Excessive acid was employed at one time in foundation making. If soft water cannot be obtained the use of some acid is desirable (**390**), but only sufficient to prevent waste of wax.

The best comparative test for foundation is to mount samples side by side in the same frame and observe the way the bees work on it.

TABLE XX

NUMBER OF SHEETS OF FOUNDATION PER POUND

Frame.	Makes without Wires.		Dadant Wired.	
	Dimensions.	Number.	Dimensions.	Number.
British Standard Brood .	13¼ × 7¾	7–8½ medium	13⅜ × 8	8 medium
British Deep Brood . .	13¼ × 11¼	6 medium	13⅜ × 11½	5 medium
British Standard Shallow	13¼ × 4⅞	14 thin	13⅜ × 5	12 thin
Langstroth Brood . .	16¾ × 7½	7 medium	16⅝ × 8⅜	5 medium
Langstroth Shallow . .	16¾ × 4½	12 thin	16⅝ × 5⅝	{ 8 medium, 10 thin
M.D. Brood	—	—	16⅝ × 10¾	4 medium
M.D. Shallow . . .	—	—	16⅝ × 5⅝	{ 8 medium, 10 thin
Australian Shallow . .	16⅝ × 3⅜	18 thin	—	—
Sections . . .	about 100 per lb.			

835. Foundation exposed to frost becomes brittle, opaque and whitish in appearance. It becomes tougher on the temperature rising and may be restored in appearance and freshened by careful warming. Foundation as purchased is 30 to 50 per cent. stronger in length than width, but this difference largely disappears when the bees have worked it. It can be modified by annealing by exposure to the sun, with care to avoid overheating.

Foundation Making in General

836. The use of foundation is dealt with above and in **425**. We have to deal here with the product and its manufacture. The earliest foundation was made by hand in plaster-of-Paris moulds. The product was improved by the substitution of metal moulds. The Rietsche press, still used in remote parts, will give a serviceable article with enough wax at the base of the walls to give the bees a good start, but some skill is required to avoid excessive thickness in the cell bases, even when closely following the instructions sold with the press. The bees are apt to draw out the walls, leaving the thick base. It generally pays the small beekeeper to send his wax to the foundation maker who will allow him full value and return an equivalent weight of foundation for a small sum. It is impossible to make foundation for sections save in machines of the roller type.

837. Machine-made foundation is much tougher than the hand- made article and much less likely to warp and bulge. The cheap machines have rollers made of type metal. The best product is made with rollers of steel, built up of numerous dies, each forming one cell bottom. A much higher pressure can be used with steel rollers than with those of type metal, and this results in a tough product of good appearance, being transparent and with a thin base.

Preparing Wax Sheets for Foundation

838. In large plants, continuous sheets of wax are prepared. In a smaller way, sheets are prepared on dipping boards. The dipping board should measure, say, 1 inch way each larger than the finished sheet required. Knot-free mahogany or pine is suitable for making dipping boards. The dipping vessel (for materials, see **391**) should be, say, 6 inches deeper than the length of the boards, and is preferably a double vessel, thus obtaining better control of temperature. The inner vessel contains the molten wax floating on a body of water. The water is put in first and allowed to boil to drive off air, and cooled to at least 175° F. (80° C.) before wax is added. The working temperature for dipping should be about 160⁰ F. (70 C.), or higher for the thinnest sheets. The useful range is about 155⁰ to 175⁰ F. (68⁰ to 80⁰ C.), and it is undesirable to exceed the higher figure. The temperature selected should be one found to yield readily sheets giving foundation of the desired weight in the hands of the operator. The contraction during solidification depends upon the molten temperature.

839. Two other vessels are required, one to hold the dipping boards and another in which to cool the dipped sheets. The former should be worked at a steady temperature of 120⁰ F. (50 C.), the boards being inserted previously and the water well boiled to drive out air. The cooling vessel should be, maintained at a temperature between 90° and 100° F. (72 to 78° C.).

840. For normal working each dipping board is dipped twice from opposite ends alternately, being cooled off by dipping in the cooling tank between each dip.

841. When the wax has set, it should be cut round the edge and peeled off and the dipping board returned to the water tank for use again.

842. To retain a uniform temperature in the dipping tank it is undesirable to add solid wax during working. Wax should be melted in a separate water bath, replenished as required and the molten wax added to the dipping tank as required at about the temperature of the tank. All scrap from the dipping boards goes straight to the melter. The dipped sheets are brittle but become pliable during rolling and retain their pliability.

The edges of the sheets having been trimmed straight and to a width just exceeding the finished size required, the sheets are ready for the roller press.

Rolling Foundotion

843. A warm room is required with a temperature between 70 and 80° F. (21° and 27° C.). The rollers should be raised to blood heat before commencing rolling and will be maintained at about that temperature, the wax sheets being warmed in water at a temperature of 110° to 120° F. (43° to 48° C.) before rolling. The higher temperature is necessary for the thinnest sheets. The rollers are lubricated with paste or soap, the latter preferred, but all soap must be washed off all trimmings and scraps before they are returned to the melter.

844. The thinner end of the sheet should preferably be inserted first. A spring clip furnished with large wooden faces is useful for drawing off the sheets. They may be trimmed to size forthwith and piled with insertions of thin paper to prevent sticking. Wood pulp tissue paper is used.

845. On no account should type metal rollers (and presses) be cleaned with metal picks. Use a quill pick where necessary and a good wax solvent (**384**).

SECTION IX

APPLIANCES AND THEIR USE

Protection of the Person

846. Bees differ much in temper, but the best are apt to be at least on the defensive when queenless, or during a dearth, or during robbing, or when disturbed by accident, shock or damage. The wise bee-keeper, with the best of bees, will have a veil handy, though, on occasion, it may be worn open and minor examinations may be made without it. The bee-keeper may be called on, however, to deal with an intractable lot, and will then have to see to it that his or her clothing allows no entry for angry bees. The ankles and wrists must then be protected, as well as face and hands, and a long bee-tight, light-coloured coat or overall is desirable.

847. Bees are more disposed to object to dark clothing than to light-coloured. Even good-tempered bees are liable to become annoyed by clothing in which their feet become entangled. Bees are quick to detect and object to animal odours, and all should be sweet and clean. If other animals have been handled the hands should be washed before handling bees. Under normal conditions it is best to have the coat closed at the wrists, either with rubber bands or by cotton gauntlets fitted, with elastic at both ends to close on the wrists and sleeves. Gaiters are useful, serving to protect the ankles and as a protection against crawling bees. Bees always tend to climb upwards. Cyclists' clips may be used on the trousers. Veils and gloves are treated in greater detail below.

848. It should be remembered that most manipulations are carried out in the warmest part of the day, so that the tendency is to discard clothing. A loose-fitting white duck overall with large pockets is cooler and more acceptable than coat, waistcoat and trousers, and may be fitted permanently with anklets and wristlets and even with veil and hat.

Veils

849. It is impossible to see clearly through a light-coloured veil. Black is, therefore, generally used, either for the whole veil or, at least, as a window in the front. A screen of wire mesh gives a clearer view even than cotton netting. A cylinder of wire netting is sometimes used, which, hanging below the hat, keeps clear of the face and neck, but its place may be taken by two rings of wire sewn to a fabric veil, serving the same purpose and being more portable. Sometimes the metal gauze is arranged in panels hinged at corners. The veil will then fold up, but only if taken off the hat. Generally an old hat is kept, such as a discarded straw hat, to which the veil may be stitched, saving time in adjustment. Any wire gauze used should preferably be non-rusting.

850. The lower end of the veil is made long enough to tuck in at the neck. The "Langdon" bee veil is cut away at both sides to fit over the shoulders and is tied round under the arms, being comfortable and leaving the neck free, but bees are apt to enter through a fold in the coat, especially at the back, and a better plan, leaving the neck free, is to have a band or collar fitted to the bottom of the veil, fitting around and below the coat collar with a V front secured to a button by an elastic loop at the point of the V.

851. The simplest veil is a plain cylinder of black netting with tie or elastic at top to fit the hat, and is a very serviceable article if put on under the coat and waistcoat and then drawn up and out to stand clear of the face and neck.

Gloves

852. Bees can be manipulated with gloves on, but with more satisfaction to both bees and bee-keeper if gloves are discarded. The only absolutely bee-proof glove is a rubber glove, but rubber gloves cause excessive perspiration and are most uncomfortable in hot weather. The next best is a glove of soft leather, such as the "Birkett," but they are not easy to clean and soon become soiled with propolis from the lugs of the frames. An alternative is to use a knitted woollen glove with a white cotton glove worn over it (see also Sting Preventives below, **870**). All gloves should have gauntlets reaching over the sleeves and furnished with elastic bands or other fastening to make them bee-proof, and any stings should be removed before they work their way through. Gloves should be dipped occasionally in vinegar and water, allowed to dry on, and cotton coverings may be used damp. Propolis is removable with hot soda or bleaching liquor and is soluble in alcohol and petrol, but leather gloves require treating after washing with a leather soap or with lanolin to keep them flexible.

Subduing Bees

853. Bees are apt to be bad-tempered if there is robbing going on, or if queenless, or during a dearth, and after removal of harvest, and can be made bad-tempered at other times unless quietly handled and with consideration. Avoid all sudden movements and anything that jars the combs, and of course avoid crushing bees. The bees over a fortnight old are the more disposed to be troublesome, and when they are out flying in the middle of the day the hive may be examined with greater ease, but an extensive examination during a honey flow may so disorganize the labours of the hive as to result in a dead loss of 5 lb. or more of nectar. The operator should stand behind or beside and never in the line of flight.

854. When using the smoker, put in a puff or two of smoke at the entrance and wait a minute or two (not a moment or two, but 60 to 120 seconds) for it to take effect, before opening the top of the hive; then use a little smoke at the top, the less the better. If bees in a normal condition do not answer readily to a little smoke and deliberate handling they should be re-queened.

Use the smoker to clear the top of the hive before replacing cover or supers.

855. Always use as little smoke as will serve to keep the bees from getting out of hand. The more docile strains do not need smoking at the entrance, whereas some strains of black bee require thorough subduing, and opportunity to gorge themselves, before they can be handled comfortably. Obviously such procedure seriously upsets the economy of the colony; such stocks are best examined in favourable conditions.

856. It is no use trying to subdue bees that have not access to open stores and, in such a case, they should be sprinkled with syrup and left for a few minutes to absorb it.

The Smoker

857. There are many patterns of smoker. The American Bingham pattern is one of the most practical. A little fuel may be burned in a large smoker, but a small smoker cannot be kept going for a long time without replenishing. The Bingham pattern may be opened and closed when hot with a hive-tool, for recharging.

A smoker, well cared for, should last for many years. It should be kept under cover when not in use. The leather should be treated with neatsfoot or collan oil or castor oil at the end of the season, and metal parts rubbed over with vaseline. All deposit within should be frequently removed. Never work a smoker so hard as to produce hot smoke. The fuel, after ignition, should be inserted into the smoker, burning end first.

858. The fuels that give the coolest smoke and least tar to foul the smoker are dry decayed wood and stringy bark. These are rarely obtainable, and old corduroy, cotton rag, old soft sacking, dried grass cuttings and old hay are good substitutes. Corrugated paper is a favourite, being so readily prepared in rolls and burning well, but it produces considerable deposit. The rolls of paper may be cut more quickly with a hand-saw than scissors, the material being rolled before cutting, and the loose flaps may be secured quickly with gummed strip paper sold in rolls for fastening parcels. For easy ignition just dip one end in a saturated solution of saltpetre coloured with red ink and then dry well. Sacking may be rolled up between corrugated paper and gives better results than the paper alone.

859. A little tobacco smoke is an effective subduer, and smokers frequently use it for minor examinations made with no veil and with quiet bees, or a roomy veil with metallic (non-inflammable) front.

860. After use, the smoker should be emptied forthwith, unless there remains useful fuel in it, in which case quench the fire by stuffing grass into the nozzle of the smoker near the bees. If hot particles are seen to emerge, the smoker requires recharging.

Use of Carbolic Cloth

861.a I have never seen this in operation and would certainly not recommend trying it. If nothing else all your honey would taste and smell of TCP! JC

Carbolic acid (Phenol).

H301 Toxic if swallowed.
H311 Toxic in contact with the skin.
H314 Causes severe skin burns and eye damage.
H331 Toxic if inhaled.
H341 May cause genetic defects.
H373 STOT RE2 Prolonged exposure may cause damage to body organs.

---

861. *Some prefer the carbolic cloth to the smoker, but with some bad-tempered bees it serves only to infuriate. Moreover, unless used with caution, there is some risk of contaminating the honey. Nevertheless a carbolic cloth, large enough to cover any hive body, hanging over on two sides and with 1/2-inch round rods thrust through the hems on the corresponding edges, is a very serviceable subduer and temporary cover for bodies, especially at any time when robbing is likely to occur. If one at least of the sticks is round and the cover is kept rolled up on that one the material will keep flat and last for years. The devotee of the carbolic cloth uses a long cylindrical box of tin with tight-fitting lid to hold one or more cloths when not in use. The cloths are damped with carbolic solution, a suitable solution being 2 parts (ounces) of Calvert's NO·5 Carbolic and 1 part (ounce) glycerine in 20 parts (1 British pint) water.*

Use of Chloroform

862.a If you have "really vicious bees" this is not the way to deal with them! Try re-queening. JC

Chloroform (Chloromethane).

H 302 Harmful if swallowed.
H315 Causes skin irritation.
H351 May cause cancer/genetic defects.
H373 STOT RE2 Prolonged exposure may cause damage to body organs.

---

862. *Chloroform can be used for rendering bees insensible, for certain operations (105), or for subduing really vicious bees. It should be kept in the dark and in a bottle with stopper and cap sold as a "chloroform bottle." For use, a wide-mouthed bottle holding two ounces is convenient, having a cork with two holes; one carries a bent glass tube with one limb reaching* *nearly to the bottom and fitted at the outer end with a rubber bulb attachment, as used for a scent spray; the other carries a short tube also bent at right angles, the inner end flush with the cork and the outer end serving as a nozzle. For subduing, apply vapour to the hive entrance for about 15 seconds.*

Use of Ethyl Chloride

863.a This method is not to be used. JC

Ethyl chloride (Chloroethane).

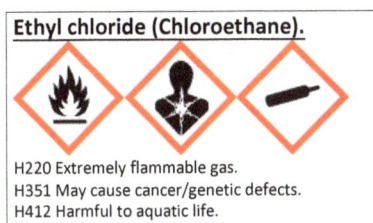

H220 Extremely flammable gas.
H351 May cause cancer/genetic defects.
H412 Harmful to aquatic life.

863. *This anaesthetic can be purchased in a small glass container fitted with a spring controlled valve and lever whereby a discharge of gas may be obtained or of liquid in drops. The bees can be stupefied without danger or risk of injury. This anaesthetic has been used very successfully for queen introduction, queen marking, and the combined operation.*

Use of Temporary Covering

864. When opening up a hive it is convenient to have handy two or three sheets of material, which can be used as temporary covers for bodies set aside or for covering part of a chamber during examination. Some use cotton squares furnished with wood rods or laths inserted in two opposite hems as for carbolic cloths. Thin sheet rubber mats may be purchased for a few pence; they last for many years, keep clean and are equally stable in a wind.

Remedies for Stings

865.a Honey bee venom is a colourless liquid where the active components are a mixture of proteins causing local inflammation and pain. The solution is acidic with a pH of 4.5 to 5.5. a bee is able to inject about 1 mg venom. ID

865. *The poison of the bee sting is not formic acid. It consists of acid and an alkaline substance combined, but supplied from separate glands. Neither substance alone produces the effective symptoms. There is little or no justification for the use of the blue bag, soda, ammonia or other alkalies as a remedy. Prompt application of Milton or of Eau de Javal, both of them bleaching liquids, has utility, also the application at a later stage of tincture of iodine or of aloes to the swollen parts. Some say that the application of any of these remedies is mainly psychological in its effects and a kiss would do as well; I leave this to the experimentalists under expert statistical guidance.*

866. As a rule, the bee is unable to remove its sting from the wound before leaving. Owing to the peculiar structure and properties of the sting, the sting continues to penetrate and empty its poison sac into the wound after the bee has left. It is important that the poison sac should not be squeezed in an attempt to remove the sting. Fine forceps are most effective, but it is common to remove the sting by pushing the thumb-nail along the skin towards the sting, so as to scoop it off. The antidote should be applied at once to get the best result, but tinctures applied to the wound and neighbourhood 5 or 10 minutes later and after the poison has spread may still be beneficial in allaying local inflammation.

Treatment of Serious Cases

867.a Approximately one quarter of people will develop a large local reaction comprising swelling, redness and itch that is larger than 10cm. Generalised reactions (anaphylaxis) can occur in up to 7.5% of people who are stung. There is some evidence that the group of medications known as ACE inhibitors, usually prescribed for high blood pressure, increase the risk of a generalised reaction. Generalised reactions may be fatal. Features suggesting a generalised reaction are the rapid onset of widespread red rash, swelling of the face, throat or tongue, problems breathing and collapse.

Local reactions can be treated with over the counter antihistamine tablets. Prednisolone tablets may be prescribed for large local reactions.

A generalised reaction is a medical emergency. Lie the patient flat, remove the sting carefully and get immediate medical help. RP

867. Most bee-keepers are but little affected by stings and become hardened to them, so as scarcely to notice them after the first few moments. On the other hand, some persons suffer very severely from a single sting, the heart being affected, so that the patient is giddy for some hours, hearing strange noises, the sight becoming affected and the skin becoming yellow with purple and blue marks under the eyes, and considerable swelling appearing in the neighbourhood of the sting. In a severe case the patient should remain prone and a doctor should be called in. Artificial respiration may be applied and massage of the limbs to assist the circulation. The specific remedy is adrenalin 1 : 1,000, 3 to 10 drops being injected into the blood of an adult of normal weight.

Internal Remedies

868.a Anyone who has experienced a generalised reaction after a bee sting should be referred by their GP to an allergy specialist. Local reactions do not warrant further investigation or management.

The treatment plan for those who have had a generalised reaction will include provision of an adrenalin auto-injector, instruction in its use and recommendation to wear a MedicAlert bracelet. Venom immunotherapy (hyposensitization) is also recommended after generalised reactions and is around 80% effective. RP

868. Those who suffer badly from stings can be immunized by the use of "Ostocalcium," a preparation of the Glaxo laboratories. For a fortnight before the season opens a tablet should be taken after each of the three principal meals each day, followed later by the use of one per day for several weeks.

Some claim relief after being stung by the use of the homeopathic remedy "Apis melifica," in strength 6x, taking two or three, pillules at once and repeating in a few hours if desired.

Destruction of Toxins by Heat

869. The toxin causing inflammation looses its poisonous properties at about 50° C. (122° F.). This is a temperature uncomfortably high, but the prompt application of the hot nose of the smoker to the wound is said to destroy the venom without creating a burn, as also a hot compress made up to a temperature above the limit given above and of course applied promptly, but only for a few moments.

Anti-Sting Ointment

870.a Whilst working with bare hands may sound attractive, popular advice is to wear disposable non sterile surgical gloves or fine nitrile gloves. These will not prevent bees from stinging your hands but provides a barrier between the bees and your hands. This will prevent the bees being affected by human pheromones (sweat) and, should any contagious disease be found in a colony, the gloves can be easily replaced before continuing to inspect any other colonies in the apiary. Nervous or beginner beekeepers who are concerned about being stung can wear plastic washing-up gloves. Although these will reduce the sensitivity of touch they will provide some protection from bee stings. These gloves can be washed in a 5% washing soda solution to remove any contamination before moving to another colony or apiary. ID

870. The beginner likes to use gloves but should be urged to abandon their use, as the bare fingers are so much more sensitive and are as acceptable to the bees. A nervous beginner can obtain help by applying a small quantity of extra strong methyl salicylate ointment, rubbed all over the hands and especially on the exposed parts around the left thumb and first finger. The bees may be seen to pitch with intent to attack and then change their attitude. The use of the ointment has the additional advantage that it hinders propolis from adhering to the fingers. Oil of wintergreen (i.e. methyl salicylate) alone, may be used, but it evaporates too quickly to be effective in hot weather for more than a few minutes.

Sting Preventives

871.a Bees can be antagonised by strong smells as well. Wearing perfumes or aftershave lotions when handling bees can cause bees to become defensive and encourage stinging. Similarly, protective clothing and beekeeping tools should be kept clean to prevent the build up of pheromones and other chemicals that can affect the temper of the colony. ID

871. *The most sure preventive of stings is, however, the cultivation of slow and deliberate movements, like those in a slow-motion picture, when the hands are over the hive, and, above all, to avoid crushing bees. In hot weather the hands should be washed and dried immediately before use. A little vinegar rubbed in keeps the hands in good condition for working, acting as a sting and perspiration preventer.*

Bee Stings and Rheumatism

872.a There have been many claims about the therapeutic properties of bee stings. There is also anecdotal evidence that some of these claims have been validated. It is recommended that any use of bee venom should only be administered by a qualified doctor. ID

872. *There are many cases of rheumatism, severe articular rheumatism, being relieved and even cured by bee-stings. The matter is one for the medical profession, as there are contra-indications to the use of this remedy in some cases, and it is not a universal remedy for rheumatic affections.*

Sundry Tools and Devices

873. Super clearers, queen excluders and the like, fitted to the hive, are described under Hives in Section VII. The frames for comb building and sections are dealt with in Section VIII. Queen-raising appliances and queen cages will be found described in Section II under Queen

Raising. Appliances for rendering wax, cappings, etc., and for making foundation, will be found under Wax in Section III. Appliances pertaining to the apiary will be found in Section VI. Other minor tools and appliances are dealt with below.

Hive-tool

874. Some bee-keepers are content to use a large screw-driver as a lever for separating hive bodies and a painter's scraper for removal of wax, propolis, and the like; but, while a wide scraper is particularly useful in cleaning hive floors and for its normal purpose in removing paint, it will be found that a hive-tool of the American pattern is by far the most convenient for general use. The best form has one end thinned to a wedge shape for use as a prise or lever, and the other broadened out, sharpened and bent over, for use as a scraper. This form is particularly useful in scraping top bars of frames and in removing burr comb, etc., as well as in opening up hive bodies with a minimum of damage to the wood, provided the end is not too small or the taper too sudden. The straight end should be 3/4 inch wide with a taper of 1 in 8 and an end nearly sharp. The scraper end, bent at a right angle, should be 1 1/4 to 1 1/2 inches wide. For American frames, however, with short lugs, the narrow end should not exceed 7/16 inch, but the taper must be right, as the narrow tool is more likely to injure the hive bodies.

Bee Brush

875. A large goose feather is just right for the job of brushing bees from combs, the bees being scooped off by the soft edge, approached as near as may be by way of their heads, the bees being gently brushed downwards off the combs; but a useful long-haired, thin brush is sold for the purpose. If bees are brushed from behind they may show much annoyance, but if properly used a feather creates less disturbance than shaking or thumping.

Vaseline Pot

876. A pot of petroleum jelly (vaseline} should be kept handy, containing a 1 inch sash brush, for putting the thinnest of coats of vaseline on the tops and edges of frames, hive bodies, sides of division boards and elsewhere, where comb building is to be discouraged, or the breakage of adhesions and the removal of propolis may be necessary. Petroleum jelly is also useful as a rust preventative and rust remover.

Indicators and Cards for Notes

877. Bee-keepers are strongly recommended to keep notes. The least that should be done is to have a card for every hive and a pencil handy, so that dated notes may be made of important matters and operations completed or required. Elaborate indicators have been sold for attachment to hives, giving indications of the commonly recurring features, such as queenlessness, signs of disease, lack of stores, etc., but they fail through giving insufficient information. The note-card hardly serves to give details of particular combs, but these can be marked by the use of a drawing-pin. The author uses his own comb-indicators, which are metal strips bent to fit over the cross-bar and removed with the hive tool. They are coloured, but coloured drawing-pins serve the same purpose. A red indicator serves to show the presence and approximate position of queen cells, green indicates drone cells, black, a defective comb and plain, anything special. By the use of such indicators the state of the brood chamber is visible as soon as it is uncovered and without moving anything.

Apiary Barrow

878. A useful appliance in the apiary is a light barrow with one or two wheels at the front and with two legs and handles at the back. It should be built low and flat to receive a heavy hive, but it is useful to have fitted to it a removable hive body for use as a frame tank, with a canvas cover fixed at one edge to lie over it, having a heavy bar at the other edge. Combs with bees or honey may be carried in this. It serves also as a temporary support for a super or hive body during manipulations. An open box or tray between the handles serves to hold smoker, hive-tool, vaseline tin, etc.) and veil or gloves when out of use. Projections from the tank or body to receive a frame by its lugs are useful, giving access to the comb face for caging a queen or cutting out queen cells, etc. There may also be narrow projections below the supports to keep the hanging frame clear of the body, top and bottom, so that the frame is firmly supported when placed against them.

Comb Holders

879. In the absence of a barrow a comb removed from the hive can be leaned against the side of the hive, but a metal comb-holder may be used; it consists of a metal frame to hang on the side of the hive, having projections to receive two or three combs and stops to prevent slipping.

If deep covers are used, a frame with a queen on the comb can be stood safely inside the inverted cover; failing which it is well to use a spare floor and brood chamber, or a nucleus or travelling box, or the queen may be lost.

Scale Hive

880. This is a valuable adjunct, indicating as it does the income and expenditure of the hive, the prosperity of the stock and the general conditions of the honey flow. The hive selected for weighing should be one in normal condition, so that its indications may be representative of what is going on generally in the apiary. Its weight should be taken always at the same time of day and -preferably at an hour when in the height of summer flying has ceased for the day. When nectar is coming in fast there will be a considerable reduction of weight by night due to evaporation. During the day the weight will be changing by bees coming and going. A heavy storm of rain makes a temporary increase of weight, larger in the case of unpainted hives. For this reason some prefer to keep the scale hive under cover. Another reason for doing so is protection of the weighing apparatus, if this is permanently in position. It is not essential to provide for weighing the hive when piled with supers during a good harvest, as there is at that time plenty of evidence of what is happening. It is valuable to get the weight to the nearest half-pound or pound, and this cannot be done so readily if a great range of weight must be provided for.

Cheap Alternative

881. It is not essential to use a weighing machine. A spring balance capable of carrying half the weight may be used by applying the hook to one side of the base and lifting until that side just rises off the supports so that part of the weight is on the remote edge of the floor board and the remainder on the edge being lifted. The operation is repeated at the opposite side, and the sum of the two weights so observed is the true weight, even if the weight is not central. It is obvious that the figures will not be correct, however, if the point of the hook is not on the exact edge, and to attain this it is best to fit a small metal plate to receive the hook, projecting somewhat and having a groove in it to take the hook of the balance (see Fig. 24, opposite). It is important also that the hive should be only just lifted, as if tilted appreciably part of the weight is transferred from the balance to the opposite side.

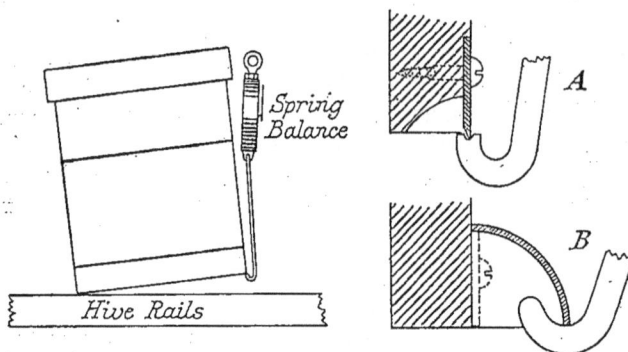

FIG. 24.—WEIGHING A HIVE WITH A SPRING BALANCE.

The hive is weighed from both sides and the readings added. The hive should be lifted only just clear of its base, not as far as shown.

A and *B* show alternative constructions for applying the balance hook, method *B* requiring a correction as described for the extension piece.

FIG. 24.-WEIGHING A HIVE WITH A SPRING BALANCE.

882. Another way of applying the balance is to fix handles which may be of strip iron on either side of the floor board opposite the centre of the hive body and have them project 1/16 inch for each inch in the width of the floor board, e.g. 1 1/8 inch for an 18-inch board. The hook of the balance engages the handle, but the reading it gives is low and must be corrected by adding 1 oz. to every pound.

883. The common patterns of spring balance have a dial on the side, and even with a 6-inch dial it is difficult to get a good reading, unless the line of vision is perpendicular to the dial. It is useful, therefore, to add a chain extension which will bring the dial level with the face of the observer, and to put a bar through the ring at the top of the balance, so that the balance may be lifted with both hands.

Feeders

884. These will be found described and illustrated in the catalogues of the appliance makers, save for the simplest and commonest of all, which is a plain tin, such as a honey tin with patent lid, pierced with a few holes. The lid is a push fit and, being hollow, provides a bee-way for the bees when inverted over a feed hole. For slow feeding two or three holes, the smaller the better, are punched from the outside with a short length of stout needle driven in with a hammer or with an awl or sharp nail. For rapid feeding a number of holes are punched all over the end, but they must be small in diameter to keep the syrup from leaking. A shallow, wide tin is better than a tall one, as it is more easily covered to keep it warm.

The tin should be filled practically full and then inverted quickly over a bucket or other receptacle, as on inversion some syrup will fall through until sufficient vacuum is created. Tins carrying only one or two pounds can be inverted over the open hive.

885. If metal feeders are bought from the appliance dealer it is worth while paying the extra first cost of aluminium feeders, as they will not rust and will outlast two or three of the tin variety. Rapid feeders may be purchased, made of wood, with or without a metal lining. All feeders require thorough washing and drying after use, and those of wood should be filled with water to tighten the joints before filling with syrup, and allowed to dry slowly after use.

886. A new, simple and effective type of top feeder has been introduced in recent years, of which Brother Adam's design was the original. It takes the form of a tray of wax-impregnated wood of the same outside dimensions as the hive body, on which it sits. It is capable of holding about 10 lb. of syrup, to which the bees obtain access by climbing a central chimney into a diving-bell-like cover. The bees obtain access to a small or large surface of syrup according to the way the cover is put on. As generally supplied the bees can escape from the cover. The author sees to the clearance and secures the cover with a screw from above. The top is closed with a crown-board or metal cover.

887. The dummy- or frame-feeder is useful, especially with small lots of bees in cold weather. The dummy-feeder is constructed of wax-impregnated wood, or metal-lined. If metal-lined, the bees cannot maintain a foothold and many are drowned, a horrible sight, unless the metal is wax-coated. If the float is rectangular in section it is unstable and tilts, leading to more bees being drowned. If, however, the underside is rounded, or the under edges well beveled, this defect is completely remedied. If the top is open it is easier to refill than if a so-called filling hole is provided. In any case the sides are cut away to give the bees access.

On the writer's suggestion dummy-feeders are now made like a true dummy, i.e. with bee space at both ends. The usual full-width pattern frequently cannot be inserted when wanted or removed when done with.

Construction of Travelling-boxes

888. A hive body may be converted into a travelling-box by furnishing covers for top and bottom, each having a large hole covered with perforated zinc or wire gauze, well secured with tin-tacks. These covers may be made of ply-wood or match-boarding, and two battens of 2 inches x 1 inch or 1 1/2inch x 1 3/4 inch nailed across the grain with the latter, the battens running parallel with the edges and 2 inches from the edges (see Fig. 25). The boards should be so used that the battens run one way across the top and at right angles across the bottom cover, as they pile better thus, leaving ventilation free. Ventilation at both top and bottom is very desirable.

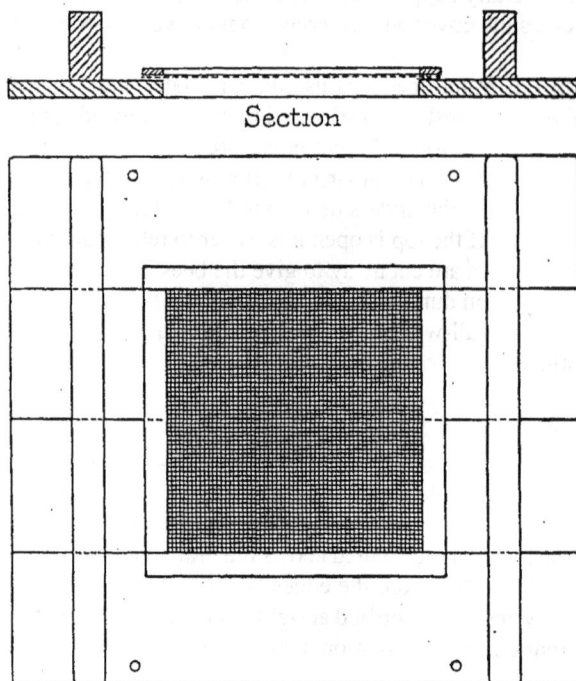

FIG. 25.—VENTILATING COVERS FOR TRAVELLING-BOX.

Temporary covers and bottoms constructed of ¼″ tongued and grooved or matched boarding with 2″ × 1″ cross battens. Centre opening covered with wire gauze sheet. Screw holes made to cardboard template to secure interchangeability. Outside dimensions to match brood only.

889. The covers should be secured each with four screws near the corners, so that when a covered body is placed to stand diagonally across an empty body, the four screws are accessible from beneath, at the corners. The cover should secure all frames in place. In case, then, there is a bee space above the frames, the cover must have suitable strips on the underside to fit down on to the frames near their ends and hold all secure. With deep frames, unless provided with Hoffman ends, it is desirable to have some sort of rack at the bottom to support them; for example, a row of wood pegs with rounded tops of suitable diameter and properly spaced.

Use of Travelling-boxes

890. On arrival at its destination a stock so traveled can occupy its permanent site, a temporary entrance being made with a wedge after removing the bottom screws, the bottom and cover being removed at leisure after the bees have settled down. Frames travel best upside-down if well built out and not overloaded with nectar. They should, or course, be inverted endwise, not sideways.

891. The travelling-boxes sold by appliance makers are not, as a rule, interchangeable with hive bodies, owing to the wide range of sizes employed, but have temporary flight boards and entrances. Those made to carry four to six combs can be used also as nucleus hives and should be furnished with weather-proof covers overall, ventilated at the sides and secured at the ends for transit.

Shipping Bees in Skeps

892. Stocks in straw skeps have to be sent by rail sometimes. They should have one or two sticks transverse to the combs, and if not already in place in the hive these should be inserted several days before the stock is to be moved, so that the bees may make all secure. A piece of stout canvas with a window of wire mesh should be tied securely over the bottom of the skep and the skep inverted for travelling, being labelled "this side up, live bees." If the skep does not stand securely inverted, it must be crated.

Shipping Swarms

893. A travelling-box, as indicated above for a nucleus hive, may be used for carrying a swarm, or a straw skep treated as above, but swarms and package bees are generally forwarded in light, strong, wooden boxes, having large openings in at least two sides, covered with wire gauze, some provision being made for feeding for long journeys (see package Bees, **1332** *et seq.*).

Drone and Queen Traps

894. It is uneconomical to breed drones not required and then to destroy them. The worker bees must gather several pounds of pollen and honey to produce a comb of drone brood, and the drones produced are themselves consumers. Drone comb should be almost suppressed in stocks showing undesirable characteristics, but some drones will be raised, and it may be desired to suppress these also at a time when queens are being mated. The drone trap may be used to advantage for this purpose. It consists of a box having an entrance covering the hive entrance and a large exit through "queen excluder," preferably of the wire variety. A horizontal division cuts off the upper part and has in it one or two conical bee escapes. These admit the drones into the upper part, from which they cannot return.

895. Any queen emerging from the hive while the trap is in place will be caught with the drones. The device may be used for catching an undesirable virgin on her first exit from the hive.

These traps greatly impede the traffic of the hive, impede ventilation, and encourage preparations for swarming.

Robber Screen

896. Some strains are particularly troublesome as robbers. Such should not be tolerated. Where they have to be coped with it is helpful to have a folding frame covered with 1 inch wire mesh, about 3 feet each way and surrounding three sides of a hive. The operator puts this in place, working through the open side. The robbers are bothered by the screen, but the rightful bees fly out readily and can return after operations are over.

Care of Appliances and Cleaning Metal Parts

897. The end of the season is a busy time, but the work must not be considered done until all appliances have been looked over and cleaned where necessary. Extractors, metal feeders and any other containers for honey or syrup cannot be cleaned too soon after use. Extractors never wear out, except the inadequate bearings of cheap extractors, but through neglect they will rust out in a few seasons. Worn bearings can be renewed.

898. Metal parts should be well washed, dried and rubbed over with vaseline before storing. The outsides of extractors and ripeners are frequently painted, for which purpose an air-drying japan, preferably with a bituminous base, is suitable. It is good to rub over the bright metal with strong vinegar before painting. An aluminium paint with a varnish base is also good and is sometimes used on internal metal parts as well.

899. Metal parts soiled with wax or propolis, such as metal ends for frames, may be readily cleaned by boiling with bleaching powder in the water, or Fels-naptha soap. A solution of caustic soda, 1 lb. to 2 gallons of water, is good, but highly caustic. It must be used with care and removed with a liberal supply of water.

Coating Wooden and Metal Vessels with Wax or .Lacquer

900. Wooden and metal vessels, even rusty ones, can be coated with hot wax and may then be used to contain honey or syrup. Paraffin wax is employed, but bees-wax can be used. The wax must be smoking hot and the operation should therefore be carried on out of doors. The vessel to be coated should be partly filled and then turned in all directions and the wax returned to the heater. Wood will absorb a considerable quantity, according to its nature. The vessel to be coated must be thoroughly dry before coating.

901. Cold lacquer suitable for metal parts may be purchased from a good oil and colour shop, or a solution of ½ ounce of shellac in 1 pint of methylated spirit may be used, the articles being dipped and drained, or the lacquer applied with a brush. To obtain the best results the metal parts should be warmed, and it is, of course, essential that they should be clean. Give a last wipe with methylated spirit before lacquering.

SECTION X
SEASONAL MANAGEMENT
Introduction

The Problem of Presentation

902. A writer setting out to give the best-known methods of managing bees, applicable to the small and to the large bee-keeper, to all varieties, all conditions of honey flow and climate and to all types of bee-keeper, has undertaken a large task, but one not impracticable if approached in an orderly manner. Management is applied to secure certain objects, readily understood, which are to be secured by methods adapted to the particular circumstances; methods based on simple principles of procedure which must be grasped, and on observations of bee behaviour. Each season has its problems; then there are problems peculiar to bees, arising from their swarming and robbing propensities; and, again, problems pertaining to bee-keepers, who have to move hives and, indeed, whole apiaries, and do many other things to suit their own convenience. Finally, there are a large number of manipulations entering into all systems of management which must be conducted in an effective and efficient manner, or difficulties will be made faster than they are solved.

The Problem Treated

903. The treatment of management has been greatly simplified by relegating to a later section all details of the manipulations involved. Management is first treated in broad outlines associated with the progress of the seasons, cross-references being given to those parts in which particular details will be found fully worked out. Next comes a novel section showing how the general system, of management must be modified to suit local circumstances. This is followed by sections on swarming, swarm prevention and honey getting, accompanied and followed by instructions for conducting several score of manipulations required from time to time, according to plan and circumstance.

In this way the whole subject is brought under review in such a way that the whole complex business is fully presented, yet the beginner need not lose his way.

Systems of Management

904. Many well-known bee-keepers have set their mark on certain systems of management, developed successfully by them to meet the conditions with which they were faced. Some of these systems are now obsolete in certain particulars, others have many features in common and most are applicable to particular conditions not universally obtaining. The more valuable and distinctive contributions to the art made by these men will be found in their appropriate place in this book, but as explained in the preface, no attempt is made to set out systems in detail against the names of individuals.

905. Every bee-keeper, however, should have a system of his own as the foundation of his management and procedure, adapted to the conditions under which he works and by which he is surrounded, or there will be much confusion, waste of time and energy and disappointment. The system should, moreover, be a simple one, involving a minimum of well-considered manipulations, so as to reduce labour and avoid unnecessary disturbance of the bees.

Single v. Double Brood Chambers

906. Something has been said about management in the section dealing with hives. The bee-keeper will have to make his choice, at the commencement, between the systems involving the use of single and double brood chambers. This is not, as might first appear, simply a matter of hive size or frame size. The question is, Are all the brood frames to be in one box, or in two or more? For either system the size and number of frames must be selected to suit the prolificacy of the queens (**50**). With very prolific strains even a 12-frame M.D. body is not always large enough, while for certain hardy long-lived but non-prolific strains of black bee, for example, one 10-frame body of British Standard frames may be enough.

907. For single-body working the M.D. body is largely used, but single bodies of 11 or 12 British Deep 12 inch frames are sufficient for average prolific queens. For multiple-brood bodies, the Langstroth frames are most used, but the British Standard frame is equally well adapted and is, of course, in general use in Great Britain.

908. The advantages of single-body management are, that there are fewer frames to handle, so that time is saved; there is no horizontal division in the brood nest, so that the queen has free natural use of the combs; the tendency to swarm is reduced (provided a large enough body is in use, not to cramp the queen); and the bees winter well on large frames.

909. The advantages of working with double-brood chambers are associated with the greater flexibility provided; the double body admits of manipulations impossible with single-brood bodies, such, for example, as Demareeing (**1226-48**); the frames are more convenient in size for the formation of nuclei (an advantage only partly met by the use of divided frames for nuclei in single-chamber working); and, where there is any considerable harvest to be secured, the same frames may be used for extracting and for brood raising.

910. Manipulations suitable only for single-body working are marked **S**, and those for double-body working only are marked **D**. All manipulations not so marked are suitable for either system. In certain cases, with single-brood chambers, doubling is necessary as a temporary measure, for example, in uniting by the newspaper method (**1409-12**), the bodies being re-arranged later to remove the superfluous one. Such manipulations are not marked **D**.

Seasonal Management

911. Under the headings following there will be found an outline of the important features of management that belong to the different seasons of the year. The features dealt with are of universal application and form the skeleton of all systems of bee-keeping.

912. Management may be defined as the planning of operations to secure maximum return with minimum effort under given circumstances. Without management bees would rapidly deteriorate, stray and eventually be lost, becoming a source of trouble to other bee-keepers in the district.

In pursuing a system of management and carrying it into effect many manipulations have to be performed, such as finding the queen, doubling, dividing, increasing, uniting, transferring, etc. Each manipulation has its proper technique. There are certain features which must be observed to secure success. These manipulations are detailed individually in later sections devoted to them.

913. Progress in the spring is so dependent upon preparations made in the autumn that it has become fashionable to commence seasonal instructions with those belonging to the autumn. The author has elected to follow the older fashion, for the active season commences in the spring, and the beginner is well advised to start in springtime.

Spring

Springtime Management in General

914. Where winter conditions are so severe that the bees have to be wintered in cellars, the bees remain quiescent until brought out of the cellars, and this is not done until the frosts are over and conditions propitious for rapid development, i.e. with ample honey in the hive and ample pollen and water available, with weather suitable for frequent flights(**109-12**).Where the bees are wintered outside with outer covers and more or less packing this covering should not be disturbed until the period of frosts and cold nights is past, but many flights will take place before this stage is reached. Where little or no external winter packing is used flights will occur on mild days even in mid-winter if the sun is strong and the shade temperature is high enough (**27** and **411**).

Early Flights

915. The earliest flights are for the purpose of evacuating the bowel and must not be mistaken for a sign of prosperity, for it should be noted that at this time a weak lot will be more restless and fly more than a strong lot, using more food per head to maintain the correct internal temperature. Again, breeding commences earlier in a small lot than in a large one in the same surroundings, as the temperature at the centre of the cluster has to be kept higher, and this leads also to early activity outside the hive. Again, clusters located near the entrance begin to move sooner under a temporary stimulus from sunshine than do those farther back or located higher up in the hive. Bees in a thin-walled hive will respond more quickly to a temporary sunny spell and rise of temperature than will those more fully protected. There is not much to be learned from observation of early activity, and, indeed, early activity may be merely the omen of early loss of a weak stock. The stocks of an experienced bee-keeper will, however, have gone into winter with a full complement of young bees and stores, and there should be no reason to diagnose weakness in the early spring, unless numbers of dead bees have been found near or within the entrance, a sign of disease or other untoward circumstance.

External Examination and Re-arrangement

916. In the early spring the hives should be looked over to see that the foundations have not shifted in winter, and any out of plumb should be corrected. Hives may be moved and re-arranged at this period when the bees are still confined for a fortnight or more at a time. Two persons are required to lift the hives (**632**) so that they may be lifted and replaced with a minimum of disturbance and without jar.

917. About the time that early pollen is available and breeding is well started in all the hives, a general external examination may be made to advantage and particularly for signs of queenlessness. The carrying in of pollen is a good sign but not a certain sign of the presence of a queen. The presence of drones of the previous year is almost a sure sign of queenlessness, but a very strong stock with ample stores will sometimes though rarely, tolerate a few drones through the winter. Each hive may be rapped smartly, preferably towards evening, and a characteristic sound is heard on applying the ear close to the hive. If all is well the bees are put at once on the alert and, nothing following, they settle down at once, save perhaps for a few guards appearing at the entrance. The sound heard is a short sharp hiss; the largest stock in best fettle making the loudest and sharpest hiss. A queeniess stock, on the other hand, suffers from uneasiness and the hiss of alarm is followed by a sort of roar, indicating that all is not well in the commonwealth. Bees seriously short of food lie as quiet as circumstances permit and give a bad response on the hive being tapped. Any stocks not showing a favourable response should be marked for early treatment, but an early examination of stocks made before remedial action is possible is likely to accentuate the trouble.

918. The relative strength of colonies having ample or similar top-packing, may be ascertained by lifting off the packing and laying the hand on the exposed quilt or crown board, finding the warmest spot. The degree of warmth gives an indication of colony strength. The food supply can be roughly gauged by lifting the back or side of the hive gently.

It is most undesirable to subject the bees to any unnecessary disturbance in early spring. Too early examination, even without touching the brood combs, may lead to the queen being balled. The bees seem to be particularly liable to do this if disturbed at a time they cannot replace her. The best time to peep in is on a still, cold day when the bees are still.

919. Those who use a wooden inner cover with a glass panel in it can, however, make an early examination without disturbing the colony, as it is only necessary to lift the cushion or other covering off the glass cover for a few moments when some observation may be made of the position and approximate size of the cluster and of the state of the stores.

Use of Candy and Spring Feeding

920. Starvation should never occur at this early stage, except through most exceptional neglect in the autumn, as very little food is consumed in winter, the bulk of the stores left in the autumn being intended for sustaining the stock through the period of heavy brood production which sets in when milder conditions supervene, accompanied by stimulation from a light early honey flow and a good supply of fresh pollen. During such a period the daily demand for food is very heavy, and if the weather turns colder for a spell, there is a greater risk of food shortage then than at any other time.

Nevertheless some bee-keepers habitually rob their bees of honey so severely at the end of the season as to feel in doubt, quite rightly, as to their condition in the winter. They then supply a box of candy as a precautionary measure, renewing it later when found empty. Now this course may save the bees alive, but it is fatal to rapid development in the spring. The use of candy in winter should be abandoned. Its use in spring should be discouraged. If stores are short it is better to feed thick syrup on a warm day, periodically, and as spring opens, leaving on a feeder until the bees won't take more or are known to have ample (**934**).

First Internal Examination

921. On account of the danger of balling the queen (**56-8**) the first internal examination should be deferred as long as possible; preferably until supplies are coming in and the colony should be actively expanding, until drone brood is likely to be present in the colony. This examination is primarily to verify the state of prosperity, i.e. the presence of ample stores, a laying queen and conditions in general. The presence of sealed brood, indicating the presence and activity of the queen, can be seen by looking down between the combs, as also the presence of sufficient stores, but the beginner may not see all he wants. If he feels he must disturb the combs, let him first remove an outer frame, then lift and examine the next, proceeding only until he is satisfied that all is well. Note should be taken of the number of combs covered by the bees. A beginner should not open up the brood chamber if the shade temperature is below 60° F.

922. At this time, in similar hives having good queens, the amount of brood is determined mainly by the number of bees able to cover it and by the amount of pollen and honey present or available. A stock weak in brood and bees should be regarded as a weak one, but not necessarily as one having a poor queen. If, however, the amount of brood is small compared with that in other stocks *with a similar history*, the queen should be suspect (see **54** for signs of queen failing). Such a stock may be de-queened and united to another small colony believed to have a good queen by the newspaper method (**1409-12**), which involves minimum disturbance.

923. If neither small colony has a good queen, and if a ripe queen cell is likely to be available shortly for re-queening, there is no occasion to hunt for either queen. Just unite and let them fight it out. If, however, re-queening at an early date is not convenient, it pays to unite such small colonies to stronger ones. One may appear to lose the output of a laying queen, but the same bees, if used to strengthen a colony of moderate size, will so assist that colony to build up, that one will have in fact more bees in a short time. Indeed this colony may be used later on for increase, so that one ends up with more bees and the same number of colonies.

Any colony found queenless at the time of the first examination should be united at once to one of medium strength.

924. Before uniting a weak or doubtful lot to a stronger one make sure there are no signs of brood disease in the weak lot, and if in doubt, wait and see. If the queen is wanted, make a nucleus for her arranged as in **1472**, but if stocks have all entered winter well-housed, strong and well-fed, weakness in spring is a sign of defect and the queen can well be sacrificed.

925. Any hive found to have only dead bees should be closed at once to prevent robbing and possible spread of disease. Any weak stock which cannot be dealt with at once should have its entrance contracted to an inch or less to prevent robbing and conserve heat.

The amount of contraction depends upon the strength of the colony, which requires to use, as well as defend, the entrance, and upon the amount of top ventilation. If a ventilated entrance grid has been used, or is available, ventilation is well cared for, and one only has to consider defence and the amount of traffic. A weak stock is not helped by a damp or foul atmosphere.

Further Examination

926. If, as is best, the first internal examination has been deferred until the development of the colony is well advanced, the following can also receive attention, failing which they await a second examination, which would be made at the same time as, or before, it is necessary to give additional

room (**935** and **1394**). Spring cleaning is done by means of the hive-tool or scraper. On uncovering the frames the top bars should be scraped from end to end with firm pressure and steady motion, the bees being kept down with a little smoke. The debris will fall between combs. Those who use vaseline will now brush over the top bars with the thinnest of coats. The covering is then replaced, being also scraped if of wood, and renewed where necessary if of fabric. The body is then lifted and placed on a temporary stand, such as an inverted flat-topped roof or on an empty body placed on a board or newspaper so that the queen may readily be seen in case she reaches the ground. The body is placed somewhat diagonally on its support so as to touch only at four places, thus reducing the risk of crushing bees. The body being removed, the floor board is quickly scraped over a box or newspaper, so that wax may be saved and rubbish collected and burned.

If the above is the first examination, then queenless and weak stocks will be attended to as previously described, and later when some stocks have built up to full strength a comb or two of brood may be taken from the strongest and given, one each, to any of medium strength, helping the strongest of them first.

927. At about this time any defective or undesirable combs, moved to the outside in the autumn for removal in the spring, should receive attention. This may determine the proper time for examination, as any such combs should be removed before the bees occupy them afresh.

In a large brood chamber empty combs may be inserted in the centre of the brood nest, in good weather, one at a time and a week apart, but this is not the best way to get defective combs repaired (see **350-1**).

Some have advised removing all unoccupied combs and closing up with a division board with the idea of conserving heat. Very little heat is in fact conserved, and a serious risk is incurred of overcrowding. Close watch has to be kept and there is unnecessary disturbance of the colony. Good bees don't have to be coddled. The bee-keeper who doesn't coddle his bees has time to keep more bees.

928. *The combs should be looked over for signs of brood disease* (**1497**) and to be sure the queen is right (**54**). Much may be learned of what has been going on and what may be expected by noting the relative amounts of eggs, unsealed and sealed brood (**5**).

The activity of the bees in expanding the brood nest can be estimated by noting the relative area of comb, or number of combs, occupied by eggs and unsealed brood compared with that occupied by sealed brood. If the ratio is about 1 : 1½ no expansion is occurring. A ratio of 1½ : 1 would indicate rapid expansion and 1 : 3 rapid contraction.

929. It is good to have a spare body and floor board, and when cleaning the frames to remove them to the spare clean body, examining them in passing. The emptied body may then be scraped and used for the next stock, but this is undesirable if there are indications of brood disease. Any bodies requiring new paint or creosote or other attention may be sorted out at this time.

It is better, however, to see to the soundness and suitability of all parts in the autumn, so avoiding trouble in the spring. With bees that are good house cleaners and hives with a full entrance it is rarely necessary to clean the floor board by hand.

Attention to Stock of Combs

930. All spare brood and all super combs should be examined and graded before they will be required, the worst being set aside for melting down, others for repair and the rest for use, new ones being separated from old ones.

Old brood combs are quite suitable, however, for use in supers even for pale honey. The bees have no liking for fouled honey and see to it that the combs are suitably cleaned and conditioned. The bee-keeper, however, will find a difficulty in judging the colour of honey stored in old combs. He will have to sample it. It is also easier to ascertain the presence of pollen in combs that have not been rendered opaque.

Stimulating Brood Production

931. The best stimulation for early brood production is autumn stimulation. Early spring stimulation is liable to overtax old bees and lead to spring dwindling, with consequent loss of brood and increased danger of disease. Some bee-keepers endeavour to enlarge the brood nest by inserting empty frames and by other re-arrangements. If, however, conditions are favourable, the bees will, at this time, have arranged for all the eggs they can care for. Egg-laying and brood rearing is encouraged by the presence of a young queen and of plenty of stores, and especially of uncapped stores. Thus slow feeding or a steady small flow of nectar, or the uncapping of some stores, all stimulate enlargement of the brood nest. The danger of artificial spreading of the brood nest is that the brood may become chilled and weakened, if not lost.

932. A strong and well-established stock will expand naturally, emptying the combs of honey and pollen, turning the stores into brood as reached, one comb of honey with good patches of sealed pollen being used in producing one comb of brood. If, however, the activities of the last season and of the bee-keeper are such that the brood nest is confined by the presence of large slabs of pollen, especially old pollen, or its expansion is hampered by slabs of drone comb, help should be given, and good combs inserted in the nest as in **927** above, but if combs are so inserted, or if for lack of comb, foundation is so inserted, be sure that there are at least two fully occupied combs on either side of each such comb. There is no risk involved in inserting empty combs between the combs containing brood and the outer combs containing the pollen supply only. Always keep pollen combs on the outer margin.

933. Egg-laying is controlled by temperature and food. If the entrance is opened wide and a queen excluder prevents the brood nest being extended upwards, egg-laying will be checked. During a spell of bad weather, keeping the bees indoors, egg-laying will be increased unless there is a serious shortage of stores. In the latter case preparations may be made by the bees for hunger swarming. Egg-laying is increased also by uncapping some of the stores.

Importance of Maintaining Stores

934. Sometimes a colony, built up late in a warm season with a young queen, may continue breeding heavily into the late autumn, so reducing to danger-point stores which had been considered ample. At other times and with other bees, they will take particular note of supplies; if stores are low and supplies not coming in, breeding will be curtailed and there may be a definite loss of morale. The bee-keeper should aim to have stores maintained at above 10 lb. minimum at all times. If allowed to fall to five, discouragement and consequent loss will occur; if reduced much below this the colony will suffer a serious loss of heart and will need serious attention. Fortunately a good supply of syrup promptly administered will provide the necessary encouragement, but it will not compensate for the set-back that has already occurred.

Preparations for Swarming

935. Under good management, where there is an early flow, a colony should reach swarming strength in late spring, if not before. It will make preparations for swarming unless hindered by bad weather or by appropriate management on the part of the bee-keeper who does not desire swarming. The swarming problem and swarm control are fully discussed in Sections XII and XIII. The outstanding feature is that crowding of the brood nest is the principal cause of preparations for swarming. Now the state of development in relation to crowding may generally be ascertained by a superficial examination of the brood nest, and if the colony is allowed to expand until the outer faces of the outermost combs are occupied, the bees will be already aware that they are on the point of being seriously overcrowded by the emergence of many bees from the recently rapidly expanded brood nest. Room should be given *before* the outer faces are so occupied. Indeed if much nectar is coming in the colony may be threatened with overcrowding before this stage is reached.

Onset of Early Harvest

936. The first considerable flow generally comes from fruit trees and other flowering trees. On its commencement the bees begin to draw out the food cells above the brood nest by adding new wax, but this is a sign that the bees are already coping with a flow substantially in excess of daily requirements, and are already crowding the queen with incoming nectar. Additional room should be given before this stage is reached. The need should be judged by watching the development of the trees and of the colony and the weather conditions as well.

The beginner in fear of chilling his bees and inexperienced in the rapidity of expansion of the brood nest at about this time, nearly always errs by giving more room too late. See **1394-9**.

Summer

Summer Management-Principal Object

937. From the earliest spring to the close of the season the prime object is to secure at the lowest cost the largest number of bees of harvesting age (**20**) at the times that nectar, is flowing freely and to divert their activity to honey-gathering. The general management must be varied to suit special local conditions, as dealt with in Section XI. The general scheme of management at this time is bound up with that of swarm control and is dealt with in Section XIII. The following instructions are of general application.

938. Whatever the conditions may be, the bee-keeper should observe what is going on; the state of the weather and forecasts and especially the shade temperature and the flowering of the principal sources of nectar. The indications of the scale hive should be studied, if one is used. The unobservant bee-keeper will miss the great opportunity of the unusually heavy flow which occurs now and again, and which to the more observant becomes a source of substantial gain. Bees cannot store honey in supers which the beekeeper fails to provide and cannot store so quickly where they have to draw out comb as they require it. See, therefore, that sufficient supers are available and ready before they are likely to be wanted. The removal of surplus and the treatment of the honey is dealt with in detail in Section V. The observations following are of a general character.

Removal of Surplus

939. It is desirable to remove most of the surplus before the flow ceases, even if the combs are not wanted again for refilling the same season, as the bees are good-tempered while the flow is on and disinclined to rob, whereas, when the flow ceases, their mood may be the reverse.

The honey should not be removed, however, until well ripened, a process which may continue after sealing, in warm weather.

940. It is as well to remove the surplus in the evening, or early in the morning, after or before the bees are on the move, super clearers having been inserted one to two days before. If the honey is to be extracted this may be done while the combs are warm. It is found that bees leave the supers more readily through the clearer if it is of the ventilated type, but it must be of correct design with double screen, not one through which bees can hand down honey readily. The two screens should be at least half an inch apart.

941. In the event of a sudden cessation of nectar flow, entrances should be reduced to hinder robbing, especially those of nuclei.

942. In Great Britain the main crop is practically over by the end of July or first week in August, except for those who are near the heather or near enough to move some hives to the heather (**634**).

Making Increase

943. Summer-time is also the time for making increase (**1424-74**). If the crop is a bad failure and feeding is necessary, the time is favourable for making bees, i.e. increasing the number of stocks, as there will not be much else to do, and breeding can be stimulated by steady feeding (see **1299-1302**, also **931-5** and **1014**).

Merits of Extracted, Comb and Chunk Honey Production

944. In some parts the principal demand is still for comb honey, but, per pound of honey, comb honey sold in sections is more expensive to produce than extracted honey. It commands a higher price but must remain a luxury article. Comb honey in small quantities for home use can be produced at practically the same cost as extracted honey (**1384**) by fitting a few marked shallow frames with super foundation or starters only and giving them to be drawn out, filled and sealed with others used for extraction, but such combs are not convenient to store (see, however, footnote to **538**).

945. The demand for chunk honey is local. Chunk honey consists of cut pieces of comb honey inserted to fill a vessel, the spaces being filled with similar honey extracted.

946. On an average 50 per cent. more surplus can be got as extracted than as section comb honey, and 33 per cent. more chunk honey than section comb. The relative outputs are therefore:

Extracted honey	100
Chunk honey	90
Section honey	67
and, ignoring labour, the prices should be in the inverse ratio, or:	
Extracted honey	67
Chunk honey	75
Section honey	100

The labour figures are higher, however, for preparing and handling comb honey, and the relative equivalent prices would be more nearly represented by the following:

Extracted honey	55
Chunk honey	65
Section honey	100

947. After allowing that the section contains less than 1 lb. of honey it would appear that, to be equally profitable, sections must be sold at a price nearly half as much again as the price per pound of extracted honey.

Management for Extracted, Comb and Chunk Honey

948. It is commonly held that the principal difference in management for extracted and comb honey is that with the former the bees may be given plenty of room, whereas for the latter they require crowding into the supers. Both statements are only half-truths and liable to lead away from success.

949. In the first place, before a honey flow and while there is danger of swarming, the bees for either system may be given plenty of room, but when the flow commences the aim should be to divert their activities from brood rearing to gathering. Manipulations are necessary to prevent choking of the brood nest with stores and sometimes to check breeding (see **1376-83**). It is generally advantageous to reach the maximum brood peak early by stimulating early brood rearing, so that there is a natural diminution of breeding coinciding with the onset of the flow. This is assisted by pressure of the arch of honey over the brood nest extending downwards as honey comes in, unless unlimited room is given below. If breeding is not checked and if the nectar flow is in two periods far apart as, for example, mainly from fruit and from clover, there is danger, in a large hive, of the bees using the early harvest for unprofitable production of brood which does not live long enough to do full duty during the second harvest. This does not apply to long-lived bees (**20**).

950. In working for sections, it is customary to restrict the brood chamber at the commencement of a harvest, but a strong stock will warm and enter the section racks without being forced to. In an early season or in late summer the section rack must be warmly covered. The bees may be hindered from going upwards by cold, i.e. lack of ample packing around and above the rack. They may be attracted upwards by the use of a special shallow frame at one or both ends of the rack and by a filled or partly filled super above the empty rack.

951. In the particular case of a white clover flow, weather being suitable, the supers should be required about 10 days after the first few blossoms are observed, and the maximum flow should be experienced perhaps 10 days later.

952. When an early flow is brought to an end by bad weather, the bees stay at home and there is fresh risk of swarming and of stores being converted into bees at an uneconomical period. For this reason it pays then to remove surplus and give foundation in supers and feed slowly. When working for section honey it pays during a temporary bad spell to feed back thin honey and extracted (disease free) honey diluted with an equal amount of water, so as to keep the bees busy working on the sections at a rapid rate. Sections so made are well finished and kept clean, and the bees kept in training for storing until the flow re-opens. With long-lived, hardy strains there is no harm, however, in encouraging breeding at such times, and the same is true in cases where there are no long breaks between flows.

953. For good work in sections it is essential to have a heavy sustained flow and plenty of bees to handle it. If the heavy reliable flow is a late one, use the early flow to build up the stock and for extracted surplus. It is undesirable to use sections for dark honeys, or for those that granulate quickly unless the latter can be disposed of quickly for early consumption.

954. The comb honey for chunk honey (**504-6**) is produced in shallow frames furnished with starters only. To get the starters started treat as for section honey and preferably put combs of honey or brood above them, thus forming gaps in continuity of comb which the bees will tend to fill.

Close of Nectar Flow

955. Breeding is generally reduced during the principal nectar flow, but if a long warm spell continues after the flow has ceased and before the onset of cold weather slows down all activity, some strains are liable to turn a considerable part of the surplus into bees. Those thus raised early are not so valuable as those raised a few months later. The bees would be better occupied storing pollen and any nectar yet to be had.

Some strains, generally of the darker and long-lived bees, have a good habit of packing the stores at this time around the margins of the brood nest, thus restricting it, carrying down the honey and sealing it early. Others, generally the more prolific and short-lived bees, are best deliberately restrained by the use of an excluder, limiting the queen to one brood chamber during and following the flow. The excluder is removed when preparing for wintering. This limitation of breeding space is an important feature, and in England can generally be carried out round about mid-summer day.

Autumn

Early Autumn Examination and Manipulations

956. The bees being easy tempered while the nectar flow still lasts, the novice, who is not quick to know the condition of his stocks, would do well to examine them all when inserting the super clearers. He should make notes of the condition of every stock, noting particularly that there is a laying queen and having a note of her age, also the approximate strength of the stock, as it will be necessary shortly to unite any weak stocks and in doing so to depose any remaining old queens. A weak stock may be requeened and strengthened by adding a nucleus containing a young queen. Note also the amount of stores.

957. Take any opportunities of removing any defective combs, especially those with any considerable patches of drone cells. Any such combs, when noted, if not empty should be placed against the outer sides of the hive on the coldest side, so that they may be the first emptied by the bees. They may be removed later, probably in the spring, before they are brought into use again.

958. This is the season for robbing and not for swarming. Swarming is unlikely, and one may ignore the precautions which, in the spring, are so important to prevent swarming. Entrances should be reduced. Any queen excluders will be removed with the last of the surplus. The bees in most districts will still gather some winter stores and, to keep breeding going, it is important that they have room below. If any considerable late flow is expected and a double or multiple brood body is used, put the body with the laying queen above and others below. The bees are generally disinclined to store at this time in empty combs above the brood-nest combs if the latter contain sealed stores near the top edges, but will extend the stores downward in these combs readily, making a compact body of sealed stores for wintering.

Autumn Feeding

959.a   Autumn feeding should be done whilst the temperatures are still high enough for the bees to be able to convert the feed into stores for winter use. In general it is recommended that a full colony will require about 20 Kg of stores to have sufficient to last through winter. The bees in the colony in autumn should be as healthy as possible to ensure a long life through winter.   ID

959. *Autumn feeding is best done quickly and finished at least 3 weeks before the first killing frost may be expected. The excitement of rapid feeding assists in the conversion, storing and sealing of the syrup. In Great Britain, feeding should be finished in September and as early as may be (see, however, 920). Autumn feeding is dealt with in Section XIV and should be practised if there is not a steady supply of nectar from minor but reliable sources after the cessation of the main harvest and not interfered with by bad weather. Bees bred at this time will be young (18) in the spring and ready for the heavy duty of rebuilding the stock. Furthermore, by forced breeding, the old bees can be worked out and will die before winter, reducing the risk of disease, especially acarine disease.*

960. Some persons object to sugar feeding on the grounds that honey is the natural and better food. Honey contains valuable salts in addition to the hydrocarbons, whereas sugar provides hydrocarbons alone. On the other hand, pure inverted sugar is totally absorbed by the bee without residue, and if fed last is used first in the winter, i.e. it is used during the coldest period, almost exclusively for the purpose of maintaining temperature, not for raising bees, and this is done without adding to the load the bowel has to carry, a load which in turn determines the number of weeks the bee can live in the hive without taking a flight. If during this period there are only the darker honeys available, which leave much residue in the bowel, there is serious danger of winter loss and spring illness.

It remains to be proved by comparative trial on an adequate scale under controlled conditions how far, if at all, bees suffer from sugar feeding. There is no doubt that the deficiency due to pollen shortage, too frequently experienced, is a greater evil than can be brought about by sugar feeding.

Stores Suitable for Winter Use

961. Honey from the later sources, golden rod, asters, heather, etc., is not the most suitable for wintering, and where sugar is not supplied, it is a good plan to keep some of the clover honey, if any, or other honey suitable for the purpose (**1303-6**), removing combs of the late honey to make room. Sometimes this late honey is of poor colour or otherwise not so saleable, but it will be excellent for brood rearing in the spring. Honeydew is definitely unsuitable and should be removed. If sugar is used, feed 10 to 15 lb. quickly last thing before closing down for winter. The combs must not be re-arranged after this final storage has been made.

962. The feeder should be well covered and packed round where nights are apt to be cool, and feeding should be completed before there is any risk of frosts causing the bees to cluster before stores have been sealed. I t is useful to note that a strong colony will take down and seal over large quantities of sugar syrup very economically if fed during a lull in the nectar flow in the summer, e.g. in early June in most parts of Great Britain. These store combs may be removed when preparing for the main flow, put away in a dry store and distributed later as required.

963. During rapid feeding in autumn it is important to see that the combs do not become so filled that there is insufficient room for egg-laying. It is important to remember that bees cannot form the winter cluster without some empty cells to cluster in. It is a good plan, before feeding, to insert an empty super below the brood chamber. The bees can then pack more stores round the brood nest and start the winter cluster hanging partly below the combs. The cluster will work up on to the frames before the severe cold sets in. This raising of the combs from the floor also helps to protect the cluster later on from the impact of winds blowing about the entrance.

Weak and Queenless Stocks

964. Before the final feeding (if any) and packing down, all weak stocks should be united and any found queenless provided with queens. Stocks queenless at this season will be short of young bees and even though strong in numbers, are best united to weak queen-right stocks.

With some strains, on the sudden cessation of a late flow, egg-laying is liable to be brought to a standstill. This should be remedied by feeding. An inexperienced bee-keeper, not having attended to this, and examining the stock 3 weeks later, finding no brood whatever, is liable to assume queenlessness.

965. If there are small stocks with good queens and well found with stores, there is an alternative to uniting, which saves the necessity of sacrificing queens, and that is, to place two stocks in one hive body. In the spring they may be united and the spare queen used elsewhere, or, if conditions are favourable and they build up well, they may be put in separate hives. The moved lot should be moved in the evening and other steps taken to secure that it does not lose its flying bees (**621**). Put the bees on either side of a bee-tight division up the middle of the hive and extending outside the entrance.

Old Combs for Wintering

966. Some make a practice of giving a frame or two of foundation to the stock while feeding, to ensure room for egg-laying, but old combs are best for wintering, and conditions are not favourable for building out the new combs to the best advantage (**315** and **318**) unless carried out early, as in **962**.

Amount of Stores for Wintering

967. The provision in the autumn of ample stores for winter and early spring is one of the most important factors in securing success, and it is well to err on the safe side, remembering that excess stores will not deteriorate, whereas a shortage of stores spells certain loss, as, although there may be no actual starvation, the bees will accommodate their expansion in the springtime to the stores available and prospective. They will be hindered from adequate expansion at the most critical period if ample stores are not available.

968. The amount required for early spring use varies mainly with the strength and vigour of the individual colony and, paradoxically, the most successful colony may be in the greatest danger. The amount varies also with the extent to which early supplies are available by the time the temperature is suitable for rapid breeding.

969. The amount required for wintering only, i.e. for consumption during the period of inactivity, depends mainly upon the length and severity of the winter and but little upon the size of the colony, as a weak lot will consume more stores per head in maintaining temperature than will a strong stock, the cluster being smaller, offering less resistance to cold, and being also more easily disturbed and more frequently disturbed by the duty of maintaining temperature (**23-5**).

970. Varying with circumstances above indicated, a supply of 10 to 25 lb. should be given for winter use and 15 to 35 lb. for spring, or a total of 25 to 60 lb. In Great Britain 30 to 45 lb. is ample, the larger figure for the strongest and more prolific stocks. In Michigan 45 to 60 lb. is recommended. In Victoria (Australia) 20 to 30 lb. is customary, but rather on the low side.

971. Where a stock is wintered in a single body, it is a good plan to keep back a shallow super of good honey and place on the body when packing down for winter.

Arrangements for Wintering

972. It is important to guard against the possibility of the bees being starved in the inactive and cold season through the stores becoming divided. The part on which the cluster is located may become exhausted, and the bees be unable to reach the other part. There should be plenty of stores at the top or back of the central combs.

Strong lots can be wintered with an inner cover giving access over the tops of all the combs, provided they have ample top packing.

973. Where quilts are used, a winter passage-way over the top is also sometimes provided. Place two bars of wood ¼ inch thick and ½ inch or more apart across all the frames in the centre, thus forming a tunnel under the packing for passage from frame to frame in the warmest part of the hive. The ends should be pared down so that the quilt lies snugly.

974. Instead of an overhead passage some prefer to make pop-holes through the combs near the top bars, and, indeed, on the continent of Europe tubular pieces are sold for insertion in the combs, which are built into place by the bees, but kept open as passages.

975. Where a shallow super of food is placed on the brood nest there is no need for a winter passage as one will be found beneath the super.

976. The last operation in the season with each stock is packing down for winter. This should be done early enough to give the bees ample opportunity to seal all crevices and adjust the ventilation to their liking. This cannot be done by the bees at a low temperature and is best done while the propolis is available, that is, while the temperature is still moderate and well before first frosts. (For Winter Packing, see Section VII). If the hives are to stand out of doors, make sure that they are weather-proof and mouse-proof and secured against times of storm and stress. Covers may be blown off unless of the deep variety, and should be secured with weights, say, by a brick hung on the end of a rope, the rope passing over the top and being secured to a stake the other side. The entrance must be screened from high wind, and it is bad to have hives so exposed that there is any risk of their being overturned.

977. The preparation of the last lot of honey or wax for market should be completed, after which the last operation of the season consists in cleaning, overhauling and careful storing of all appliances, empty combs and the like (see **357-60** and **897-9**).

Winter Conditions

978. In sub-tropical regions, breeding and even the gathering of nectar may continue without omission throughout the winter. Some hints for management under such circumstances will be found in **1052**, but in temperate and colder regions there is a period· of quiescence in the winter, with close clustering, subject in the cold period to periodic disturbance for internal re-arrangement and the raising of the temperature. When the surrounding atmosphere in the hive is low enough to cause risk to the life of the bees, then sufficient heat must be generated and pass from within the cluster to maintain the bees on the outside at a temperature above the minimum to sustain life.

Generation of Heat and Size of Cluster

979. The winter cluster is formed on the store combs and when first formed will include and cover the last of the brood. When this brood has all emerged the bees cluster very closely, filling the empty cells and the space between. The closer the cluster, the better the heat is retained and the smaller the external surface. The mass of bees and heavy old comb and of any stores within the cluster has considerable heat-storage capacity, but the heat has to be maintained by the consumption or combustion (oxidation) of food within the bodies of the bees within the cluster. If the temperature (**27**) falls too low, the bees will die. If it is too high, the activity and consequent consumption of stores increases, with risk of the bowel becoming overcharged (**1305-6**). Between the two extremes there is a small range of temperature within which the food required is a minimum and also the risk to life and health.

980. Now a strong lot of bees offers a smaller ratio of external surface to volume than a small lot. If the number of bees is halved the ratio of external surface of the cluster to the number of bees is increased by 30 per cent. and more heat must be generated per bee to keep up the outside temperature. In a small lot the rate of consumption of food per bee and the rate of loss of heat are both increased and one finds a higher average temperature within the duster, and more frequent disturbances of the cluster to save the lives of the outside bees. Calculations based on outer surface alone, do not seem quite adequate to account for the marked differences observed as between small and large clusters. The air will pass more readily through a small cluster than a large one, and it seems likely that this considerably intensifies the burden of maintaining temperature in a small cluster. In any case, the saying, now classic, that the best packing for bees is bees, has its justification in practice and in theory.

Causes mid Prevention of Winter Losses

981. In preparation for winter the first object is to conserve the bees and the second to conserve the stores. When one reads, as one may occasionally, of an average of 40 per cent. of stocks lost when wintered out of doors and about half the number when wintered under cover, it is evident that the problem of successful wintering, to some people, is no light matter. With plenty of young bees, *disease free*, a young queen, plenty of suitable stores and appropriate protection one may reasonably expect all stocks to come through successfully, barring an occasional mischance. One has heard of bees of a hardy strain successfully wintered in the open, in the Hebrides, in small hives without external packing and others showing high losses when wintered under heavy packing in a mild short winter in the south of England. Taken the world over, probably disease and damp are the greatest enemies in winter, but lack of observance of important details set out below, and in autumn preparation, is probably responsible for as much loss.

982. Cold alone will not kill bees that have access to stores, but excessive consumption of stores, and the use of unsuitable stores (**1305-6**) in maintaining the temperature of the cluster, leads to rapid accumulation of waste matter in the bowel, and puts a limit to the number of weeks the cluster can withstand the cold; a period short with weak stocks but longer with strong stocks for reasons already given (**980**). If such conditions are prolonged therefore, some packing is a necessity. How much is desirable is a matter of compromise, both stores and packing costing money, in material, or labour, or both. Bees, well packed, in the open, have withstood temperatures down to -35⁰ F., but with relatively heavy consumption of stores and serious risk of dysentery. It is sometimes economical to provide winter cellars in districts where the mean temperature falls only to 25⁰ F. for long periods. Winter cellars are almost necessary for commercial work where the mean temperature reaches 15⁰ F. Heavy packing is necessary where these limits are approached. It is used successfully under more severe conditions and this may be justified where only a few hives are kept, but for fifty or more colonies it is a question whether a permanent cellar would not be the more economical proposition in the long run.

Keeping out Damp

983. In milder regions, damp is more to be feared than cold, detrimental as it is, not only to combs and to stores but to the health of the bees. On the other hand, as already indicated in **711**, the practice of wintering bees out of doors and with less and less packing is on the increase, even in regions where cellar wintering or expensive packing was previously considered essential. The problem is seen to be largely a question of economics in which cost of labour, of materials and of stores consumed all figure.

984. Waterproofed packing paper may be used in a similar manner. A covering of roofing felt with 4 to 6 inch overlap, enclosing an air film, adds appreciably to the thermal protection, as well as to the moisture resisting quality of the hive. A loose outer cover as employed in the W.B.C. and Kootenay hives, used in Great Britain and Canada respectively, serves to keep out damp, but see **658**. Damp air being lighter than dry air at the same temperature, tends to rise and will leave by a top or middle entrance (**758-61**).

Production of Damp

985. If the temperature of the hive walls is below that of the dewpoint of the air within, condensation occurs on the walls. The prime cause of condensation is the charging of the relatively warm air in the hive with moisture through the activity of the bees, the combustion of their food causing the production of much water vapour with carbon dioxide. This moisture condenses on a fall of temperature taking place. Conditions are much worse in the presence of unsealed stores, as the moist surface exposed tends to saturate the air with water vapour. However charged, the warm moist air circulating in the hive is cooled as it passes down the outer walls and over the outer combs and discharges its moisture over those surfaces. Excessive humidity is bad, apart from direct moisture. The cure is adequate ventilation.

986. Heavy packing tends to equalize temperature and thus promotes a healthy atmosphere. With heavily packed large hives, as used for example in Canada, a ½ inch tunnel entrance provides sufficient ventilation, but where packing is inadequate increased ventilation is essential to cause the air to change more frequently, both on account of increased activity of the bees and of greater and more rapid temperature changes occurring.

Freedom from Disturbance

987. It is important that the bees should not be disturbed during the quiescent period by straying animals, vibration, water dropping off tall trees or other causes.

Moving Hives in winter

988. Advantage may be taken of the quiescent period to do any necessary ground clearing and any necessary re-arrangement of the apiary. The hives may be moved with but little fear of the bees losing their way, as they will note their location anew after a few weeks of quiescence, but there will be some drifting and loss of bees. The hives must be lifted, carried and replaced gently and carefully, so as to make only a momentary disturbance. On no account use a wheeled vehicle. The best time for moving, however, is in the summer (Section VI).

Cellar wintering

Conditions

989. The conditions under which cellar wintering is desirable are discussed in **982**. Outside wintering with heavy packing is making headway in regions in which cellar wintering might have been considered essential. Cellar wintering is not so desirable where bees can have an occasional flight during breaks in the cold period. Hardy strains naturally winter better out of doors than the softer strains. The use of no cellar is better than the use of cellaring imperfectly carried out. To secure success with a cellar it must be so constructed and worked, that the temperature is properly controlled and the bees protected from damp.

Construction of Cellar

990. The cellar should normally be really dark, and it is best to use an artificial light for inspection purposes.

991. The cellar may be built below a dwelling-house or below a bee-house or workshop, or into the ground with earth over the top, the side of a small hill affording convenient entry. All parts of the cellar must be below the frost-level. This generally necessitates having the ceiling about 30 inches below the ground level. If below a building, the ceiling should be packed to a depth of, say, 18 inches, with heat resisting material, say, sawdust; but any brick arch counts as part of this. If there is a risk of heat entering, but more especially if there is a risk of heat escaping, a double door will be necessary and, indeed, in very cold regions the doors must be separated by an antechamber, itself packed so that the inner door is protected from frost at all times.

992. The cellar may conveniently have a height of 6 to 7 feet and a capacity of 12 to 15 cubic feet per colony. Thus a hundred colonies can be wintered in a cellar with a floor area 10 feet by 20 feet.

993. If the temperature is kept within proper limits very little ventilation is required. A 2-inch pipe entering at bottom and one or two exits at top, which may be smaller if there are two, say, 1½ inch, should be ample, but there must be means for partially closing them. If more room is provided per hive than above indicated there may be difficulty in keeping up the temperature.

Cellar Temperature

994. A temperature of 45° F. (7° C.) used to be recommended, with two or three degrees higher at the commencement and two or· three degrees lower at the end. Experience has shown that temperatures several degrees lower give better results in the coldest regions and some increase is permissible in the warmer. The best temperature is that at which the bees make least noise, and experience will lead the bee-keeper to judge how things are going by the amount of noise made. The temperature may be observed by a thermometer freely suspended about level with the eye. Periodic inspection is important, and once a week is not too often.

Moving Bees into the Cellar

995. The most favourable time to move the bees into the cellar is on a day after they have had a warm spell and good flight (to clear the bowel) just preceding a cold spell. All brood should have emerged and the cluster have formed. This points to late October and early November in the northern hemisphere, according to location. All hives should be marked and a record kept of their position, so that they may be restored to their original stands. It is true that bees tend to locate their position afresh after long confinement, but if the hives are mixed some undesirable drifting will occur and there may be some unnecessary loss of bees.

996. The quality of stores is a vital factor in cellar wintering, and the supply of sugar feed is recommended last of all before moving (see **1303-6**).

997. If the bees are cellared in single bodies it is essential to have in reserve (**533-6**) a second body for each hive well stocked with food. This should be placed on when the hive is first put out, so that breeding may be stimulated and ample provision made for rapid increase.

998. There should be not less than 25 lb. in the body taken to the cellar.

Conditions in the Cellar

999. Before moving to cellars, hive entrances should be lightly plugged, but all entrances should be opened as soon as the bees are in the cellar. Some top ventilation is desirable, obtained by the use of porous packing. If inner boards are used a somewhat wide entrance is desirable. The hives must be lifted, carried and set down as quietly as possible.

1000. The temperature of the incoming air being raised after entry, is above dew-point, and is in fact capable of absorbing and carrying off a certain amount of moisture. If the temperature is correctly regulated so that food consumption and moisture production is small, all will be well and the slow ventilation required will be sufficient to carry away the moisture produced. It is important, however, that this moisture be not added to by damp walls and floor or by the storage of vegetables in the same cellar.

1001. The floor, though dry, will nevertheless be cold, and the bottom row of hives should be raised several inches off the floor on bricks or creosoted wood runners. The hives may be piled in two or more tiers up to the ceiling.

When visiting for inspection see that all entrances are clear of dead bees, using a wire hook for the purpose and reaching well in.

Removal from Cellar

1002. Bees of some strains wintered outdoors will fly in the sunshine with a shade temperature of 50^0 F. $(10^0$ C.) or less, but there is considerable risk, as if the sun is hidden for a while the bees become benumbed and cannot return. They will then die by exposure in the cold night following. Flights are frequent with a shade temperature of 55^0 to 57^0 F. $(13^0$ to 14^0 C.). Thus the time to bring the bees out of the cellar is the time when such temperatures are reached as maximum shade temperatures. Bring out the bees preferably when the weather forecasts predict a rise of temperature to follow. The period in question is generally in late March to April in the northern hemisphere, according to location and season. The blooming of the willows in most places at this time ensures a supply of fresh pollen (see **1275-6**).

SECTION XI

SPECIAL MANAGEMENT

Introduction

1003. Instructions for general management are usually based on average conditions, and but little guidance is given for coping with the peculiar conditions of particular districts. Only too frequently writers in the journals, giving advice or experience on management, omit to state for what precise conditions they are catering. In Great Britain alone almost all kinds of conditions are to be met with, except a winter flow. There are main harvests to be secured only in late autumn, early harvests mainly from fruit, irregular, long-continued flows and the characteristic early and late crop with a gap between. In America the clover, buckwheat and tulip tree regions require different management, and appropriate methods have been worked out in detail. In Australia the forest regions present unique problems. In all parts, however, every district has some peculiarity, and, to secure success, it is necessary to recognize what the conditions are and, to know how to adapt management to fit them.

1004. General guidance for seasonal management is given in the preceding section. The technique of particular manipulations is given in later sections. Here we have to deal only with what has to be done, not how to do it. It is thus possible to present in small compass an account of the secrets of profitable bee-keeping applicable to almost all circumstances in which English-speaking bee-keepers may find themselves, presented in such a way that the intelligent bee-keeper may grasp the problem as a whole and adapt the details to his particular need.

1005. In attempting this problem for the first time, the author wishes it to be understood that he himself has had personal experience of only a few of the conditions enumerated and respectfully submits that if there are flaws in this first attempt they are mainly due to the fact that his predecessors have evaded the problem. He hopes that those of his readers who can criticize from personal experience will not hesitate to communicate with him, as well as those who can suggest useful additions to this section.

Relation of Management to Flow Periods

1006. In deciding upon a system of management for a particular district, the first and principal considerations are (*a*) what are the principal sources of honey, (*b*) when do the flows occur, and (*c*) what is their relative magnitude? To take a representative case, there may be an early flow from various fruit-trees in April and May (northern hemisphere), or October and November (southern hemisphere), continued by some later flowering trees, followed by a gap in early summer, when notwithstanding fine growing weather there are no good nectar-giving flowers in sufficient quantity to give a surplus, perhaps even insufficient to provide for the daily need; then comes a heavy flow from clover and lime, weather permitting, and practically a cessation in the beginning of August (northern hemisphere), February (southern hemisphere).

This case is representative of conditions prevailing in a large part of Great Britain and in many other parts of the world in temperate climates. The instructions given for seasonal management in the preceding section are all appropriate, but some of them require modification where other conditions prevail, as indicated in what follows here.

1007. The more important sources of nectar are listed in Section IV, **455** *et seq*, but sometimes it is as important to have adequate knowledge of the pollen supplies as of honey flows. Fortunately in most parts there is no lack of good pollen when wanted. Bees do not readily fly long distances for pollen, especially in cool weather (**583**), and in places where the principal sources of nectar are also the principal flowering plants in the district, there may be lack of pollen at the time when bees should be being raised, ready for those flows. The need is felt especially in Australia, where nectar-bearing forest trees cover large areas frequently unsupported by the numerous earlier sources of pollen found in most other places. The more important sources of pollen are listed in **455** *et seq.*

1008. Management includes all those things that appertain to bee-farming, such as queen raising or purchase, increasing stock, stimulating breeding, checking swarming, and, above all, honey-getting.

1009. *Now, in all special systems of management - that is, in all adjustments of method to the particular local conditions - the great secret of honey production is to have a large force of bees of the harvesting age (23-5) available throughout the time each flow is on. Without the harvesters the harvest will not be gathered. If large forces of bees are raised at any other time, it will be at the expense of the honey stores. It is of no benefit to raise bees in numbers, except when they can he utilized for harvesting at the harvesting age, unless one intends to make one's profit out of bee-breeding and not out of honey.*

1010. If harvesters do not predominate at the time of harvest there is, furthermore, a probability that much of the honey gathered will be utilized forthwith for brood raising, which brood will mature too late to be profitable. If, however, the large majority of the bees are of the flying age at harvest-time, then the stores they bring in will keep the home workers busy storing, thus actually checking breeding. Hence, the benefit of manipulations aimed to increase the proportion of flying bees in a particular hive during the honey flow.

To produce honey one must therefore plan the whole procedure so that the periods of activity and of inactivity in brood raising are properly related to the periods of heavy flow of nectar.

Influence of Longevity of Bees

1011. Some bees, having very prolific queens, are short-lived, their life when hard-worked extending only to about 5 weeks, first as nurses, then as harvesters. With such bees not flying for the first

fortnight (**23-5**) and gathering only for about 3 weeks, it will be seen that considerable thought must be given to securing an adequate force of them at the critical time. For example, if the main flow lasts from June 21 to July 15 (December 21 to January 15, southern hemisphere) with but little before and a rapid falling off after, it will be seen that no bee emerging before May 17 (November 17, southern hemisphere), 5 weeks before the harvest commences, can be counted on as a harvester, and no bee emerging later than July 1 (January 1, southern hemisphere), 2 weeks before the harvest finishes, will be of much service. Put another way, eggs laid before April 26 (October 26 southern hemisphere) cannot give bees for the main harvest even for a day's work, and scarcely any work in the profitable period will be got out of bees emerging from eggs laid after June 10 (December 10, southern hemisphere). In such a case, earlier laying is useful only to build up the colony so that egg-laying is in full strength about or soon after April 26 (October 26, southern hemisphere), and egg-laying may be slowed down to advantage from a little before harvest until, say, the latter half of August (February, southern hemisphere), when it should be stimulated again for a few weeks to produce new blood for wintering.

1012. If the main harvest is uncertain as to date, as it is in a variable climate, some margin of days must be allowed in the programme to provide against this.

1013. Consider now, however, the same flow, but to be harvested by bees having a life approximating to 8 weeks in the busiest period, even though of a less prolific strain. It will be seen at once that there is a 6-weeks' harvesting period in the life of each bee instead of 3 weeks, giving a greater latitude in fitting the bees to the period of nectar flow. For a given rate of egg-laying, the proportion of harvesters in the hive tends to be doubled, and for any given period of flow a larger proportion are able to put in a long period of work.

Where nectar flows are short and uncertain, long-lived bees are much easier to manage profitably than short-lived bees, other things being equal.

Working to Produce Bees v. Honey

1014. Very weak stocks can scarcely maintain themselves even in warm weather during a flow. Stocks very strong in bees tend to check brood rearing during a honey flow. 'There is an intermediate size not overstrong in harvesters able to expand its brood nest rapidly and which will do so at the expense of a maintained supply of food, whether artificially fed or coming in during a natural flow. Such stocks, especially with young queens, produce bees, not honey. Thus, to produce bees one may work with stocks occupying, say, five to seven frames, and see to the food supply. Old bees may be removed and used elsewhere (**1418**).

To check breeding, keep supplies short and remove combs of emerging bees, or make nuclei for some of the queens and unite their bees to other stocks.

Importance of Keeping Stocks Strong and Timing Development

1015. During a flow early in the year a small stock of good strain still engaged in building up, if below a certain size, may nevertheless actually loose weight, for it must keep most of its bees at home for the necessary minimum of home duties, including maintaining the temperature of the brood, and it will be turning stores rapidly into bees, say 1 lb. of stores for every 1,000 bees. On the other hand, a larger colony that has attained its maximum expansion can put out a larger proportion of its workers and a much larger total number to gather the harvest, for not only does it make a smaller relative demand for nursing, but the larger volume of brood nest presents a relatively smaller surface to be covered to maintain temperature.

Special Case of Large Apiaries

1016. Other important details will be brought out in discussing particular cases which follow, but it is convenient to mention here a relationship between the size of the apiary and the prospects of a profitable harvest. The matter is touched upon in Section VI, where it is shown that apiaries may be profitable in which, say, 100 colonies are located in one place, and indeed for certain very heavy flows much larger numbers may be profitably managed, but, by the same token, when the flow is less a smaller number may represent the profitable limit. Now, if there are substantial minor sources of flow as well as the main source, it may be found that, while 100 colonies in one place are profitable during the main flow, the same lot divided into two suitably placed lots a few miles distant may be equally profitable during the main flow and give surplus during minor flows as well. For example, in a clover and lime district, with orchards few and far apart, a double harvest cannot be got from one large apiary. It may pay, however, to place out hives in individual orchards in the spring and collect them for heavier work later, moving them in the slack period between harvests. An alternative, where it is not convenient to move hives, is to work only a portion of them for honey during the early harvest and the remainder at that time for breeding and increase.

1017. With simplified management and convenient means of transport, it will pay a bee-keeper owning, say, only 50 hives to place them out in several stations, and to move some of them from time to time to where a flow may be found.

Influence of Disease

1018.a It is generally accepted that the best bees for any region are those that naturally occur in that area. They will have acclimatised to the local weather conditions and forage. Improving the quality of stocks (healthy and productive) is better performed by selectively breeding from local stocks rather than importing queens from other countries or regions of the country. ID

1018. *The importance of providing flying bees for the harvesting period has been stressed.*

Next in general importance is the business of ensuring strong stocks not only for the rigour of winter but to stand the strain of a rapidly growing brood nest in the spring. This is especially important where there is an early surplus to secure; but where the main flow is late and the earlier flows precarious it is advantageous to avoid too early breeding such that the bees bred will be a source of expenditure with no hope of a corresponding return. Bees accustomed to luxurious regions are particularly wasteful in these circumstances. What is wanted in such a district are strains accustomed to lying quiet without swarming when times are bad, and working furiously when there is a ready flow.

1019. Conditions are, however, altered for the worse in any district where disease is prevalent, and it becomes of first importance to secure that stocks are not depleted by disease at a time when rapid access of strength is essential to a good harvest.

1020.a   European Foul Brood (EFB) is a notifiable disease in the UK and must be reported. Uniting weak colonies in areas where EFB is prevalent should only be united once it has been confirmed that neither shows any sign of infection otherwise there is a danger that the disease will be spread to other colonies. ID

1020. *European foul brood is a spring disease seriously hindering spring development and tending to shorten life. While attending to the disease itself (1515) it is essential to make up for the wastage by special attention to all that appertains to rapid expansion in early spring, and, by uniting, to avoid carrying weak lots into the harvesting period.*

1021. Acarine infestation (**1529**) is another disease apt to be devastating in effect and to hinder spring development. Stocks suffering from acarine should be treated during the off season (**1537-44**). Management tending to separate emerging brood from old bees is helpful.

1022.a   Any colony with suspected American Foul Brood (AFB) must be reported to DEFRA via the Appointed Bee Inspector (ABI). If the disease is confirmed the colony will be destroyed. ID

1022. *American foul brood comes late in the season and must be dealt with to prevent serious autumn and winter loss.*

Classification of Districts

1023. Diagrams of nectar flows are liable to be misleading. A considerable minimum flow is necessary to maintain a stock at an even level. Under the same conditions, however, one beekeeper will snatch a surplus while another is advancing backwards. The most useful classification of districts is one according to the period of the principal flow giving surplus; such a flow as that from clover, heather, buckwheat, orange-blossom, etc. The general management must be planned to take full advantage of the principal source of income, and minor adjustments made to fit the minor flows and nature of the district.

1024. The classification of types of district is indicated by the side headings in what follows. No attempt is made to classify by name of locality as this is only possible for important areas such as the buckwheat area in U.S.A., or heather districts in Scotland; but in such cases special leaflets or bulletins are generally obtainable from the local departments of agriculture.

Bees Suited to the Conditions

1025. Different strains of bees have different habits. It is desirable, if possible, to obtain bees from a strain developed under and giving good results under the conditions in which they will have to work (see **118-20** and **125**).

Principal Flow Early

1026. This is the case, for example, in the Tulip-tree region in the U.S.A. and in some large fruit-growing areas. The problems are how to get a strong enough colony early enough and what to do for the rest of the year.

1027. I t is essential to go into winter with the strongest stocks of young bees and young queens, with ample stores to encourage early and rapid expansion of the brood nest. Pollen stores carried over winter and early pollen supply are important. Stimulative feeding (**1294-1300**) is useful in early spring, and the use of pollen substitutes (**1275-1281**) where there is an inadequate early supply. The stocks should be packed to winter quietly so that no wasteful premature excitement is encouraged. In regions where packing is practised, stocks should be unpacked, say, 10 days after the last killing frost in spring.

If swarms should issue before unpacking, treat as in **1193** to avoid increase, giving ample super room before the flow is expected.

1028. When the flow is due to commence, or previously if queen cells are started, it is essential to hinder swarming. See **1156-76**.

1029. If further surplus is expected at a later period the method selected must be one which can be timed to secure a supply of harvesters at such period. It may pay to transport the bees to another district where a later flow occurs.

1030. Queen raising in late July is favoured, and replacement of all queens annually each autumn, as the flow arrives too soon to enable stocks found with failing queens to be dealt with in time to save them for the early flow. The making of increase is best practised in early autumn.

1031. If there is a moderate late flow, not enough to harvest, it is a good plan to work for increase and unite for winter under the best queens, thus going into winter with the strongest possible stocks.

In districts with a cold spring and early flow, the purchase should be considered of package bees from a warmer region to assist in harvesting (**1332-48**).

1032. Naturally a very substantial amount of honey is required to maintain the colony in such a region after the principal flow is over. It may well pay to purchase a strong packet in time to build up to harvest the flow and then destroy the bees, thus getting at the honey.

Principal Flow Late

1033. There are districts where a heavy flow occurs in autumn, and but little earlier in the year. Such late flows occur where buckwheat is grown extensively, in good heather districts, and in some where asters, golden rod and other autumn flowers flourish in abundance, and perhaps beans.

1034. Conditions being difficult in the early part of the year, trouble from disease must be watched for, and where prevalent it is essential to carry over ample stores from the previous year to secure an early start (**1303-6**). The problem is how to avoid excessive consumption in spring and summer and yet have hives full of bees for the late harvest.

1035. The special management appropriate depends upon the possibility of securing also some surplus in the summer, say, from the clovers.

1036. If no summer surplus is obtainable, divide the colonies as soon as strong enough in the spring (**1446-52**), and re-unite before the flow, under the best queens. With stocks re-queened, see that the queen will be, and is, laying at least 6 weeks before the main flow commences. This necessitates raising queens early and dividing early if the divided lots are to be used for the emergence and fertilization of their own queens (Section II). Increase, when required, from divided lots. Never raise queen cells except in strong stocks (see Section II).

1037. If some surplus from clover is obtainable, doubling without increase (**1219-25**), and separate queen raising, or artificial swarming with increase (**1453**) and re-uniting under new queens for the autumn harvest, or a convenient method from those in **1226-64** may be used. Increase is best made in the autumn if inferior honey constitutes the main flow.

Early and Late Flow with Gap Between

1038. This is characteristic of regions having a good flow from fruit-blossom and early blossoming trees followed later by a heavier flow from the clovers and perhaps bass (lime). If there is also a still later flow from heather or beans, asters, golden rod, etc., the bee-keeper lives in a sort of paradise and does not need much special management, save to feed, if there is interruption, at important breeding periods, and to avoid breeding harvesters (**1009-10**) for the slack period. Generally the autumn flow is not more than enough to keep the bees supplied from day to day, and the main problems are then, how to secure that the old bees are used on the early flow; and how best to tide over the dull period between the spring and summer flows.

1039. To secure the early flow the stocks must go into winter with ample young bees and ample provisions to secure an early start and rapid build up. The older bees should be used for early surplus by such methods as **1213-18** and **1219-25**, which also check swarming.

1040. If the main flow is late, there being no early summer surplus available (alsike), and if the spring flow is good, some stocks may be divided and worked for breeding, then united on the eve of the main flow. The stocks should be strong enough for division preferably 8 weeks before the main flow. If, say, in swarming condition in early May, this method could be worked where the main flow does not commence until the beginning of July; but, for a June flow, strong April stocks would be required and can only be got in mild districts.

1041. Increase may be made in spring or autumn and the various methods of checking swarming practised, provided always that the bees are kept breeding most heavily 6 to 7 weeks before the height of each flow, and any manipulations which check breeding carried out at such a time that the check operates 6 to 7 weeks before the middle of a slack period (**1009-10**).

1042. Queens may be raised on the early or late flow, according to the general plan adopted, but it is most safe to build up with young queens in autumn, ready for the spring.

Prolonged Heavy Flow

1043. This occurs in a few favoured districts where a summer and an autumn flow are extended and meet; where, for example, alsike and white clovers are supported by late limes, numerous wild sources and continued by beans, buckwheat, early asters, etc.

1044. This makes a bee-keeper's paradise, but has its special problems, and the lazy or unwatchful bee-keeper will miss the best results and be troubled by swarming. The problems are, how to secure ample harvesters for the continued flow and how to avoid swarming.

1045. It is important to avoid the use, at the beginning of the prolonged flow, of checking methods, such as **1211-12**, which hinder brood development, as this late brood is needed to provide bees for the latter part of the flow. Such methods may be used later in the season, as a second check, desirable especially if a check in the weather and flow causes congestion in the hives likely to lead to swarming.

For the beginning of the flow it is better to use Demaree or the like, **1226** to **1264**, but one can form nuclei to be re-united later (**1216-18**).

1046. It is well to recognize that it is difficult to work stocks to full advantage for a long spell hard at harvesting, and better to note differences in stocks, those ready early to be applied to immediate harvesting and the others to breeding for harvesting later (**1025**). Swarming may be checked by manipulating these weaker stocks so that their flyers join the early harvesters (**1418-23**) and then building them up in the early part of the flow to work them hard in the later part.

1047. If all stocks are strong early, some may be divided (**1448-52**) to be united for the later part of the flow. It is desirable to re-queen late, so as to avoid queen failures in early spring. Increase may be made early or late according to the minor flows.

Irregular and Uncertain Flows from Spring to Autumn

1048. Where the flow is irregular and uncertain and also limited in period, bees cannot be kept to show a profit. If the condition exists throughout the year, as in many urban districts, the bee-keeper in a small way can yet get some return, but he will need to study the weather and weather forecasts and to use snatch-crop methods. He should know the more promising sources and their relationship to the preceding weather. A wet May in the northern hemisphere, for instance, gives promise of a clover harvest. He must continually watch for signs of increasing flow; and when a flow is on or pending he should use manipulations such as **1219-25** and **1418-23** to concentrate the force of the harvesters.

With this condition to meet, it is peculiarly valuable to use the long lived strains, as with such, the bees will manage their own affairs to advantage, and will not miss any opportunity to store.

1049. Queen raising in July with forced autumn increase (Section II and **1464-7**) are suitable to such conditions, but both queen raising and increase may be carried out at any period timed to secure that the hives are not lacking harvesters at the probable harvest times (**1009-10**).

1050. If, during the best flow, an attempt is made to produce sections, the bee-keeper should be prepared to feed honey (**1285**) during any temporary break in the weather.

1051. Comb-building presents a difficulty; new combs are best provided by a stock especially worked for the purpose in warm weather (**318-22** and **1431-4**).

Flow Continued in Winter

1052. This condition exists only in tropical and sub-tropical regions. For example, in parts of Australia, iron-bark, white box and "cider" eucalyptus may give nectar in the winter months. There will be no winter surplus, but also no winter rest period. The special problems are, how to secure a rapid build-up in the spring with but few unworked bees, and when to re-queen.

1053. A good plan is to raise queens in late summer or early autumn, bring these through the winter in nuclei of four or five combs, which will be self-supporting but will not drain the energy of the queen; then in early spring re-queen the main stock by uniting with these nuclei.

If the weather is relatively cool a whole nucleus can be united by insertion at one side of a stock with a minimum of disturbance, making room by removing empty combs, the old queen having been removed previously.

In other respects the management will depend upon the time, of the principal flow or flows.

Week-end Bee-keeping

1054. The weekend bee-keeper generally keeps bees as much for pleasure as for serious profit, and enjoys manipulating. He may increase his profits by cutting down his manipulations and so keeping more bees with the same labour, but his principal problems are to plan his management so that manipulation is not required during the week, and so that emergencies do not arise in his absence. Bees have been kept profitably by a man who could visit them only once a year.

1155. The week-end beekeeper has to remember that bad weather will necessitate occasional postponement of a manipulation to the next week-end, but should know that, in fact, manipulations can be carried out during rain, preferably under a big umbrella and avoiding all hurry. The bees are bad to handle in bad weather, but better if they have been fed overnight. The week-end bee-keeper is generally in an urban district, and for this reason also it is important that he should have a docile strain.

1056. Re-queening annually used to be advocated, to avoid losses due to queens failing, but the cost and labour is reduced and the difficulty remedied by the use of strains that re-queen by supersedure without swarming, and by the use of long-lived queens, with which there is no justification for annual replacement, and which indeed frequently give their best performance in their second full year.

Section honey may be worked for during a rapid flow, but generally it pays at least to start and finish with combs for extracting, the bees taking readily to a rack of sections inserted below a half-sealed rack of shallow combs. More section racks are inserted as required, followed by the use of extracting combs to finish, put on top, thus avoiding unfinished sections. See also **1384-6**.

1057. The system of management will depend upon the flow, as indicated under preceding headings, modified as described above. In selecting methods, it is useful to note that methods involving the use of nucleus hives add flexibility to the time-table, and are therefore of special value to the week-end bee-keeper, and not much less so in the out apiary. The most useful size is probably one of four combs in a six-comb body. It tends to increase in size and strength and will withstand exchange of combs or sudden additions in mild weather. Its queen can be robbed with or without the comb it is on, and spare combs distributed to other nuclei, or used, with help, to form a new nucleus if a queen cell is to be had. A nucleus-raised queen is of no potential value (see, however, **190**), but the nucleus can care for ripe queen cells if its queen is removed. Swarming can be checked by the prompt removal of brood to a nucleus, the bees or brood being available for use in the same stock at a later period (**1213-8**). Preparations for swarming can be postponed also by exchange of brood with a nucleus, putting emerging brood in the nucleus in exchange for unsealed brood and eggs. *Almost every emergency of management can be met forthwith by putting something into or taking something out of a nucleus, while nuclei themselves seldom present emergencies.*

For honey getting and swarm control, see Section XIII, the methods referred to in **1166** and **1177** being suitable for weekend bee-keeping. For queen rearing, see **178**.

SECTION XII
SWARMS AND SWARMING
Theory of Swarming

Introduction

1058. In the first edition of this Manual the author discussed at considerable length the evolution of the habit of swarming, of the construction of honey comb, drone and queen cells and other related matters "all capable of verification or refutation in time, by observation and research." Some verification of the novel views put forward has since been obtained, but no refutation. Those interested in theory should refer to the full statement. What follows here is a highly abridged statement of that part bearing directly on the evolution of the swarming habit, showing, *inter alia*, the close relationship that must exist between the evolution of mating and swarming. A good test of any theory of swarming is to try its fit with the evolution of the colony to the stage with which we are familiar today, and of colonies of nearly related insects.

Evolution of the Colony

1059. It is obvious that a fertile queen, dependent upon a worker caste for her food, could not be developed before the worker had been differentiated. It is reasonable to assume, therefore, that co-operative colonies of honey bees, male and female, were developed before there were distinct castes of females, and that in the active season batches of males and females were raised to leave singly or in swarms on suitable days, to mate and found new colonies. In any large colony a large number would be ready to, and would, leave at the same propitious time.

1060. It is important to note that at this stage the fertility of the individual queen would be small, limited by her ability to provide for her early offspring unaided, plus later offspring with such help as might be got from the early ones before they leave the parent nest.

Development of Modified Females

1061. We have to remember that we are dealing with a race of insects which, as later experience has shown, have the characteristic that a reduction of gland food given in the larval stage leads to imperfect development of the ovaries.

It seems likely that the first imperfectly developed females appeared among the earliest brood raised. With the development of the cooperative habit, the mother bees would find themselves in possession of larger and larger nests, with more and more help in bringing up their offspring, and so develop increased fertility. But, in the spring, before the first lot of brood had hatched, the business of providing royal jelly would lie with the mother bee alone. She would have ample stores of honey and pollen, but her heavy duties and limited gland capacity would lead to some of the bees getting a scant supply of royal jelly and emerging with imperfectly developed ovaries.

1062. The balance is now set so that some loss of fertility on the part of this first batch, or even only delay of fertility, would lead to development of even larger nests, even more fertile mothers, and a more rapid production of new bees and new colonies per colony, but at the start the imperfect females would be not far from perfect development.

1063. A sudden diminution of nursing duties, by the filling up of the brood space, would lead to an excess of brood food, which, be it noted, could not be used up by filling queen cells because at this stage every cell is a queen cell and all or most are full; but must be absorbed and would result, as it does in our time, in increased ovarian development in the imperfectly developed females. Thus, the hitherto imperfect female becomes complete or sufficiently complete, and a swarm of females ensues each intent upon founding a new colony after the manner of her ancestors.

Advancement of the Imperfect Female

1064. Thus far we have some females with delayed or somewhat imperfect fertility and others produced under rather more favourable conditions with better ovarian·development, also a tendency for this contrast to increase. The tendency is cumulative, for the less perfect females cannot hope to compete with the more fertile ones in the race for supremacy. The advantage in every way lies with those colonies having the most fertile mothers and most successful in delaying or suppressing fertility of a large part of her offspring. This again leads to increased differentiation of function as between the fertile and less fertile females and ultimately the production of the queen mother, as we know her, and the female worker.

The queen is sometimes referred to as the perfect female, but the worker lacks only the ability to lay, whereas the queen retains only the ability to lay eggs and slay her rivals; she can do no other work in or out of the hive for the benefit of the colony. She has become the most imperfect female in the hive, though the most essential.

1065. In the extreme case today we have workers not given to swarming and not given to the development of laying workers. In less extreme cases, we have bees still given to the production of large numbers of queens, frequent swarming, the ready production of laying workers and even of laying workers able to be fertilized by the drone.

A queen for supersedure may be raised from a cell with a queen-cell base or a worker base. It is difficult to account for the discovery and development of the latter device. Possibly the alternative method developed side by side, starting from a time when there was no difference and then but little difference between the two kinds of cell and their contents.

Co-operative Flight of Workers and Queen

1066. We have seen in **1062-3** above that the workers probably did not learn to swarm, but rather had never left off doing so. In the early stages, we have premised in **1064** that the inferior females were able to function in starting new colonies, but were at a disadvantage. With increasing differentiation between queen mother and modified daughter, we have also premised increased dependence of the mother on her helpers, and we may assume increased desire of the helpers to be with and assist the mother, as this is a development which undoubtedly emerged. I suggest it is here that we may find the genesis of the habit of swarming of queen plus workers and their co-operation in founding a new colony. It is certain that when this stage of differentiation is reached the queen can no longer found a new colony unaided.

The Trigger

1067. Other things than the making and filling of queen cells happen when a sudden check occurs in the call for brood food. We can hardly have a more extreme case than that of removal of, or accidental death of, the queen. It is known that this leads, especially with certain races, not only to the production of laying workers, but that large numbers of the workers, though not reaching the laying stage, are found with expanded ovaries.

1068. Observation has shown that a temporary development of ovaries occurs if the proportion of young brood is artificially reduced, and without removal of the queen, or by the artificial increase of the proportion of nurse bees. It has been suggested in 1063 that this ovarian development is a contributory factor in swarming, and it may be the trigger that starts the final act, leading to swarming.

1069. Preparations for swarming actually commenced may be checked by adding frames of eggs and young brood and removing emerging brood.

Drone Raising, the First Act

1070. There is another important factor in determining the act of swarming, and that is the way of raising drones, a feature disclosing yet another link between the evolution of mating and of swarming.

1071. Drones take much longer to raise and also to become potent, than do queens, the difference amounting to about nineteen days. Colonies normally commence to raise drones not less than this period before they commence queen raising. Furthermore, if a colony is deprived of all drones and drone brood, it will recommence drone raising before it will start queen raising. (See also **313**.) Incidentally, this does not afford a practical way of stopping swarming, as a colony so treated is badly demoralized.

1072. Again, a colony losing its queen early in the spring will probably not start raising another if it has no drone brood. Incidentally, colonies have been known to miss the chance of doing so while there were eggs and young larvae available even after drones were present; the explanation remains to be discovered by more detailed and extensive observation.

1073. The raising of drones, then, is the first act in the series of steps that eventuates in the issue of a swarm. The appearance of flying drones is indeed the first external sign that preparations to swarm may be in train. On the other hand, if preparations for swarming are advanced, a change of circumstance may lead to their abandonment. Methods of control of swarming, based on bringing about this change of circumstance, are discussed in Section XIII.

Conclusions on Swarming

1074. From what has been presented above, it will be seen that the author's view of swarming is that it is a complex act, arising out of habits developed and practised in the earliest stages of the evolution of the honey bee, intended primarily to secure the continuation of the species, but adaptable to meet many hazards and emergencies in the life of the bee. In its various stages it depends upon conditions of prosperity and upon the stirrings of sexual impulse, and may be hindered, or prevented, by conditions antagonistic to prosperity, or which interfere with the normal development and progress of the sexual impulse, in which all the bees in the hive are concerned.

Formation, Flight and Settlement of Swarms Preparation for Swarming

1075. The first sign of preparation for swarming is the breeding of drones, and the next, the raising of young queens. The conditions controlling the commencement of drone breeding have not been sufficiently closely observed. When, with expansion of the brood nest, started in late winter or early spring, the queen reaches drone comb, drone eggs are likely to be laid, but this will not occur until the brood nest is well established, and occurs, generally, on the opening of spring, enabling the close winter formation to be broken up. The production of some drone brood continues throughout the season unless stopped by a serious shortage of stores when, not only the drone brood, but drones also, may be cast out. Drones are conserved, however, in any queenless hive.

A few weeks after the first drone eggs are laid, if the colony is prosperous, queen cells may be formed and the queen will lay in them.

1076. It is generally safe to reckon that a colony in which queen raising has not been started, i.e. one in which any queen cells in the acorn-cup stage, are empty, will not swarm for at least nine days. Certain prolific strains of Italian origin are apt to swarm before queen cells are sealed, especially if overcrowded, but this is unusual in modern practice or with hybrids.

1077. When preparing for swarming, comb building ceases and egg-laying is diminished to prepare the queen for flying, as shown by a relatively small amount of eggs and young brood in relation to sealed brood, notwithstanding that the conditions as to weather and nectar flow are good for brood raising. If egg-laying is neither increasing nor diminishing the proportion of eggs to unsealed brood to sealed brood is approximately 3 : 5 : 13.

1078. When queen cells are being raised a prosperous colony previously showing many flyers will be found to show considerably reduced activity in comparison with other colonies of similar strength but which are not preparing to swarm, and particularly before 7 a.m. G.M.T. (or

9 a.m. in double summer time).

1079. Wax production is active before a swarm issues. On the evening before issue workers may be observed clinging to the top-bars of the frames in serried ranks.

1080. Expansion of the brood nest hindered by the presence of drone cells and pollen combs as

detailed in **932** and **1174** may result in; preparations for swarming.

1081. A day or two before a swarm issues, scouts are sent out to find a place for the starting of a new colony. Such scouts may be observed prying into any empty hive, and shortly afterwards a swarm will be found in possession if the scouts are satisfied.

Queen Cells for Swarming and Supersedure

1082. Before accepting the starting of queen cells as a sign of intention to swarm, it is necessary to make sure that the intention is not merely supersedure. This possibility is too frequently overlooked, leading to wrong procedure, unnecessary disturbance and loss.

For discussion of supersedure and the distinguishing signs of preparation for it, see **157-69**, especially **168-9**.

Issue of the Swarm

1083. The swarm will emerge as soon as weather permits, that is to say, as soon as the shade temperature is suitable for free flying and there is absence of rain and high wind. The flight board and the surroundings of the entrance become covered with bees and the bees issue in clouds into the air. The cloud remains somewhat dispersed until the queen issues, when the cloud concentrates about the queen and. shows a general movement tending in one direction.

1084. The swarm most generally issues between 10.30 and 2 p.m. (11.30 and 3 p.m. Summer Time), during a sunny spell, with barometer above normal. This issue is stimulated by sunshine falling directly on the hive entrance, but about 5 per cent. of swarms issue before 10 and an equal number after 2.30.

1085. Many young bees go with the swarm and many bees of harvesting age are left behind. The idea, commonly held, that only old, bees fly with the swarm, is erroneous.

Bees too young to fly may sometimes be found on the ground in front of a hive that has swarmed.

Issue of Casts

1086. As a rule, the first cast may be expected 8 or 9 days after the swarm. It will issue sooner if the issue of the swarm has been delayed by unsuitable weather. It will later only if the swarm flies before queen cells are sealed (**1076**).

The cast is, of course, accompanied by a virgin queen, and may be followed by a second cast, two days later, and even a third, fourth and fifth, etc., at shorter intervals where the bees are bad swarmers. Casts are sometimes accompanied by more than one virgin.

External Signs of Issue

1087. Careful observation of what is taking place on the alighting board before a swarm issues will disclose signs that something unusual is in progress. Instead of the steady traffic outwards and inwards, under the observation of a few active guides, moving about the entrance, we find many bees resting or moving about on the alighting board and communicating with each other. The amount of regular business is greatly reduced, returning bees joining those on the, alighting board and about the entrance. These bees enter the hive, probably to load up, shortly before the: swarm issues, and there is considerable excitement within and much noise emitted.

Delayed Swarms

1088. If the weather turns cold or stormy about the time a swarm is due to issue, the issue is delayed. The delay may extend for several days. If the bad weather continues, the idea of swarming may be abandoned for a time, or altogether. In the latter case, all queen cells will be destroyed.

1089. If swarming is delayed, the virgin queens may become ripe for emergence before the swarm issues. In this case they are prevented from doing so by the workers, who omit to cut away the wax around the point of exit, but the imprisoned queens are then fed through an orifice, large enough for the passage of the tongue.

1090. If the swarming is delayed so that one or more queens approach maturity in their cells, the queens may be heard at night calling to and challenging each other (**41**).

Settlement of the Swarm

1091. The swarm, or cast, having issued, the cloud of bees finally gathers around some suitable settling-place. A swarm with an old queen, especially a heavy one, will fly low and make the best of any possible settling place in the vicinity of the hive, but with young queens, and delayed exit, and especially with virgin queens, swarms and casts may fly considerable distances and to high places before settlement, although the first flight and settlement is generally confined to a radius of 20 to 100 feet.

1092. It has been suggested that swarms prefer to fly in a westerly direction before clustering. The author's limited experience of apiaries on several sites supports this finding.

1093. The swarm having issued, and having found a suitable settling-place, one somewhat shaded, and affording, not only a firm hold for the cluster of bees, but preferably also a clear space below the support for the cluster to hang free, the formation of the cluster is started, the bees and queen gradually settling in the well-known manner. Sometimes, one or more secondary clusters are formed, the bees tending to settle on anything that looks like the beginning of the cluster proper. Sooner or later the clusters containing no queen, break up, the bees joining the main cluster. If the queen has been lost the main cluster breaks up and the bees return to the hive, spreading over the front and sides in search of the queen.

Bee Bobs

1094.a It is now known that the settling of a swarm is organized by some of the worker bees in the swarm alighting on tree branches and other objects and releasing assembly pheromone produced by the Nasonov's gland. This attracts other worker bees to settle, and many of them join in the release of the pheromone. Eventually, the queen lands at one of these incipient clusters, and when this happens the workers greatly intensify their release of the assembly pheromone. It may be possible, therefore, to guide emerging swarms to settle in a handy location by mounting a commercial "Swarm Lure" at this spot. These swarm lures are plastic vials that contain the blend of chemical compounds produced by the Nasonov gland. The vials slowly release their contents, and so last for several months. TS

1094. *To persuade the swarm to settle in a place convenient to the bee-keeper a bee bob is sometimes used. A ball of worsted made to resemble a small cluster, or a small fragrant piece of old comb, is secured in a situation near the hive suited to the convenience both of the bees and of the bee-keeper. It may be noted that there is a tendency for a swarm to settle where a previous swarm has settled, not only because the spot is found convenient but probably through an odour remaining and sometimes wax deposited on the branch.*

A swarm remaining at one spot for some hours may commence comb building and actually found a brood nest in the open, following the practice of bees living in warm climes.

1095. A good form of bee bob consists of a board of a size to cover the top of a brood chamber, with hooks on its edges for suspension with cord from a convenient bough and having the bob proper fixed on its under side. This may consist of one or two short strips of comb, glued to the board, or a piece of rough oak bark giving a good foothold, the bark being smeared with a solution, made as in the next paragraph, from old comb. If wax is used, the board should be stored when not in use with the wax freely exposed to the light to save it from the wax moth. When the swarm has settled, the board is unhooked and the swarm carried away for hiving. If the board has a large hole near the bob it may be placed with the swarm on an empty hive and a body with frames of foundation placed on top and covered over. The bees will soon run up.

Attracting a Swarm

1096. The part of a comb next the top bar contains propolis added to strengthen the comb where the strain on it is greatest. This part is the most aromatic and smells also of honey and of brood. Some such old comb should be broken up and boiled in a little water. This preparation gives an odour highly attractive to bees and if smeared over the inside of skeps or box hives fitted with foundation serves to attract swarms and sometimes prevents the loss of unnoted swarms.

1097. Essential oil of orange or of lemon, especially the latter, also citronella oil, which is lemon-scented, are attractive to swarms. Lemon-scented Balm can be used, by rubbing the leaves on the inside of a skep.

1098. A decoy hive should be given an entrance of moderate size, not too large.

Final Flight of Swarm and Legal Ownership

1099. A swarm will remain in its first place of settlement only until the flying bees have settled and the scouts have found a place suitable for founding a new colony. The scouts will then lead the swarm to its permanent abode, and this last flight may be for a mile or more if no suitable place is near at hand. A flying swarm has been met with no less than ½ mile above the earth.

1100 A The case of *Kearry -v- Pattinson* [1939] 1 K.B. 471 established that if a swarm of bees entered the land of another, even if the beekeeper had kept it in sight and could be confident that the bees were from his hive, he had no right to go on the other's land without permission. The swarm would belong to anybody who collected it. Although the beekeeper could claim that the bees were his, he lost his property in the bees as soon as they were reduced into possession by another. D.S.

1100. *Among themselves, bee-keepers recognize that the owner of a hive from which a swarm issues has a claim to the swarm; but the legal position is that, if a swarm leaves the bee-keeper's land and enters the land of another, the bee-keeper can claim it only if he has kept it in sight.*

Obviously a decoy hive should not be used to steal one's neighbour's swarms. Its legitimate use is to prevent the loss of swarms, for which purpose it has considerable utility. A stray swarm reverts to the wild and becomes a nuisance and a menace.

Under a recent legal decision in England, failing to recognize that the bee produced by the modern bee-breeder is very different from and a much more valuable article than a wild honey bee, a bee-keeper can no longer claim a right to enter the land of another to recover a swarm, but unless he has previously annoyed said other party he may expect permission and will probably be welcomed.

A tabby cat, if allowed to stray beyond control may take to the wild in the woods and remain the legal property of its original owner, but an equally bred, almost equally domesticated and more expensive bee, becomes another's property immediately it strays by swarming, although the original owner has not in fact yet lost control.

Taking and Hiving Swarms Settlement of the Cluster

1101. Before taking a swarm, the bees should be allowed ample time to settle, as sometimes more than one cluster is started, and the queen may join one of the small clusters. If the clusters are sufficiently separated they may be dislodged in turn and collected in the same skep. If the clusters are near together, those outstanding should be disturbed with smoke or a twig and the bees persuaded to join the main cluster. If the main cluster is much spread, the same means may be used to consolidate it. A fall of temperature or light shower of rain accelerates the formation of a good cluster; thus, on occasion, a syringe may be used to produce an artificial shower.

1102. After a swarm has been taken any old bees missed will eventually return to the parent hive, but frequently a number of bees will gather on the same spot, later making a small and persistent cluster. There may be a virgin queen present, but see also **1131**. Before hiving the swarm this cluster should be taken in the same way that the swarm was taken, and hived with it.

Taking a Swarm in a Skep

1103. In taking a swarm, as in most other acts of the bee-keeper, patience is a great asset, but a swarm should be secured as soon as may be convenient after the cluster has formed and the bees have settled down quietly. If the scouts have found a suitable home, the swarm may move off a quarter of an hour after it started its first settlement, especially if it does not contain a mated queen; it may fly and cluster a second time or it may fly right away to its new site. Most first swarms hang for an hour or two, especially if they have a good foothold and place to hang suspended and in the shade.

1104. A swarm should not be shaken before being taken, as if the bees are given notice they will obtain a more secure foothold. If, however, the branch from which the swarm rests is struck sharply with the fist or a wooden bar or mallet, the sudden shock takes the bees unawares and practically the whole swarm will fall.

1105. It is usual to dislodge a swarm into a skep, this being light and handy for the purpose, and serving as a temporary home for the bees. The skep should be held just below the cluster. If the bees have a long drop, the queen may be injured. Immediately the swarm falls into the skep, the skep is covered with a piece of canvas or other strong porous material. The covering may be tied round to secure it in place, unless the swarm is to be carried only a short distance. For a long journey it is desirable to employ, as a cover, a piece of canvas having a window of wire gauze sewn in it, say 3 inches (7½ cm.) square, to secure adequate ventilation. For shipment the skep may be stood in a box with the opening upwards.

A fine sieve makes a good cover for a skep provided any entrance cut in the edge of the skep is blocked before use so that the sieve makes a bee-tight joint. The sieve should be used, bottom upwards, unless for transit a long distance, as it is then easier to provide an exit for the bees later on (**1113-14**).

Swarms in Inconvenient Places

1106. Occasionally, however, a swarm is located where it cannot be dislodged in this simple manner. The cluster can be caused to move by the use of a little smoke, remembering always that the tendency of the bees, on being disturbed, and especially on being jarred, is to move upwards.

1107. If the swarm cannot be taken from below the bees may be caused to run up into a skep or box so placed over them that one edge and inner face offers a ready path. The surest way to make them run is to tap sharply the support on which they have clustered, using the smoker gently, to discourage wanderers. The operation may take ten minutes, or it may take an hour, according to the skill of the bee-keeper in getting the cluster on the run.

1108. A used comb, but especially one containing unsealed brood, affords a great attraction to a swarm. If the cluster is in a high tree, in an inaccessible position, the surest way to secure it is to fasten a box, or skep, containing a brood comb on to a pole, so that it may be held, or rest, just over the swarm and touching the branch, so as to provide a footpath. The cluster will gather on a brood comb, without box or skep, and it is easier to manipulate a single comb on a pole, than a box or skep, and to bring it in contact with the cluster.

Use of Bag Net

1109. Where there are many tall trees near the apiary it is highly desirable to use bee bobs at a convenient level and a bag net for swarm taking should be kept at hand. This consists of a bag like a butterfly net with a stiff rim, secured to a pole, or with a hinged rim, like that of a purse, with a cord for closing it. The cluster is swept up into the bag and the bag closed. If there is no hinge the net is closed by a turn of the wrist the same way as a butterfly net. Flying bees may be allowed to cluster on the bag if the swarm is a stray one, but if from a known apiary it is more convenient to let the stray bees find their way to the apiary so that the bag is taken with but few bees outside it. See, however, **1102**. The swarm is easier to handle later if the pole can be detached readily from the bag.

Box for Taking Swarms

1110. It is good to have a light box, of a size to receive six or seven combs, on runners as in a brood chamber; it should have a ventilated removable cover and no bottom and carry fittings such as large staples and one or more sockets, so that it may be held up, tied up, hung up or held on a pole. If filled with combs or foundation, including one brood comb with some brood in it, and secured over but touching the cluster, the bees will run up and take possession and are then already temporarily hived. If the cluster is awkwardly situated first insert the brood comb into the cluster so as to collect bees on it, put it back into the middle of the box and then place above and as near the cluster as possible, where the bees can follow those already within and which will signal to them.

Temporary Resting-place

1111. On securing a swarm the receptacle should be put as soon as possible in a cool place so arranged that ventilation is not impeded and preferably so that the cluster may form again without difficulty.

1112. If possible, the swarm should be taken to the place where it is to be hived and placed temporarily on the new stand, an exit being provided by inserting wedges or stones under the edge of the container. The container must be well shaded. Time should be allowed for the cluster to form before an exit is made, and if the swarm is shaded, ventilated and confined until towards evening, the danger of a further flight is minimized.

It is, however, possible to hive a swarm immediately it is taken, and it is best to do so if the swarm is in a bag, on a comb, or on a bee bob.

Sending Swarms Long Distances

1113.a Before a colony is sent long distances it should be inspected by a qualified person to ensure it was free of disease. The colony should be packed in an appropriate manner to ensure that the bees and any handlers are fully protected. Beekeeping suppliers are able to provide suitable travelling containers and packaging. ID

1113.b The shipping of nuclei by rail ceased probably 35 years ago. Most colonies of bees are now collected by the beekeeper from their chosen supplier or delivered by an overnight carrier. The use of skeps to supply swarms is also no more than an historic anecdote. If nuclei are shipped long distances by road by either the carrier or the customer they must have sufficient stores, ventilation and water for the journey. It is a good idea to advise the purchaser that they spray them with water occasionally, through the roof vents if the weather is particularly warm.   PS

1113. If a swarm is to be sent a long distance by rail, it is best to transfer it to a swarm box(see makers' lists), or to a box as used for package bees (1339), and to provide syrup, as with package bees. In hot weather this will greatly add to the comfort of the bees by enabling them to regulate the temperature. A swarm may be sent in a skep, secured as in 1105, to anyone able to deal promptly with it. It should be clearly labelled "Live Bees. Keep in a cool place."

1114. In modern management a swarm is usually hived on the parent stand (**1193-99**), but the following instructions for getting a swarm into a hive are of general application.

1115. To hive a swarm taken in a skep or box, it is customary to shake out the bees in front of the hive, on a temporary extension of the alighting board. The hive should be provided with frames, fitted with foundation, or with combs, but some foundation should be furnished, as a swarm makes wax, and the bees will desire to utilize it. If the bee-keeper has had trouble with swarms decamping he may give a comb of unsealed brood, as the bees will not desert such. It is most unusual for a swarm to refuse a good home, but they may refuse one which is too hot or lacks ample ventilation.

1116. To receive the swarm a board, say, 3 feet x 2 feet (say, 100 cm. by 60 or more) is propped up so as to form an extension of the alighting board. If it is not a good fit so that the bees will not get an unimpeded path from board to entrance, a sheet may be thrown over the boards secured with stones, or a piece of newspaper may be used fastened down with pins or stones, leaving no point higher than the entrance as the bees like to climb all the way.

1117. The swarm is thrown from its temporary container onto the board in front of the hive. The bees will soon be seen running over one another in a steady march to the new home (Plate VII). When this stage has set in the queen is frequently seen to run rapidly over the backs of the bees and make for the entrance. To dislodge the bees from their temporary receptacle, it must be held over the board and shaken down sharply, being brought to a sudden stop, and this process repeated. Any bees remaining will find their way from the empty receptacle, but they should be looked over in case the queen is with them.

Two small swarms may be shaken together if it is a matter of indifference which queen survives.

1118. If the swarm is on a comb or in a bag it may be placed in an empty body on the floor board and the body containing combs put in position on top. It is convenient to open the bag from the top after the upper box is in place, temporarily removing the combs for the purpose. See that the bag is so disposed and opened that all the bees have free exit from it and that the main entrance is not blocked by it.

1119. It is good to see a swarm running into a hive, and to make sure, by looking over the mass of bees, that the queen is among them, but a swarm may be hived more quickly and with less preparation, by shaking the bees into an empty brood chamber, or empty super, placed on the permanent stand, and promptly placing the chamber of frames, or combs, on top, with the quilt or inner cover in position. The empty chamber is ,removed later.

Manipulation of Supers

1120. It always pays to feed a swarm unless perhaps when there is a good nectar flow on. Do not start feeding, however, until the second day after hiving, and continue until ample food is coming in. If fed too soon the swarm is more likely to decamp. If there is a honey flow on and if it is desired to hive the swarm on the site of the parent stock, some drawn comb should be given as well, then the supers can be put on at once. If the supers are put over foundation, pollen will go into the supers. If no comb is given, do not put in the supers until 3 days later. An early swarm may be given some comb to assist in building up quickly. Do not use combs wet with honey for hiving swarms. Do not give eggs to a swarm, especially if hiving on the old stand.

Notes on Swarming

Clipped Queens

1121. If the queen's wing or wings have been clipped she will fall to the ground on attempting to fly with the swarm and may be lost or may return to the parent hive and crawl in. The bees finding that the queen is lost may return and fly later with the first virgin, flying further afield, so that clipping does not afford a guarantee against loss of swarms; moreover, it may lead to supersedure first.

If the issue of the swarm is noted the queen should be found and caged. She is generally on the ground with a small group of bees. Action depends upon what the bee-keeper wants. He may return the queen to the hive and deal with the lot when the swarm has returned by procedure described in **1188** and others following.

Swarm Returning to Parent Hive

1122. Sometimes a swarm will issue and return to the parent hive without settling. This is probably due to the queen failing to follow, which may in turn be due to her being crippled in some way. After one or two failures the queen is likely to be superseded and thrown out. Always examine a hive that has thrown out a queen. Always attend to a hive from which a swarm has issued and returned and deal with it according to requirements, but it may be desirable that the stock should be given a new queen.

1123. Sometimes on a very hot day a large cloud of bees, behaving like a swarm, may fly out and return again. Hives should not be too freely exposed to the sun in very hot weather.

Signs of Cessation of Swarming

1124. In a stock which has not built queen cells, the only sign of abandonment of all intention of swarming is the casting out of drones which occurs late in the season, but a stock hardly ever swarms after the main flow.

A stock which has built queen cells and decides not to swarm, or not to swarm again, will tear down any remaining occupied cells and cast out the young immature queens. When a queen cell has been torn open, a large gap will be found in the side, while a cell from which a queen has emerged will have a circular opening at the tip and sometimes a cap hanging from one edge of the opening. When a cell is done with, the wax is soon used elsewhere and the cell reduced to its original dimensions.

Factors tending to Encourage Swarming

1125. The following factors tend to encourage swarming:

1. The presence of queen cells raised under the swarming impulse;
2. Swarming tendency in the race or strain;
3. Lack of shade. Flat top roof board too close on top of brood nest (**714-15**) in hot weather;
4. Expansion of brood nest ceasing and nothing much to do. Light and irregular honey flow after
    colony is fully developed;
5. Insufficient room for egg-laying in worker cells; brood nest cluttered up with incoming nectar
    for lack of room elsewhere;
6. Presence of excess queen cells raised during supersedure or replacement of lost queen, with
    conditions otherwise suitable for swarming;

7. Presence of drones;

8. Lack of sufficient ventilation;

9. No room for comb building;

10. Lack of food. Starvation swarm;

11. Obstacles to expansion of brood nest such as brood combs with honey at top in lower

of

two brood chambers, hindering passage of queen.

1126. It is not usual to manipulate to encourage swarming, but if desired, an early swarm may be encouraged by warm packing, steady stimulative feeding, crowding, having drone comb available for early use, and by inserting combs of emerging brood in exchange for unsealed young brood.

Factors tending to Hinder Swarming

1127.   The following factors tend to hinder swarming (for Swarm Control, see Section XIII) :

1. Necessity for other activities;

2. Young queen keeping workers busy up to harvest-time;

3. Plenty of room for egg-laying, sometimes effective even when given after queen cells have been started in a crowded chamber.

4. Absence of drones and advanced drone brood;

5. Lack of pollen and stores;

6. Ample ventilation;

7. Sufficient shade;

8. Undivided brood nest;

9. Combs not built near entrance. Vacant comb below brood;

10. Absence of much drone comb;

11. Having swarmed (conditions in swarm itself);

12. Having given a swarm (conditions in parent hive), provided casts are prevented (**1188-92**).

1128. Any steps securing the above are helpful. In a large hive, through ventilation is useful, between the brood combs, as well as outside them. In the hottest weather the entrance cannot be too large during a flow, and additional ventilation at top is helpful, such as is obtained with porous covering, or a quite small gap around the inner cover, to be closed as soon as the nights are cool again. Sufficient shade is essential (**586**).

Natural and Artificial Swarms

Characteristics of Natural Swarms

1129. The conditions underlying the making of artificial swarms are considered here; the procedure for making them is considered in later paragraphs.

1130. A natural swarm consists of (*a*) bees of all ages, (*b*) other than those too young to fly, (*c*) most of them well loaded with honey, and (*d*) many having already started the making of wax, a process which takes about 36 hours, (*e*) collectively bent on starting a new colony and accompanied by (*f*) an old queen or (*g*) a young virgin queen.

Furthermore, the nursing activities have been reduced, owing in case (*f*) to the reduction of egg-laying, and still more so in case (*g*), as egg-laying will have ceased for some days and, unless the issue of the prime swarm was delayed, all brood will have been sealed by the time the virgin issues.

Finally (*h*) issuing bees commence to form themselves into a new colony; they collect somewhere where the queen and more issuing bees join them prior to the final flight to their new home. Seeing that the new home has frequently been chosen in advance by scouts, although the bees never go to it without first gathering in a cluster at some spot near the hive they have left, it would seem that one object of this clustering is to secure that the queen is with the swarm. The cluster, once settled with the queen, the bees seem to establish themselves as a new colony and will not return to the old home.

1131. Rarely, a swarm will start to build where it first settles. If the swarm moves off, all the bees go with it, but if the swarm is taken, any bees missed will cluster again, and if the swarm had been hanging for some time these bees behave as though there were a settled colony admitting no other home; they refuse to return to the parent hive.

Making an Artificial Swarm

1132. An artificial swarm is usually made at a time when there has been rapid expansion of a colony; the average age will be low, so there will be ample nurse bees to cope with brood in the new colony. Thus the absence of a reduction of laying prior to making the swarm is not of consequence save that the queen will require immediate room for laying, and careful handling, being in full lay. The other conditions of a natural swarm require consideration.

1133. There is no difficulty in obtaining (*a*) bees of all ages. It is not important to avoid (*b*) taking bees too young to fly, but in fact if the workers are shaken off the combs gently the youngest bees will be left clinging to them.

1134. It is, however, important to note condition (*c*) and its relation to (*d*) and (*e*). If the bees are well loaded with honey they are not only protected from starvation, but are in a condition to produce wax. They will be disposed to cluster for wax formation and ready to start comb building in the new home.

Condition (*c*) can be met by causing the bees to load up, by giving them a good smoking and by drumming on the hive, but there must be honey or syrup available. If this is not attended to it is essential to feed when hiving so as to meet conditions (*d*) and (*c*), but (*d*) takes time, so some drawn comb must be provided.

1135. If there is a good nectar flow on, it is good to see that the bees are fed and given one or two combs, but hived on foundation with a super of combs over an excluder. The bees will then carry nectar into the super and draw out the brood combs as occupied by the queen. If, however, there is no comb below the excluder the bees will carry pollen into the super, for they must start gathering pollen immediately.

1136. If the bees are to stay with the queen as a new colony they should reach condition (*h*), i.e. that they should all reach the condition of organization as a new colony. This is facilitated by the drumming above referred to, as that will put the bees in the appropriate stage of preparation. It can be ensured by confining the removed bees with their queen in a well-ventilated box put for 24 hours in a cool place before putting them on their permanent site and releasing them.

1137. It has been suggested that condition (*h*) may be met by causing the required bees to cluster on a brood comb carrying the queen, supported in the open but under cover, on the original site. Flying bees and shook bees, if not too young (**1133**), will find this comb. Half-an-hour or so after the cluster has been formed the comb with cluster can be carried to the hive the bees are to occupy. See, however, that the other conditions are complied with as already described.

1138. Finally, when a natural swarm has issued, sufficient bees are left behind to look after the reduced brood, so as to build up the colony anew with the new queen in prospect. If the bees are prevented from having more than one virgin queen they will proceed to build up food supplies and a new brood nest which will take some weeks, extending probably to the next flow, the onset of which will divert activities to honey getting. Such a stock having a young queen is unlikely to swarm that year.

In making an artificial swarm, unless nights are warm, care must be taken to see that sufficient bees are left to cover the brood and keep all warm.

SECTION XIII

HONEY GETTING AND SWARM CONTROL
Introduction

General

1139. The primary economic gain from bee-keeping comes from the activities of the honey-bee in searching for pollen, which lead to greatly improved pollination and production of better and more abundant fruit, also increased seed production by the clovers and other important crops. The primary object of bee-keeping, especially to the commercial honey producer, is the production of honey.

1140. In bee-keeping, the feature in management most liable to lead to an undue demand upon labour is that of swarm control and swarm management. This feature is so intimately bound up with that of management for honey production, that the two are here dealt with as one problem.

1141. Management is based on certain methods and principles but entails also carrying out certain manipulations in precise detail, many of them common to several systems of management or being required in dealing with several situations. To avoid unnecessary repetition of detail and to simplify the presentation of methods of management, the study of method is separated from the study of common details.

1142. The ultimate object of all manipulations carried out for honey production is that of obtaining a maximum output of honey with a minimum amount of labour. All manipulation creates some disturbance of the economy of the hive, resulting in loss, which must be offset against the advantage intended. The skilled bee-keeper is known by the small number and apparent simplicity of the manipulations he employs. This apparent simplicity is attained only by employing well-thought-out methods and having everything necessary ready to hand before commencing operations. Upwards of one hundred manipulations are described in this manual. The bee-keeper will do well if he can select a dozen which he can make do all he usually requires, but he will need some of the others in emergency or after change of circumstance.

1143. The secret of good management is to work with the bees, not against them. Following the plan of the Japanese wrestler, the bee-keeper utilizes the power'of the bees but diverts it to his own advantage, and incidentally to theirs. He supplements their instincts by his intelligence and thus so reduces their difficulties that they are able to produce considerably more honey than they need for their own welfare. That excess constitutes his legitimate harvest.

How Surplus Honey is Got

1144. The importance of hardiness and longevity in bees has already been dealt with (**20**). A long-lived strain will maintain at all times and for longer periods a given supply of harvesters on a smaller brood nest. Without exception, taking the year as a whole, the larger part of the nectar collected has to be used for the production and maintenance of bees. A good colony of long-lived bees will consume, in the hive, in a season, 200 to 300 lb. of food. Short-lived bees will require and consume much larger amounts; thus their chance of producing surplus is greatly reduced. The prospect for surplus depends far more on the hardiness of the bee than upon the fertility of the queen (**1011**).

1145. A large part of the nectar gathered has to be used as fuel to stoke the engines of the harvesters themselves, and never reaches the hive. If the flowers are producing only a sparse supply, the wear and tear of harvesting may exceed the bee-producing power of the food brought in. Age-long experience has taught the bees to adjust their harvesting activities to the state of the flow. (Incidentally the adjustment acquired by bees working in one place may prove quite unsuitable to the requirements in another place where the conditions are different.) When a good flow is on, every available bee goes out and works to the limit of its strength. Taking the fuel needs of the bee into account one may expect in an average district that the nectar stored in the hive represents about half that actually gathered. Thus the amount of nectar to be gathered by a good stock before it can spare surplus is equal to that in 400 to 600 lb. of honey.

1146. Seeing that the total nectar supply in a given district is limited, it follows that the bee-keepers using that district can obtain the largest return by employing long-lived bees, as they require proportionally less for internal consumption. One can keep, profitably, a larger number of stocks of such bees in any given district.

1147. Checking breeding at a time when the bees produced cannot be used, directly or indirectly, for harvesting the more profitable flows is also clearly indicated. A hive full of brood and bees at the wrong time looks good but may prove a liability rather than an asset. Bees should be bred for the nectar flow, not on the nectar flow.

1148. What has been said also is the importance of having the colony located near the harvest. This is generally recognized in dealing with heather honey, but it is of equal importance where the flow occurs at lower temperatures involving greater wear and tear in harvesting, as from early fruit. The author anticipates that the practice of moving hives to the neighbourhood of important sources, and from less profitable to more profitable sources during the course of the year, is likely to receive increasing attention. He commends the practice to the small bee-keeper, and especially to such as can afford to capitalize a light portable honey house. The move should be made at the time the flow is just commencing, otherwise many of the bees are apt to ignore the new source at the critical period of full flow.

How Swarming is Controlled

1149. In the course of nature the continuance of the races of honey bee depends upon the production of new colonies to make good continual losses due to disease, hardship, accident and robbery by man and animals. New colonies are required also in the process of colonizing a new area. New colonies are formed by swarms. The continuation of colonies on a given site is dependent upon the renewal of the queen, following the issue of a swarm, or upon her timely supersedure without swarming.

Under the control of man, new colonies are formed at will; thus the habit of swarming is not essential to the maintenance of the race or strain in a state of domestication. It is no more necessary than that oranges should have pips. The bee-keeper, however, has to deal with the instincts and habits of the bee as he finds them.

The secrets of successful swarm control are, firstly, avoidance of conditions conducive to swarming (**1125-6**), of which, confining and overcrowding the brood nest is the most important, and secondly, satisfying the instincts connected with swarming, either by putting the colony into the condition of (*a*) a swarm, or (*b*) of one that has swarmed and cannot send out a cast, or (*c*) of one having no desire to swarm. One may aim at (*c*) and only when in doubt, or unsuccessful, utilize (*a*) and (*b*), or one may adopt a regular system of management based on utilizing (*a*) and (*b*), thus taking no avoidable chance. Procedures are discussed later.

What is a Colony?

1150. This question might be put, "when is a colony not a colony?" One must seek to understand what, constitutes a colony of bees; in particular, a separate colony, acting as such, and come to agreement with the bees about this.

A normal colony in the active season has an established brood nest, with used brood combs, drone comb, pollen and honey supplies, all arranged in a certain way. If circumstances lead a colony to swarm, the swarm issuing will start a new colony, unlikely to swarm that year. The old colony may swarm again unless assisted to alter its intention.

A colony is an organized community having a routine of activities in which individual duties are suitably apportioned. A random group of bees brought together in a hive and given comb of brood and eggs, even though it has no queen, will organize themselves in about half an hour so that all essential immediate requirements are met to good advantage. In due course they will take all the collective steps necessary to their future security.

1151. A colony acquires a distinctive smell by which the bees readily recognize their own members, but if two colonies intimately share the same atmosphere, they quickly acquire the same scent, and bees from either colony are then free to enter the other. If both colonies have queens, but are separated by two screens of queen excluder, separated by half an inch, so that the queens cannot make any contact, the two colonies will continue each on its own business, but workers will drift from one to the other and will sometimes show a marked preference for the one of them. The two communities may even share a common honey store, notwithstanding their strong instinct to defend their stores against all recognizable strangers.

1152. A colony centres around a brood nest. If the brood nest is effectively divided, two separately organized communities will be established even though they share a common scent. The insertion of two combs or of two frames of foundation in the centre of the brood nest is likely to produce such effective division, and the bees in the half where the queen is not laying will then organize themselves to raise a new queen. If a barrier of queen excluder only is introduced, the bees continue as a single colony, and although they will accept a queen cell in the queenless half, this is because the queen cannot get access to it. If a bee-proof screen is inserted between two entrances two colonies are formed.

1153. If the brood nest is divided by combs and an excluder, vertically or horizontally, in the latter case by a super, two colonies are usually formed. The bees in the upper portion are more likely to raise a queen if the two lots are put in a condition requiring very different organization, e.g. the lower lot with queen, given very little brood and much comb building. Queen raising is also stimulated by giving a separate exit. If, however, the super contains mainly empty combs in which brood has been raised, this hinders colony division.

1154. An interesting case is reliably reported in which a batch of old brood combs was put above a queen excluder and honey super and left with an extra entrance. Some of the bees occupied this portion and either raised a queen from a removed egg, or else received a straying virgin. It is a nice question whether the bees, finding themselves in possession of an empty brood nest, regarded themselves as a colony in a bad way.

1155. Recognition of the facts cited in the preceding paragraphs leads to the development of simple methods of management without swarming, as will be seen later.

Honey Production and Swarm Control

General

1156. In nature, swarming leads at least to a temporary increase of colonies and of bees, but made at the expense of food stores. The bee-keeper wishing to make increase should consult Section XV. In the present study the primary object is to avoid uncontrolled increase and to secure surplus honey. We do not yet know enough about swarming to prevent it absolutely in a practical way.

1157. Great progress in swarm control by manipulation of chambers, frames and bees has been made in recent years with hives in which two or more chambers are used to accommodate the brood nest at its maximum development. Two or more standard brood chambers may be used, or in some cases, one large chamber supplemented by a shallow one. Where one large brood chamber alone accommodates the brood throughout the season, different management is called for.

The author looks forward to the day when the swarming instinct will be modified by better utilization of hereditary factors, leading to simplification of management, and to an extension of bee-keeping on the "let alone" principle. It would be a grievous error to exaggerate the merits of elaborate methods of management so that bee-keepers come to rely upon them. The encouragement of such fashions must postpone evolution of simplicity. The extension of simple practice makes possible the control of larger numbers of profitable stocks with a given amount of available labour and reduces the number of disconcerting emergencies.

1158. Methods requiring double or multiple standard brood chambers are hereafter marked with the letter **D**. Those suitable for a single brood chamber plus shallow are marked **S +**. Those solely relating to single brood chamber management are marked **S**. Any methods having no distinguishing mark are capable of general application. It may be noted, however, that methods marked **D**, and intended for Langstroth, British or other relatively small standard chambers, may occasionally be used with large chambers such as the M.D., by those able to handle them and having occasion.

Management of Swarms

1159. The flight of swarms and hiving of them is dealt with in **1075-120**. We now have to relate swarming with management. Those who seek to minimize natural swarming are sometimes defeated and have to deal with natural swarms. If the swarm is merely hived on a new site, we have increase at the expense of surplus honey. We have still to deal with the parent stock, from which a second and unprofitable swarm is likely to follow.

1160. The natural swarm may be returned to the hive, with or without requeening. Procedure is described and discussed in **118-92**. Alternatively, the swarm may be hived on the old stand (Pagden method) and steps taken to prevent the original stock reswarming. Procedure avoiding increase is given in **1193-6**, and is applicable to single or double brood chamber management, but if bodies are used, not too heavy and large to double, one can secure the advantage of better methods without any more labour or attention. If increase is required, see **1197**. If in time for an early flow, the method of doubling without increase described in **1200-3** may be used, or at any time the use of an upper chamber as in **1204-6**.

Checking Swarming

1161. One cannot have too many bees for the harvest; every queen in full lay in the early part of the season is a producer of prospective harvesters. Methods, though simple, which involve stopping or hindering the laying of queens early in the season, are not to be recommended. With strains disinclined to swarm, or having their inclination reduced by giving room in time (**935**), swarming may be checked by removal of the queen (**1211-12**), or better still by removal of some brood (**1213-15**), which brood may be used to assist less advanced stocks; again, both queen and some brood may be removed (**1216-18**). Removal of the queen may be combined with requeening as in **1207-10**, giving an effective check. The advantages and disadvantages of these several courses are discussed in the paragraph to which reference is made.

Should such a checking fail, the resultant swarm should be dealt with as above described (**1159-60**).

Artificial Swarming

1162. If, however, the colony has commenced preparations for swarming, or if the use of more positive methods is desired, one may make an artificial swarm outright, thus anticipating and fulfilling the bees' desire, or one may employ one or other of the several procedures described later which satisfy both the bees' instincts and the owner's convenience without calling for such drastic treatment as the production of what the Americans describe as a "shook" swarm, while minimizing the risk of loss through swarms absconding, or the inconvenience of having to deal with swarms at inconvenient times. As with a natural swarm, it is generally best to use a procedure that does not involve increase, and secures maximum surplus honey.

The making of artificial swarms is discussed in **1130-7**. Where increase is desired see **1553**. Complete or partial artificial swarming enters into several plans now described.

Doubling (D)

1163. This is a method of treating two stocks whereby one is made strong with harvesters and in a condition much like that of a stock that has swarmed, whereas the other is filled with brood which the available bees can only just cover, resembling in important respects a stock that has swarmed. As first presented by Simmins (**1219-23**), the hives have to be in .pairs side by side, whereas the developments in adjacent hives are frequently not just what is wanted. This necessitates a rearrangement of stocks and consequent labour and disturbance. The author offers an alternative in **1224-5**.

1164. The principal application of doubling is to be found in districts in which there is an early and late flow with gap between, a characteristic of many districts, and to stocks which have expanded early though hardly strong enough to produce a surplus, unaided, from the spring flow. If allowed to build up on the early flow they are ready too early for the main flow and liable to swarm.

Now, by doubling, half the stocks dealt with should be in a condition to secure surplus from the early flow to advantage, and all will reach strength in good time for the main flow while being hindered from swarming. One gets the surplus at the expense of a temporary setback, which proves ultimately an advantage.

1165. Doubling carried out as described, results in building up the supply of bees of harvesting age in a hive assigned to gather surplus by transfer of flying bees from another hive. Other methods of making such a transfer, which can be used for the same purpose, will be found in **1418-23**. Any such procedure ought not to be necessary in dealing with the main flow, but can be applied where stocks are weak, which they should not be.

Utilizing an Upper Brood Chamber (**D**) and (**S +**)

1166. Since Demaree introduced the system which carries his name, whereby a sort of artificial swarm is made in the brood chamber proper, and most of the brood transferred to an upper chamber above the supers, the queen being kept below by an excluder, there have been many variants and modifications, all incorporating the use of this upper chamber idea. The great advantage of them all is that all that is done to the stock is done in the one hive. There is no moving of hives, or of brood or bees to some other hive in some state appropriate to receive them. Each stock is treated as a unit on its merits, when ready, and without regard to the condition of any other stock.

1167. The Demaree system is detailed in **1226-34**, as applied to working for extracted honey, and in **1242-7** as modified for working for sections. The system as originally planned is found to be rather too drastic for use where an unexpected cold spell may be experienced, resulting in a serious setback. Unless stocks are really strong and the prospects for warm weather are good, it is better to use the modification described in **1234**.

1168. If the colony, expanding in the spring, is given additional room as described in **1394-8**, the stock should be treated when the brood nest is well developed and before it reaches its peak and a reduction of laying sets in. Should this time correspond with the commencement of an early honey flow, the colony will get busy harvesting and at the same time continue to build up its brood nest, an excellent preparation for later main flow. If the operation is so timed it may prove sufficient and prevent preparations for swarming without further action being taken, but if queen cells are built below, the operation can be repeated as in **1235-9**. Further modifications of the method are described in **1240-1**, and other methods of utilizing an upper brood chamber in **1249-62**.

1169. If the operation is combined with queen raising it is even more effective in hindering swarming (see **181**, Method III). The upper colony, being temporarily separated from the lower by a wire screen, while the queen in the upper colony is mated, the conditions more closely resemble those of a stock that has swarmed. The new queen can be used to replace the old one, the whole lot then being reunited as one colony, or it can be used elsewhere, preferably by making up a nucleus which can be built up and added to a stock requiring it.

Again, the whole colony can be worked as one with two queens, as in **1264**.

*Transfer of Chambers (**D**) and (**S** +)*

1170. A further simplification is possible which, while retaining the merit of checking swarming, involves less labour and disturbance; moreover, it does not necessitate finding the queen, though if she is found, so much the better. Instead of rearranging combs one moves a whole chamber as found, above a super or supers, using an excluder or excluders to keep the queen where she is wanted. See **1249-62**.If done at the right time, the stock can be steered through the swarming season without swarming even if the stock has commenced preparations. The expanding colony is first dealt with by giving additional room as in **1394-8** and, preferably a little before maximum expansion of the brood nest is reached. Action may be delayed, and some preparations for swarming commenced. One must watch for this, remembering that drone brood is the first step, and endeavour to act before queen brood is started (**1075**).

If queen cells are started they are dealt with as detailed. Remember, however, that during a cold spell, while brood raising is checked, workers will continue to emerge in large numbers, so that after say a week, there will be plenty of bees to cover the divided brood and maintain the further development of the colony. It is better, then, to carry out the operation under seemingly unpromising conditions, if there is much sealed brood present, than by delaying, to allow the desire to swarm to arise. *Thus one must learn, when examining a stock, not only to ascertain its then state, but to acquire the habit of foreseeing by consideration, what its state is going to be, say, ten days hence.*

Snelgrove System

1171. The Snelgrove system is also one in which a brood chamber is placed above a super or supers but involves the use of a special floor board and entrance, whereby bees are switched from the super to the main colony. The apparatus is such that the bees must be allowed to fly out on all sides of the hive, and is thus not convenient unless there is ample room in the apiary. It incorporates the best features of the old double hive in which flying bees were switched by means of controlled passages in the main floor.

The system has given satisfaction in the hands of many who have followed the detailed procedure clearly described in a book by the inventor. For reasons given in **1157** the present author advocates simpler methods.

An important observation was made during the experimental development of the system, viz. that if the upper chamber has brood with queen cells and the queen is included, the queen cells will be destroyed. In this system the bees have a flight hole from the top chamber. It would be useful to establish just how much separation is required to bring about this end. Can it be applied to a scheme in which the colony in the upper chamber uses the main entrance through one or more excluders?

*Swarm Control with Single Brood Chambers (**S**)*

1172. There are many users of a large brood chamber who do not add a shallow chamber for wintering; this may be one taking anything up to 15 M.D. frames, for example. Management then consists mainly in ensuring that the queen has the free use of the whole chamber, avoiding those conditions scheduled in **1125** that encourage swarming, and in using a strain of bees but little inclined to swarm.

1173. To secure that the queen has the full use of a single brood chamber, it is not sufficient merely to have it full of combs, for the queen may be impeded by honey in the brood nest, or restricted by combs of pollen. While it is good to avoid disturbance of the brood nest as much as practicable, it is well, on the eve or commencement of a principal flow, to rearrange the combs so as to secure that those at the outside of the nest have young brood, and those at the centre, emerging brood, then, if ample super room is available, the bees will not crowd the queen with honey. See **1376-80**.

1174. When a new stock is formed in the summer it frequently happens that the new queen does not reach her full lay that year and the natural boundary of the nest, being combs containing pollen, does not present a full-size nest. Now in the spring, if new pollen is freely obtainable, the previous year's pollen may not be removed in time and its presence may then check the expansion of the nest. The bees are then liable to build drone comb also on the borders of the restricted nest and proceed to raise queen cells for swarming. The author suggests that this is one reason why many stocks swarm in their first full year.

1175. The danger can be avoided by inserting an old brood comb, or foundation, on the outside of the brood nest on either side but inside the pollen combs (**932**). The danger may not be removed by giving additional room below.

Management for Honey Production

1176. By adopting good methods of managing swarms and of swarm control, one has gone a long way towards securing a surplus. The principles of honey getting have been discussed in **1143-8**. It remains only to refer to certain manipulations.

The addition of brood chambers is dealt with in **1394-8**. The procedure to be adopted in adding supers is dealt with in **1388-93**, modified for heather honey by **1380-4**. When using an upper brood chamber in swarm control, this chamber is generally removed by the time the main flow is on, or has become part of the supering, for about this time it is desired to restrict the brood nest (**955**). A number of common manipulations, which are self-explanatory, will be found in Section XV.

Management of Two Queens in One Hive

1177. Two queens may be worked in one hive, thus obtaining a larger population of workers and a larger surplus. As against the use of two hives there is a saving of one floor board, stand and roof, but it is not yet established that a greater return is obtainable for a given capital expenditure plus expenditure on labour.

The queens have separate colonies as described in **1264**, and need separation only by two excluders or other queen proof screens. If the workers have common access to the two queens they are apt to favour one at the expense of the other, so the use of a screen that will not let the bees through is preferred. To employ the idea it is necessary to be able to secure young queens early in the season. These are, or may be, used for requeening later; if either queen fails to give satisfaction one can retain the other and reunite the colonies.

1178. This use of two queens may complicate swarm control, but anyone using an upper chamber system of control can raise a queen in that chamber and work the two colonies together for a while.

1179. The system may be employed also to dispose in spring of a nucleus or small stock with good queen, brought through the winter. The small stock is added to a strong one, taking the place of *C* in the routine described (**1264**) and the laying power of both queens is retained.

1180. In an alternative method suitable for use with M.D. combs, a hive body is employed, long enough to take two stocks separated by, say, half a dozen store combs placed between two excluder division boards. The colonies build up at either end and the workers unite in the middle, where the entrance is situated. The supers are placed above the centre portion, the end portions being closed with crown boards. In the author's opinion this system offers less prospect of true economy than that previously described, but it has had a considerable following.

Manipulations with Swarms

Discovering the Parent Hive

1181. If the swarm is not seen to issue or is not expected, there is frequently difficulty in discovering the parent hive. The inactivity of the remaining bees may not be very noticeable. Sometimes very young bees, carried out with the out-rushing crowd, may be noticed on the ground before the entrance, thus marking the parent hive.

1182. There is, however, a manipulation which, used intelligently, gives the answer. Have a matchbox handy when taking the swarm, having first put a little white flour into the box and shaken it well. Detach some bees from the bottom of the cluster into the box and pocket it. In the evening, when flying has practically ceased; release these bees slowly, standing at the time where the entrance can be observed. These bees, marked with flour, will fly up and then make for the entrance of the hive from which they came.

Preventing Casts

1183. Hive the swarm on the stand of the parent hive, removing the parent hive to a new site, so that its flying bees will join the swarm (see also **1120**). This manoeuvre is ample to prevent casts with a strain not disposed to swarm freely.

1184. If, however, the bees are of a strain much inclined to swarming, examine the parent hive for queen cells 7 days after the prime swarm issues, and destroy all cells except one well-developed one. if the weather has been unfavourable so that the swarm is likely to have been delayed, an earlier examination will be necessary, but in that case a second examination should be made a week later to see that no more cells have been raised around eggs, or young grubs remaining after the previous examination.

Preventing Early Swarms Re-swarming later in the Season

1185. An early swarm may build up to swarming strength before a late honey flow diverts activities and may swarm again. For such a risk the preventive is removal of brood, **1213-5**. The methods dealt with in **1170** are applicable, whether or not queen cells have actually been started. In general a stock will not swarm a second time unless the brood nest is overcrowded.

An early swarm given a new young queen is not considered likely to swarm again. It should always be remembered that an early swarm will normally contain a last-year's queen, and should be marked for consideration when requeening later in the season.

Delaying Flight of Swarm

1186. A stock may be temporarily prevented from swarming by putting a framed queen excluder below the brood nest above the floor board. The device has some utility as a means for hindering the swarming of a particular stock, say, during the absence of the bee-keeper for a few days, but while in use the flight of drones is prevented and the normal business of the hive hindered and upset, all at a time of prospective prosperity. The use of the device involves labour and planned action to an extent but little less than that required by management properly planned to control swarming and can only be regarded as an emergency device.

Stopping Swarming by Destroying Queen Cells

1187. Instructions are sometimes given to stop swarming by cutting out all queen cells, say once a week This involves much manipulation, disturbance and loss of honey, and is not the sure preventive that it might appear to be. The bees and the bee-keeper are at continual cross purposes, which is bad for both; demoralizing for the bees and frequently disappointing to the bee-keeper. Removal of all queen cells (**1364**) may not be successfully accomplished; then the unexpected swarm is probably unnoted and lost. Better methods of swarm control are described in what follows.

Returning Swarm to Parent Hive

1188. This course is generally a mistaken remedy, applied to a condition that should not have been allowed to occur. If done as an attempt to reverse the intention of the bees, it will probably fail. If the swarm and queen are to be returned to the hive from which they came, something must be done to alter the intention of the bees, and destruction of all queen cells will probably prove insufficient.

1189. Most of the brood combs can be removed and given to other stocks, being replaced by frames with foundation. Leave no unsealed brood, or eggs, or queen cells. It is better, however, either to follow the plan in **1200-3**, if before an early flow, or the plan in **1204-6**.

1190. If, however, single large brood chambers are used, the issue of a swarm shows that the chamber was not large enough, or had become clogged with honey (**1376-80**) or cramped as in **1174**, or that the bees were determined swarmers and had better be requeened. The cause must be traced and removed before returning the swarm (**1125**).

1191. A delayed swarm, or second swarm, or swarms returning after loss of a clipped queen and swarming again, or two combined swarms, may, any of them, have two or more queens. Against all these emergencies it is wise, therefore, on returning the swarm, to put excluder zinc over the entrance, or a framed excluder beneath the brood chamber for 24 hours to let the queens fight it out. If, however, a queen cell has been missed, this may not afford a remedy, as the virgin may emerge some days later.

Returning Swarm to Parent Hive, and Requeening

1192. This involves destruction of the old queen and a period during which no eggs are being laid. The old queen may be removed by the use of an excluder sieve. Place an excluder under the brood chamber when hiving, or place it on top under an empty super and hive the bees from above, covering the bees until they go down and using smoke as required. The old queen is then caught and destroyed.

Before hiving, all queen cells but one should be destroyed. The new queen will be laying three weeks later, more or less. The queen so raised is more likely to give satisfaction than one introduced.

If two or three combs are removed to a nucleus hive the old queen may be put with them and not only held in reserve until all is well, but will continue to lay, on a reduced scale.

Hiving Swarm on the Old Stand, Without Increase

1193. The swarm is to be hived on the old stand and the old stock ultimately united. The following procedure is for general use. The modification in **1199** is applicable to the issue of an early swarm from a heavily packed hive which cannot well be moved.

1194. Hive the swarm in a spare hive on the old stand, furnished mainly with foundation, placing the old hive to one side or back somewhat, with its entrance turned to one side, so that the flying bees join the swarm. This is done to prevent the issue of casts, the old stock being so depleted that it will retain the first queen emerging. In 2 or 3 days move it up beside the swarm with entrances facing the same way.

On the eighth day after the swarm was hived, or later if the weather is cold, especially the nights, remove the old stock to a new location, this again transferring flying bees.

1195. Finally, when the new queen in the parent stock is laying well, or later when her progeny are proved, unite by the newspaper method (**1409-12**), destroying the old queen.

1196. If supers were on when the swarm issued and the swarm is given combs, transfer the supers to the swarm. If the swarm was hived only on foundation, do not transfer the supers until the hives are brought together 2 or 3 days later, or pollen will be carried into the supers.

Hiving Swarm on Old Stand, with Increase

1197. Hive as in **1194-6**, but remove the old stock to a new site. It will be so depleted of bees that the colony will retain the first queen to emerge and will not send out a cast.

1198. If the swarm is early and the main honey flow hardly due, it is desirable to postpone removal of the old stock, and it may be made instead to deliver up flying bees to the swarm by removal from side to side (**1418-23**). A week after uniting, bring the two brood chambers together and examine for supersedure queen cells in the body in which the queen is not laying. This chamber is removed for increase.

1199. If the swarm issues from a hive not yet unpacked which has developed to swarming-point before the turn of the weather, hive the swarm *B* beside the parent hive *A*, pack it at top and feed. Within 7 days unpack *A* and set it to one side. Then move *B* to where *A* stood and give supers if due, then place *A* bodily upon *B*. If the construction does not admit of this, then leave *A* next *B* where it was before moving *B*. Ten days later unite *A* and *B* as above. If there is a good flow on, the newspaper may be omitted.

Manipulations for Swarm Control

Doubling without Increase, with Natural Swarm (*D*)

1200. This is applied to a pair of hives from one of which a swarm has issued. Call this hive *A*. The second, *B*, may be any stock approaching swarming strength, situated anywhere in the apiary. See Fig. 26 (a).

Move *B* next to *A*, putting a super clearer, s.c., under any supers. Hive the swarm in a spare hive *C*, on the original site of *B*, all as in Fig. 26(*b*). In a day or two destroy all queen cells in *A*; transfer the supers from *B*, now cleared of bees, to the swarm *C*; unite *B* to *A*, newspaper method, returning *A*'s supers over a second newspaper; all as in Fig. 26 (*c*).

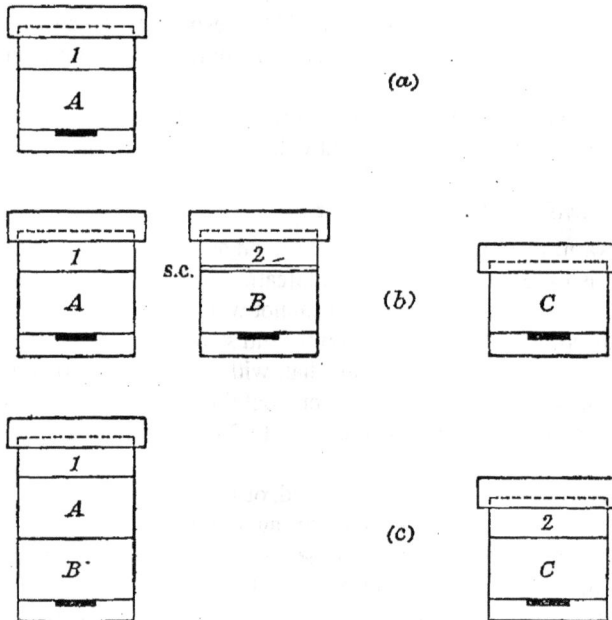

FIG. 26.-DOUBLING WITHOUT INCREASE, WITH NATURAL SWARMS.

If, however, *A*'s supers are required for the swarm, convey the super clearer with *B*, putting *A*'s supers on top, and transfer to the swarm a day or two later. Do the uniting in the evening when all bees are at home in *B*. When moving *B* put it next to *A* if there is room there.

1201. It is good to give the swarm a comb or two as well as foundation, and one of the combs may have unsealed brood or eggs. This will help to tether the swarm.

1202. If a flow commences, the united lot will divert activities to harvesting, but must have supers, and should be watched for queen cells, more especially after this early flow ceases.

1203. Stock *A* can well spare three combs for a nucleus, with ripe queen cell, or such a nucleus will be in good condition to prepare for reception of a selected queen cell from a non-swarming colony (**262**).

*Use of Upper Chamber in Hiving Natural Swarm, (**D**)*

1204. The swarm is to be hived on the old stand, as in **1193**, but the brood chamber is merely set to one side, an empty chamber being set in its place to receive the swarm as before. On hiving the swarm, put an excluder in place, then a super, using any already on the stock, then placing the original chamber on top. Before closing down, destroy all queen cells (**1364**) in the top chamber.

Examine the upper chamber again a week to ten days later for queen cells, and destroy any there.

1205. The stock may be requeened by allowing the upper colony to raise a queen from one cell left for the purpose, or preferably by substituting a selected cell from a more desirable strain. In this case, a false floor or screen board should be inserted under the upper chamber, having a flight hole; a flight hole can be provided by boring one in the hive body, to be closed with a cork when done with.

Most of the flying bees will join the swarm, and in fact it is not essential to destroy cells if the upper lot has its own floor board, excluding passage of bees from below. The colony, depleted by loss of flying bees to the swarm, will give up all idea of swarming. If, however, a selected cell is to be inserted, all other cells should first be destroyed.

1206. It is good to raise a new queen as described. Nothing is lost by it. One has the benefit of its laying power, and even if finally one queen is discarded, one unites and keeps the better.

Stopping Swarming by Destroying Queen Cells and Requeening

1207. On discovering that preparations for swarming have been commenced, remove the queen and destroy all queen cells but one ripe one, or if uncompleted, destroy all but two, returning five or six days later to select the better and destroy the other. Make sure that no other cells have been started, though this is not very likely to have occurred.

1208. There will be a gap of weeks, during which no eggs will be laid in that colony, so the method is not too good. The objection is partly met by transferring the old queen, on removal, with combs of advanced brood, to a nucleus, where she will continue to lay on a reduced scale. The nucleus must be fed, and can be re-united when all is well with the new queen, the old one being destroyed then, or the old queen may be returned if there has been any failure. Do not be in a hurry to decide that the new queen is not going to lay.

1209. If the main harvest is expected shortly, a new queen should be given, preferably with the contents of the nucleus from which she was mated. See also **1211-12**.

1210. It is of course important to make sure that the preparations made were preparations for swarming, not for supersedure (**166-70**).

Checking Swarming by Removal of Queen

1211. This method has been recommended for use just before the main flow is due. If, however, the brood nest does not become overcrowded, there is not much risk of swarming, as the diversion of activities serves as a check. For removal of the queen into a nucleus, see **1216**. If the queen is to be renewed see **1207-10** above.

The alternative to removal is to cage the queen temporarily in the hive and destroy all queen cells, also any found to have been raised 8 or 9 days later.

1212. This crude method has little to recommend it, and is liable to lead to a serious loss of morale. Bees without hope of a queen will never work hard.

Checking Swarming by Removal of Brood

1213. This method is employed under the same circumstances as the above, several combs of emerging brood being removed. The effect is a rapid diminution in nurse bees and a sudden expansion of space for breeding, the removed combs being replaced by empty combs or foundation.

This affords a good check on preparation for swarming, but removes brood which a fortnight later would be adding to the harvesters. It can be practised to advantage at the commencement of a short flow.

1214. Examine throughout for queen cells (**1364**), and while doing so, select the combs of mature brood, putting them to one side after shaking the bees off them. The removed brood may be given to a backward stock if nights are warm.

If working for section honey, the use of foundation is preferable to the use of empty combs.

1215. The removed brood has to be utilized, so it is better to use one of the methods described in **1234-58**.

Checking Swarming by Removal of Brood and Queen

1216. A nucleus is formed with combs of mainly young brood with the queen, leaving some eggs however, and filling up with combs or foundation, all queen cells being destroyed (**1364-5**). In 10 days examine for queen cells and destroy all but one (**1366**); or destroy all and introduce a new queen (**78-105**), or selected queen cell (**181**).

The nucleus should be placed beside and somewhat behind the parent hive, and when strong enough may be manipulated to furnish flying bees to the parent hive (**1418-23**). Late in the season the old queen is destroyed and the nucleus united to the parent hive. In case of any accident in requeening, the old queen is available as a spare. If a cell is kept, proceed as in **1207-21** above.

1217. If the intention is to requeen, see **1216**. If the queen is to be retained and returned, see **1207-10**, save that all queen cells in the original stock must be reduced to one every eight to nine days until the queen is returned. The methods described in **1234-58** are preferable and easier to carry out in practice.

1218. The above method is especially suitable for a prolonged flow, even in a poor district, as there is no break in the supply of harvesters and, indeed, after a few weeks the supply is doubled.

The method is good to apply when preparations for swarming have started before the flow is due.

Doubling without Increase with Artificial Swarming, Plan I (**D**)

1219. This method, due to Simmins, is applied to pairs of stocks approaching swarming strength when the early flow is due to commence. A better surplus may be got by this method than by attempting to work both stocks for surplus.

1220. Two stocks *A* and *B* are brought together, being moved into this position if necessary in advance (**627**). See Fig. 27. A spare brood chamber *C* is placed next *B* and a set of empty combs or frames, fitted with foundation, is provided. *C* receives all brood combs from *B*, all the bees being shaken or brushed back into *B*. The combs in *B* and *C* are placed to one side and the spaces filled with the spare combs or frames.

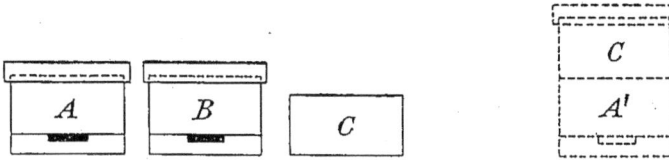

FIG. 27.—DOUBLING WITHOUT INCREASE, WITH ARTIFICIAL SWARM. PLAN 1.

If the queen is seen, she may be left in *B* in the comb she is on.

FIG. 27·-DOUBLING WITHOUT INCREASE, WITH ARTIFICIAL SWARM. PLAN 1.

1221. Hive *A* is now moved away to a new position *A'*, and the body *C* placed on top of the combs in *A*, the entrance reduced and the top well packed. It may be supered with combs two or three days later, or with foundation at once, but if supered with combs at once, pollen may be carried up.

1222. *B* will now have all its own bees and the flying bees from *A*, and will gather a full measure of surplus while building up a brood nest.

A will have the brood of both hives to care for and no flying bees. It will build up strongly for a later harvest and must be watched for queen cells. If either queen is to be superseded, the stock left with brood should be the one. Again, this stock, as soon as full of bees, is in excellent condition for queen raising.

C and *A* must be examined later for queen cells. On destroying them, note in which body the queen has been laying and put that one below.

1223. Where there is one weak stock *A* and a strong stock *B*, or where neither is up to full strength, but *B* is the stronger, or where nights are cold, it is better to leave *A* its full complement of bees, at least for a week. In this case place *C* on *A* without moving *A* until a week later.

If it be desired to requeen, do so with stock *CA* (**1207-9**).

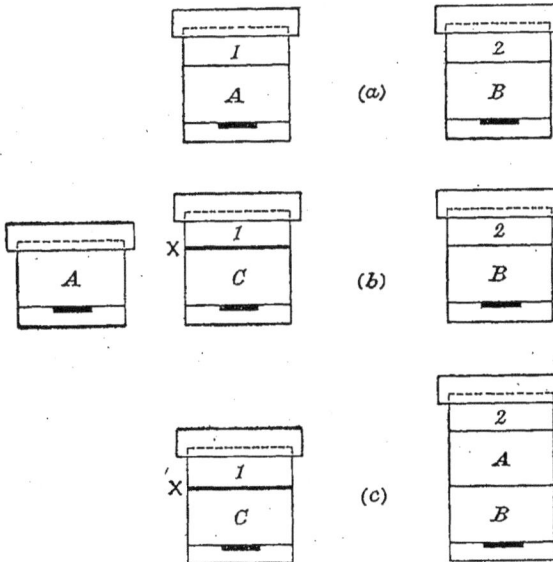

FIG. 28.—DOUBLING WITHOUT INCREASE, WITH ARTIFICIAL SWARMING, PLAN II.

FIG. 28· DOUBLING WITHOUT INCREASE, WITH ARTIFICIAL SWARMING, PLAN II (**D**)

1224. This is a modification by the author of Simmins' plan, avoiding the necessity of having or bringing the two stocks side by side, but retaining the advantage of securing a maximum of flying bees in the swarmed lot. It is suitable for use on the opening of an early flow.

Two stocks of swarming strength are chosen, *A* and *B*, in Fig. 28 (*a*), *A* being the more advanced. An artificial swarm is to be made from *A*, and *B* is to be doubled.

1225. Move *A* a little to one side, as in Fig. 28 (*b*) and smoke it well so that the bees will load up with honey. Place a spare hive *C* on the site previously occupied by *A*, as shown in the figure. Hive *C* should be furnished with foundation, or foundation plus empty combs, a space being left in the middle. Proceed to brush all the bees off the combs in *A* into *C*, looking out for the queen. Transfer the queen on the comb on which she is found, placing that comb in the space left in *C*. It is convenient to find the queen before commencing the transfer, then the frame she is on can be moved first. If the queen is not found, care must be taken that she is not left in *A*. If there, she will move on to the combs if the hive is shut for ten minutes, and a few bees left in it.

After, transferring all the bees, put any super over an excluder on top of *C* and close down. Put the box of brood *A* on stock *B*, and leave all as in Fig. 28 (*c*).

Use of Upper Chamber with Artificial Swarming (**D**) (Demaree)

1226. This plan is that widely known as the Demaree plan. The originator's own description is quoted below from the American Bee Journal, April 21, 1892. Modifications and other features are detailed later:

I begin with the strongest colonies and transfer the combs containing brood from the brood chamber to an upper story above the queen excluder. One comb containing some unsealed brood and eggs is left in the brood chamber as a start for the queen. I fill out the brood chamber with empty combs. Full frames of foundation may be used, in the absence of drawn combs.

The colony thus has all of its brood and the queen, but the queen has a new brood nest below the excluder, while the combs of brood are in the super. In twenty-one days all the brood will be hatched out of the combs above the excluder, and the bees will begin to hatch in the queen's chamber below the excluder, so a continuous succession of young bees is sustained. Usually the combs above the excluder will be filled with honey by the time all the bees are hatched, and no system is as sure of giving one set of combs full of honey for the extractor in the very poorest seasons; and if the season is propitious the yield will be enormous under proper management.

Notes on the Demaree System (D)

1227. It is disastrous to apply the Demaree system to a stock not able forthwith to expand in the bottom chamber under the conditions of weather following the manipulation. It should not be used unless warm weather is likely for several weeks after manipulating.

The method is peculiarly suited to the case in which the queen is well able to more than fill a brood chamber of the size employed and has already commenced to lay in the second chamber.

1228. The essential features are that the young brood and eggs are removed to the top, where many nurse bees follow, and the bees and queen in the lower chamber are then in a condition approximating to that of a swarm.

1229. One or two supers are generally needed about the time the operation is carried out, and these are placed above the queen excluder before putting the body of brood on top.

If supers are not yet in use, insert one, preferably filled with new combs, or with foundation, if a heavy flow is not expected immediately. If a flow is expected there should be some combs placed outside the foundation.

1230. The bees in the upper chamber are likely to raise queen cells. T'hese must be destroyed in 7 to 10 days; alternatively, at that time, a false floor or screen floor may be inserted and a new queen raised as in **1206**.

1231. In case preparations for swarming have already been commenced in a stock having two brood chambers, proceed as in **1226**, putting up all combs with queen cells, and a day later make preparation for requeening as above.

Finding the Queen **(D)**

1232. Finding the queen (**59-64**) is sometimes difficult, especially with dark bees inclined to run. If the bees have been given an additional body below as in **1394-8**, this may be left on the floor board when the queen is found to be working in it, and without further preparation, save to replace the outermost combs by foundation. The other body is raised as before.

1233. If the bees have been given an additional body above, however, as in **1394-8**, the queen is very likely to be using it, say ten to fifteen days later, and if care has been taken not to drive her down by smoke, this chamber may be placed on the floor board. Most of the combs of brood should then be removed and replaced by foundation, the bees being brushed or shaken back. The queen may be found perchance during this operation. If she is inadvertently in the chamber put above, or transferred to it, she will be found to be there when examining 7 or 8 days later. She is then more readily found, being among young bees, and may be put below, after destroying any queen cells started below, on the few brood combs left there.

Use of Upper Chamber and Transfer of Brood **(D)** *(Modified Demaree)*

1234. This plan is less drastic than full Demaree, and better suited to the British climate.

The queen and about half the brood is left below, the rest of the brood and combs removed from below being put in the upper chamber, which should receive the combs with the youngest brood other than that carrying the queen.

It will be seen that this modification resembles that in **1232-3**, which also is less drastic than full Demaree, and, as in that plan, it is not essential to find the queen. If the queen is found to be above, one merely interchanges the top and bottom chambers, searching the bottom lot for queen cells when it has been raised. While making the exchange, exchange also any food combs that were in the top lot for an equal number of frames of foundation or empty combs that were below, so that the food combs will again be found at the top and foundation below. When taking out the food combs, brush back the bees in case the queen may be on one of them.

Repeating the Operation **(D)** *(Re-Demareeing)*

1235. This is the operation of rearranging the brood in the lower and upper chambers so that combs of eggs and young brood are at the top and combs of mature brood or empty combs with the queen are below. It has the effect of securing more room for the queen to lay.

1236. The bees below are not likely to have started a supersedure queen unless, in fact, the queen is failing, but may build queen cells for swarming (**158-60**) if determined to do so. Look therefore for queen cells on the edges of the comb. The bees above, however, very generally start supersedure cells (**161**) and the upper combs must be searched throughout, although where an excluder is used some allow the virgins to hatch out above the supers, counting upon them being destroyed in trying to get through the excluder.

1237. It is convenient to operate late in the afternoon when honey gathering has nearly finished and the bees are still flying strongly, so that there are still fewer bees to handle. If done early, gathering for the day may be seriously disturbed.

1238. Provide a spare body *C* (Fig. 29) and place it by the hive. After subduing, open the top and remove the combs from the top chamber *A*, one at a time, shaking back all the bees. Combs with honey may be set aside, but combs with mature brood must be put into the spare body *C* after removal of queen cells.

Now set aside the supers and the empty top body *A*. Put *B* diagonally on an inverted cover beside the stand so that the queen shall not be lost through the bottom. Place *C* on the stand where *B* was

Now remove combs from *B*, watching particularly for the queen, as either the queen·must be shaken into *C* or the comb on which she is found must be placed in *C*. All other combs of eggs and young brood are placed in *A* and the honey combs replaced there. When all the combs are in place in *A* and *C*, replace the excluder on *C* and the supers and *A* on top. If the supers are heavy, place *A* temporarily on *B*, which is empty, until the supers are in place, and then finally on top; and remove *B*.

B is now empty and serves for manipulating the next stock. It is handy to have a cover or empty body to serve as a temporary stand for the supers.

1239. This method of working is such that, except just when *B* and *C* are being interchanged, the flying bees are free to enter normally and find combs within. Further, the combs removed to *A* mostly have bees upon them and bees shaken off first in *A* have a place to go to. If, however, it is known that the queen is one hard to find, it is desirable to make a modification. The queen should be found before any combs are removed from *B* to *C* or *A*, using a comb support to support the first two combs removed from *B*, or shaking the bees off these two combs before putting them in *C* or *A*.

It is not good to sort the combs in *B* before those in *A*, as the flying bees are then more troublesome.

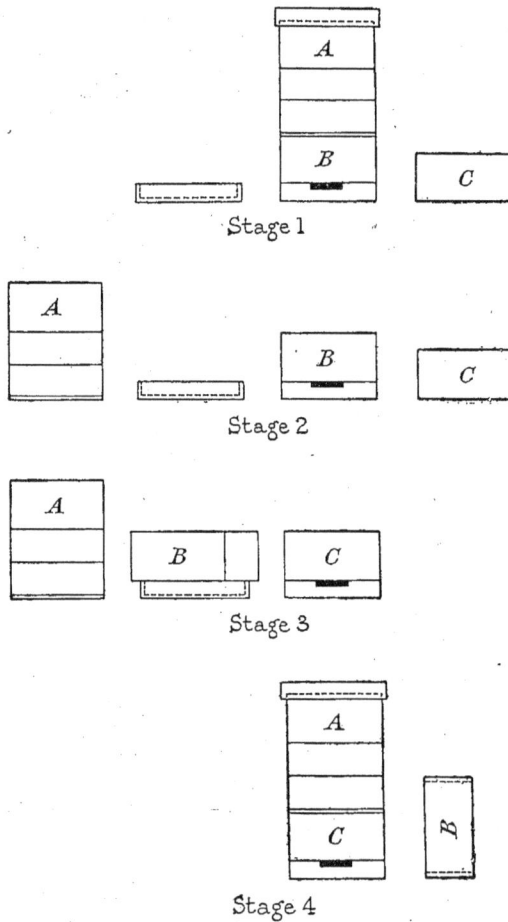

FIG. 29 - RE-DRESSING

Pseudo-Demaree (**D**)

1240. Sometimes combs of advanced brood are removed from the brood chamber to an upper story to relieve the brood chamber. This is good if done just before a honey flow, the removed combs being replaced by foundation inserted so as not to divide the brood nest (**165**).

It differs from Demareeing in that only mature brood is lifted.

The brood lifted should preferably be placed to one side and vertically above that left in the lower chamber. If the combs are arranged warm way (**647**) the brood should be at the back

Preparing a Strong Stock for Demareeing (D)

1241. When a stock already occupies two bodies with brood of all ages, or is so arranged, the introduction of a queen excluder between the two bodies secures that in 10 days' time the queenless half will contain mature brood and the other half mainly young brood. The latter is to be put on the top of the pile and the queen, with the comb she is on, transferred to the other body, in exchange for a comb of sealed brood. The older lot should of course be looked over before the queen is placed with them, and combs with honey, for example, removed, but pollen should not be taken away.

Demareeing, when Working for Section Honey (D)

1242. For good section work a copious flow and strong gathering force is necessary. If the stock is Demareed well before the flow, proceed in the usual manner, but it is important to secure that when the flow really starts there should be no room for surplus in the top chamber and a minimum in the bottom. The top chamber should preferably be full of combs of eggs and young brood. Combs unoccupied with maturing brood in the bottom chamber should be replaced with foundation.

1243. Vacancies will occur in the top combs at a period depending upon the age of the brood. If half the brood combs are in the top, brood will be emerging in 10 to 11 days; if two-thirds, then in about a week. The top lot should be examined in a week to 10 days according to the above and to general convenience. After removal of all queen cells, bring the top lot to the bottom in place of the bottom lot, which is then taken away, the queen being transferred to the top lot in its new position.

1244. The bottom lot removed will consist of young brood, probably of no value for that particular flow. It may be placed on top to advantage, provided all empty combs and unused foundation be removed and the space filled with combs of honey, or better still, brood of the same age from another colony.

1245. If the plan is being worked with a number of stocks the simplest course is to collect all the removed brood over a few selected stocks, one chamber on each stock, putting no upper chambers over the remainder. Use the strongest stocks to receive brood and shake the bees from the brood transferred into the hive from which it is taken. There will then be no uniting necessary. Alternatively, with a flow on, the bees on combs placed in the top chamber may be united by alternating after exposure, as in **1408**.

1246. It is undesirable to give room much in advance of requirements when working for section honey, and there is more risk of swarming. It is desirable to examine more frequently, say, every 7 days. This may be avoided by combining requeening.

1247. With a clover flow, the stock, if ready, may be Demareed when the first blossoms are seen, and re-arranged as above 10 days later.

Queen Supersedure and Demareeing (D)

1248. The active verb in the heading, built upon the name of Mr. Demaree, is in common use and forms a lasting recognition of his contribution to the craft.

After all that has been said about the merits of supersedure without swarming the reader will be prepared for a warning that this possibility must not be overlooked when following the

Demaree plan, or some modification of it.

It is good to apply the plan when a honey flow is pending which may lead to overcrowding and that condition of general prosperity conducive to swarming as swarm cells may then never be built where the queen is. If this procedure is followed one misses the opportunity of studying the swarming habits of that stock save to the extent that it is a good sign if the isolated portion of the colony does not build queen cells.

Room can be given for honey without dividing the brood nest, so with bees expected to supersede without swarming one can wait until queen cells are in fact built and occupied. One then operates only if it is clear that the intention is that of swarming, not supersedure (**166**). The first suspicious sign is likely to be the tearing down of good worker comb on the margins of the brood nest for the increase of the drone brood. If this is marked, any queen cells built should be regarded with suspicion. Queen cells built out on the bottom of the upper frames in a tiered brood chamber are most probably swarm cells.

Unless satisfied that supersedure is on foot one would Demaree as originally intended.

If supersedure cells are built one can still employ pseudo-Demaree as in **1240**, and encourage the bees to play according to plan.

*Use of Upper Chamber·by Transfer of Brood Chambers (**D**)*

1249. As here detailed this plan is suited to the weekend beekeeper. If the bees can receive attention at any time examination may be made, say, at intervals of nine days instead of 7 days. The remarks in **1248** above still apply. The owner of bees that may be expected to supersede without swarming will not apply the procedure unless preparations for swarming as distinct from supersedure are discovered.

1250. One starts with a strong colony already breeding in two chambers and probably supered above an excluder as in Fig. 30 (*a*). This colony is examined every seven to nine days to ascertain whether queen cells have started at the bottom of the upper chamber or near the top of the lower chamber. If cells are found, lift off the upper chamber and the super, destroy all cells in the lower chamber, then replace, but rearranged as in Fig. 30 (*b*), the super with its excluder being placed on the lower brood chamber and the other placed on top over a second excluder. The bees will rearrange themselves and the lower chamber will be relieved of congestion.

1251. Now the condition desired is that the queen should be in the lower chambers. She may have been noted when destroying cells, but there is no need to search for her, as a week later the upper chamber may be examined. If eggs and young brood are found the queen is in that chamber and the position is rectified by putting that chamber below. If there are no eggs or young brood the queen is already below as desired, but any queen cells in the upper chamber should be destroyed. Thus either by luck or by rearrangement we have reached the condition shown in (*b*), the queen *Q* being in the lower chamber.

1252. A week later examine the lower chamber, where the queen is, for queen cells. If there are none, all is well and the chamber now at the top may be placed on the lower one and the super replaced on top over an excluder. If, however, queen cells are found, they must be destroyed and the whole procedure repeated save that, as the queen is known to be below and the stock is not too ready to respond to the treatment, the next examination should be made in a week to nine days.

1253. If no queen cells were found the operation may be repeated after a fortnight as a precautionary measure. Two courses of treatment are found sufficient for most stocks, but some further features require attention as below.

1254. One object of the plan is to avoid the labour and disturbance of rearranging combs, but some intelligence and observation is necessary, for if either chamber becomes clogged with honey some relief is necessary. One can detect such a condition by feeling the weight and peering between the combs or lifting out outer ones, if in doubt.

There must be sufficient room for the bees to spread out incoming nectar, so there should be some empty combs as well as foundation.

1255. The procedure above described may be applied as a precautionary measure at the commencement of an early flow, without waiting for queen cells to be built.

1256. A shallow chamber is shown for the super in Fig. 30, so that it may be readily distinguished, but more than one may be required, and, of course, a full depth chamber may be employed.

1257. Two excluders are employed so as to prevent the queen from entering the super, but when it is known that the queen is below, one is sufficient.

Use of Upper Chamber by Transfer of Shallow Brood Chamber (S+)

1258. This plan is a modification of that just described applicable to a large brood chamber supplemented by a shallow one. It presents some simplification. The precautionary measures again apply, and if desired one can add some rearrangement of the combs in the main chamber as dealt with in **1376-80.**

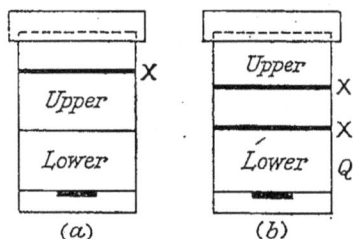

FIG. 30.—USE OF UPPER CHAMBER BY TRANSFER OF BROOD CHAMBER.

One starts then with a stock in a large chamber as shown in Fig. 31 (*a*), the brood already occupying a shallow chamber on top. There may be a super above an excluder above. One operates when queen cells appear at bottom of upper chamber and/or top of lower chamber, and after destroying them, or when about to add the upper super.

1259. The shallow brood chamber is merely exchanged with the super or placed upon it when it is inserted, the excluder being placed between the two shallow chambers. The queen can work up above the main brood chamber, but if the main chamber is of ample size and the super contains no layer of honey it is better to add a second excluder on top of it, thus effecting a more complete division of the colony.

1260. If at the end of a week the queen is found to be in the top super, examine the main chamber and destroy any queen cells, then interchange the two supers but use an excluder only below the upper one. If the queen is not there, look for and destroy any queen cells found in the upper chamber, then interchange as before.

1261. One has now returned to the condition shown in Fig. 31 (*a*) save that the position of the queen is known. Examine for queen cells as before and repeat the exchange of supers when they appear or when desired and convenient. If the queen was in the super she is likely to go below very shortly, but if below she may not come up for some days unless nights are cold.

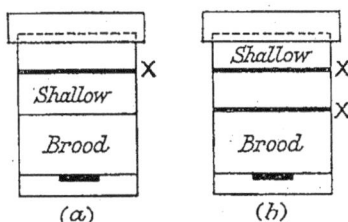

FIG. 31.-USE OF UPPER CHAMBER BY TRANSFER OF SHALLOW BROOD CHAMBER.

1262. Be sure to give ample room as required, keeping the honey supers together whatever position they occupy. Any super of sealed combs may be removed.

Rauchfuss Method of Swarm Control

1263. This method involves increase and will be found described under that head in **1448-50**.

Using Two Queens in One Hive

1264. This plan is illustrated in Fig. 32. A strong stock is shown at (*a*), occupying two brood chambers and ready for supering for the early flow. A screen floor board is placed on top of *A* as in (*b*), and a nucleus is formed in an empty chamber *C*, with 3 combs of food, 2 of emerging brood and a new queen, the space being filled up if possible with combs. The colony in *C* builds up on the spring flow and both lots may be supered as required as indicated in (*c*).

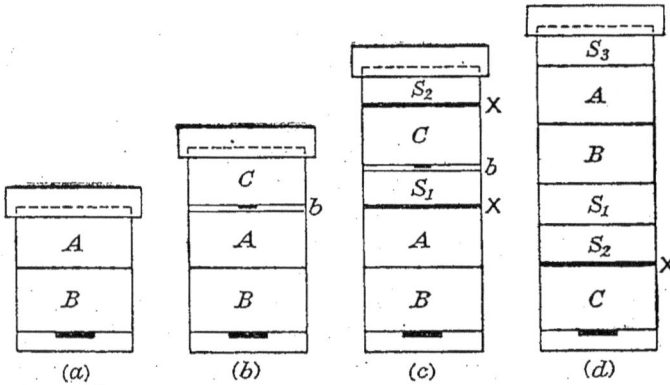

FIG. 32. - USING TWO QUEENS IN ONE HIVE.

For the main flow the new queen is put at the bottom, as in (*d*), all combs of unsealed brood and eggs in *C* save one being exchanged for combs of emerging brood from *A* and *B*. The old queen in *A* and *B* is destroyed, or housed elsewhere. Additional supers are added on top as and when required.

SECTION XIV
FEEDING-ROBBING-PACKAGE BEES

Feeding

General

1265. It must be remembered that bees, like other livestock, are dependent on food and drink, and cannot always find the food they require or can use to advantage, even in summer-time. The beginner frequently ignores the question of water supply, and even advanced beekeepers frequently have but little idea of the immense quantities of pollen required for brood raising.

Water Supply

1266. Water is required at all times, except during the quiescent period in winter. Water is required to dilute the honey to bring it to the right consistency for feeding. In the summer also water has an important use in temperature control (**734**). The evaporation of 1 oz. of water requires about the same amount of heat as is produced by the consumption of ½ oz. of honey, or the burning of, say, 1/10 oz. of coal; and, if the water be caused to evaporate, it will absorb a corresponding amount of heat from its surroundings. A nice adjustment of ventilation by the bees is required therefore, when there is much nectar to be evaporated and also much brood to be kept warm at night.

1267. A water-carrying bee is especially susceptible to cold. It requires considerably more sunshine, or the consumption of considerably more food, to raise the temperature of a water-loaded bee than of one carrying no water. The bee has no mechanism for extracting the honey from its honey sac and leaving the water behind, though it can separate out pollen. It is at least possible that heat is lost by the bee while travelling, by the evaporation of water while the bee is endeavouring to produce heat by consumption of food. In cool weather flying bees are mainly dependent upon direct sunshine for maintaining the temperature necessary to activity (see **409-12**).

1268. A colony breeding well will use about a pint per day according to size, and in the spring will readily accept half a pint or more per day, which supply is most acceptable at temperatures up to 100° F. (37° C.). Water should not be supplied until the bees are able to take regular flights, daily temperatures reaching at least 50 F. (10° C.) in the shade, but a supply will then save much bee life by reducing the hazardous occupation of water carriers. It is cheaper to heat the water by the consumption of coal or gas than to let the bees heat it by the consumption of honey.

1269. During the late winter and spring the bees uncap honey on the margins of the cluster. The increased humidity of the air and condensation of moisture is such that this exposed honey readily absorbs water (**437**), thus relieving the bees of the necessity of bringing in all the water required. The consumption of the honey itself leads to the production of water, which adds to the general humidity and hastens the process described. The complete consumption of one pound of honey releases about eleven ounces of water, including that carried in the honey as water. While the air outside the cluster may be saturated with water the humidity within the brood nest is reduced, the air being warmed as it passes through the cluster.

Giving Water

1270. Water may be given by use of an open tank or tray of a size suited to the needs of the apiary, fitted with a wooden float made of slats, large enough to cover the whole surface, thus reducing evaporation, but free at the edges and with small gaps between slats so that the bees can reach the water.

1271. The growth of fungi and formation of mould is checked by the addition to the water used for filling the tank of one part in a thousand of common salt, say one flat teaspoonful per gallon. Incidentally the bees appear to prefer the salted water to plain water.

1272. In a large home apiary it may be worth while to provide means for warming the water in early spring, but protection from wind would be required. In any case choose a warm spot, sheltered from wind and from the sun during most of the day.

1273. There is generally a building, from the roof of which water may be collected and held in an exposed vessel. If running water is available a tray full of small pebbles makes a very suitable source, or even a sloping grooved wood board or slab of concrete with ¼ inch grooves and a slope of, say, 1 in 50.

1274. Bees will take water from foul places, such as the sinks outside dwelling-houses and even off manure heaps, but this probably on account of the temperature of such sources. Water should not be given to bees for travelling. Syrup is more suitable.

Pollen Supply

1275. Honey, consisting, as it does, mainly of carbohydrates, serves principally for the production of heat and energy, but nitrogenous foods are essential for body formation and reparation. Pollen is the great source of nitrogen and is rich also in essential vitamins lacking in honey. Pollen is necessary to honey gatherers and even to bees taking down syrup, but especially to nurse bees.

1276. A large and prosperous colony will require considerably more than 1 cwt. of pollen in the season, and will harvest 2 lb. or more per day at times. About one pound of pollen is required for every 4,500 bees produced. Hence the great, importance of healthy supplies carried over to the spring for use before the pollen-bearing trees are yielding freely; and the importance of so-called pollen-clogged combs (for Sources of Pollen and their Seasons, see **459**, *List IV*).

Pollen Substitutes

1277. Owing to lack of early sources of pollen in some places, considerable attention has been given to finding substitutes to replace or, at least, to supplement the natural product.

1278. For generations past bee-keepers have furnished flour in various forms, of which pea flour has been a favourite. It is well known that the bees will carry in large quantities, especially if it is placed in the cups of early flowers, such as the crocus, or sprinkled on wood shavings exposed to the sunshine and sheltered from rain. Bees will, however, also gather coal dust and other powders of no possible use, and it is now considered doubtful if farinaceous substitutes, like pea flour, serve any useful purpose, even as a supplementary food. It has been definitely proved that it is quite impossible to produce bees on this material used entirely in substitution for pollen.

1279. Many other materials have been tried, of which the least unsatisfactory are probably fresh albumen (white of egg) or soya bean flour added to sugar syrup, also powdered, yeast. Fresh and dried milk have also been used as a supplementary (see **1297**). With these used alone it has been found possible to produce bees with developed nursing glands, but complete satisfaction in places where there are no other sources of pollen cannot yet be recorded. Yolk of egg must be avoided, and, in consequence, dried egg-powder must be avoided. Attention is now being turned to finding natural sources capable of cultivation in places lacking early supplies. Bees fed with pollen substitutes are short-lived.

1280. Satisfactory results are obtained if one quarter of the pollen supply is substituted by soya bean flour expeller processed and containing 5 to 7 per cent. of fat.

In using egg, the white of one egg is used with every ½ pint of thick syrup or honey prepared as in **1285.** Beat up the white of egg, add the boiling syrup or honey, pouring it into the egg froth, stirring well, and continue to stir until cold.

1281. On evidence so far available on pollen substitutes the author gives the following advice:

(1)   If fresh pollen is available in adequate quantities do not give any substitute as such and especially any form of flour.

(2)   If fresh pollen is available, but the supply is short, supplement it with soya bean flour, shaken over dry shavings and exposed in a shady place, protected from rain.

(3)   If the district is one in which fresh pollen is not available early, in time for building up for an early flow, look to the matter in the late summer and autumn. Large quantities of pollen are carried into hives awaiting a queen, and if removed, will be replaced forthwith by the bees in an economical manner. Combs of fresh pollen given to a strong stock, fed with syrup, will be covered and sealed in ten to fourteen days and may be distributed as required when making winter preparations. Proceed also as in (2) above, but don't expect to raise good bees without sufficient pollen.

(4)   Fresh milk, dried milk, soya bean flour or white of egg can be added to sugar syrup with honey as a supplementary stimulative food for encouraging breeding.

Honey Supply

1282. It is important to recognize that a good colony of bees will use for internal purposes at least 200 to 400 lb. of honey, which has to be found before there can be any question of surplus.

In the hive the normal position of the honey supply is around, above and behind the brood.

1283. Spare combs of honey inserted to supplement the food supply should be placed against and beside the brood nest, not in the midst, and outside any existing stores. Do not displace combs with pollen in them found next the brood combs (see also **165**).

1284. There is frequently much wastage in using granulated stores. The bees seek first the liquid portion and are apt to discard crystals in getting at it. These crystals fall to the floor. Granulated honey can·be fed from a super above a clearer board and any crystals not removed by the bees may then be saved by the bee-keepers; but partly granulated stores brought into use in a warm part of the hive in free flying weather are not so wastefully used as if distant from the brood, stored in a remote part of the hive and used in cold weather.

Feeding Honey for Storage

1285.a   It is not necessary or advisable to feed honey to a colony of bees. Once removed from a colony the honey may have been treated that would reduce its palatability for bees. It may have come from another colony and be a source of infection. It is far better to feed bees with a sugar solution or specially produced feed for honey bees. ID

1285. *Honey fed either neat or as a weak syrup is not stored without a considerable quantity being consumed in the process. The consumed portion may be partly recovered as wax. It is stored most rapidly if about one-third by weight of water be added to the honey, remembering that one pint British weighs 20 oz. . Such a solution has a density of about 1.27 at 20⁰ C. The water is, heated and the honey stirred in without boiling, the whole being stirred and heated about as hot as is comfortable to the finger held in the syrup. It may then be put warm into the feeder for immediate use. Honey should not be diluted until required for use, as it is liable to ferment. It may be kept a few days in a cool place after dilution (517-22).*

Unless the origin of honey is above suspicion it is not safe as food for the bees without sterilization.

Sugar Feeding

1286. The use of sugar for feeding is discussed in **960**, and it is shown to have value.

Candy feeding is a great labour and a common sign of mismanagement. Bees candy-fed in winter and early spring are forced into undue activity in fetching water to dissolve the crystals and in carrying the supplies to the cluster. This involves loss of life by wear and tear and hazard at a time when the colony is reduced to a minimum. Candy feeding is discussed in **920**.

The use of sugar syrup preserved against fermentation has been proposed, the addition of thymol being recommended, but this practice is now frowned upon. In general, the addition of drugs is considered undesirable.

1287. Nevertheless, after a very bad season, and even in the summer during a very bad year, auxiliary sugar feeding may have to be resorted to, and some bee-keepers will still give candy as a late spring stimulant as well as to make good for past neglect. Recipes are given in **1299** and **1308-15**.

1288. Candy is frequently stored in no-bee-way sections glazed on one side or similar glazed boxes. Cheap glass dishes can be bought at certain stores and are very suitable and practically everlasting if warmed before putting in the hot candy.

1289. It must be understood that sugar syrup is not stored by the bees as sugar syrup but as an inferior kind of honey, lacking in aroma and valuable salts, partially inverted and containing in some measure the additions made by bees to nectar in converting it. If thick syrup of the right consistency for storage is given, the bees have to dilute it to convert it, thus it is better not to feed a dense feed. The conversions make a call on the strength of the bees and they need access to pollen if their constitutions are not to suffer.

Note on Sugar and Use of Acid

1290. In making up syrup with sugar, pure white cane or a good white granulated beet sugar should be used. Never use brown sugar. In the past it has been general practice to add a certain amount of acid, such as vinegar, cream of tartar or tartaric acid, to sugar syrup when boiling same, to invert it, but the amount inverted is small. Careful investigation by more than one investigator shows that when taken down -stored by the bees- the amount inverted is less than in similar syrup made up without the acid and left to the bees to invert during storing. It is believed that the excess acid is not good for the bees. There is no occasion to use acid with sugar, even in the preparation of candy.

Feeding Sugar Syrup for Storage

1291. Sugar syrup is taken most readily if diluted to a density of about 1.27. This is obtained by using about four parts water to five sugar by weight, or 1 pint weighing 20 oz., say, 1½ lb. sugar to 1 pint water; or 9 gallons of water per hundredweight of sugar. A thicker syrup is used for winter storage if fed late, and a thinner syrup is advantageous for stimulative feeding. For winter storage use, say, 1 pint to 2 lb. Doolittle recommended adding 10 per cent. of honey. The above figures are British measure.

1292. When melting sugar be careful not to burn the sugar by overheating any part of the vessel through lack of stirring. In a large vessel sugar held in a porous container at the top of the liquid will dissolve completely without heating. For small quantities pour boiling water on the sugar in the morning, and in the evening it may be quickly dissolved completely without raising the temperature above hand heat.

1293. A syrup of about the most suitable strength for feeding for storage can be obtained in a simple manner without weights and measures. Put dry granulated sugar into any convenient vessel and shake it down to a level top. Note the height of the level top as the sugar stands in the vessel. Add boiling water enough to just cover, the sugar sinking while this is done. Later make up to the original level of the dry sugar by adding further boiling water. This will give a syrup of suitable strength.

Stimulative Feeding

1294. Stimulative feeding should not be undertaken until the bees are able to fly freely;

that is, until the maximum shade temperature reaches at least 50^0 F. (10^0 C.). This generally occurs in the northern hemisphere in the latter part of March or beginning of April, according to the progress of the season and the latitude. In the southern hemisphere the time would be September to October. When stimulative feeding is commenced in the early spring it must be continued without cessation until nectar is coming in from early sources, as shown by the daily temperature (**411-12**) and the presence of blooms or by actual inspection of combs. See recipes (**1299**), or water supply (**1268**) and uncap some honey. A good colony will use ½ pint of food per day at first, and, say, 1 pint per day later.

1295. Food coming in slowly is taken to the borders of the brood nest and, if too much is fed, may actually cause contraction instead of stimulation. The uncapping of stores on the borders of the brood nest, however, has the effect of expanding the area available for egg-laying. Early stimulative feeding must be commenced slowly and should be followed by examination of the brood nest and uncapping where indicated.

1296. Stimulation in the spring has the advantage that old and perhaps diseased bees are disposed of by being worked to a full stop, while being replaced by a greater number of vigorous young ones, but it is not good to feed syrup unless the bees can get access to adequate supplies of fresh pollen.

1297. Milk and other protein and fat-containing foods have been used for generations in parts of Europe for stimulation as additions to the spring diet, but many additions have been made on the stimulus of the bee-keeper's heart and imagination rather than on that of his intellect. Such practice has not hitherto been subjected to the critical examination which is essential to provide a basis for sound progress. The effect on the normal development, health, strength and longevity of the bee has yet to be pursued, but amongst the additions tried widely there is a strong partial case made out for milk as an addition to sugar syrup or honey syrup as a spring stimulant.

1298. Stimulative feeding is practised in the autumn just after the last considerable honey flow has ceased, with the object of deposing the old bees and providing a maximum supply of young bees for wintering and for activity in the early spring. It gives the best possible use for late gathered stores, which as a rule are of poor quality for wintering (**1306**). Half a pint every other night is generally sufficient. It is not desirable at this period to uncap sealed stores. I t is necessary to ensure that there is room in the brood nest for late breeding. Plain syrup or diluted honey is suitable, there being plenty of pollen and the weather being favourable.

Stimulative feeding is resorted to in some phases of management, especially in connection with queen raising, both for securing good queen cells and successful mating.

Stimulative Food Recipes

1299. For stimulating breeding a warm sugar solution is used of, say, equal parts of sugar and water by weight, a pint of water weighing 20 oz. in Great Britain, but 16 oz. in the U .S.A. Weaker solutions are frequently used.

1300. Nothing is more exciting than a feed of honey, which may be given with an equal measure of hot water and fed slowly, but there is some risk of starting robbing whenever honey is fed. It is safer to uncap some stores and give warm water (see also **1281** (4)).

Summer Feeding

1301. It always pays to feed a swarm until it is well established and gathering freely. Sugar feeding may be employed to supplement honey-gathering, sugar being helpful in the economical production of wax for comb building. Summer feeding may be necessary between crops, especially in a bad season. Breeding may be seriously checked in case the honey flow ceases temporarily.

1302. The casting out of drones and of immature drone brood in the summer is a sure sign of really serious shortage of food, but it pays to feed long before this stage is reached (see also **934**).

Winter Stores

1303. The provision of good and sufficient stores for the winter is important, not only for winter needs, but to secure rapid expansion of the brood in the spring and to avoid the necessity of disturbing the wintering arrangements made by the bees. The stronger and more prolific the colony, the more winter stores it will require in the spring. Twenty five to sixty pounds of honey should be present in the late autumn, according to the size of the colony, immediate and expected, and the external conditions (**967-71**).

When the bees are wintered in a cellar, under the best possible conditions, a strong colony may use only about 4 lb. of honey, while in the cellar; whereas a colony in a hive in, the open, not well packed, will use 10 to 30 lb. for keeping up the temperature, the amount depending upon the climate (**967-71**), apart from what may be used for breeding, during the same period. The large margin of stores is not required for the period of quiescence but for the period immediately following, but stores are needed also before the quiescent stage is reached.

1304. A practice widely followed consists in slow feeding during the early autumn, followed, if necessary, by a short period of rapid feeding done before the temperature is too low for evaporation, and successful sealing. In Great Britain the end of September is late enough, with average daily mean temperature of about 52° F. (11° C.). In some places the cold period comes suddenly and extra care is then needed. The last supply of food may take the form of 10 lb. of strong sugar syrup, fed warm at one operation, from a rapid feeder, or at the rate of 2 or 3 pints per evening. In very cold regions twice this amount may be given, especially if the spring is also cold. It is, however, more economical to feed a somewhat weaker feed somewhat earlier, as the bees take it down with less consumption and can evaporate it if fed early.

1305. The capacity of the colon in which the bee has to store its faeces until the weather permits of' a flight to clear the bowel is about 0.04 to 0.05 gram, and it fills up at a rate depending upon the quality and quantity of the food consumed. When the colon becomes full and flights are impossible, the contents are discharged within the hive. Experiments with good colonies of bees cellar wintered in Russia showed that the danger-point was reached on October 30 with stores mainly honey dew, on November 30 on honey containing some honey dew, on December 29 with half honey and syrup, and not until March 9 with sugar syrup alone. Without cellar wintering and in smaller colonies these limits would have been reached much sooner, owing to the large consumption of food, unless the climate permitted of occasional flights, as in Great Britain, for example.

1306. Dark honeys, any containing honey dew and honeys crystallizing rapidly, also those that jellify (**501** also **510-11**), are bad for wintering, though suitable for early spring use. It is not necessary to remove such honey if the last food given to the bees to store is sugar syrup (**1291**), as the bees will then store the sugar where it will be used first.

Outdoor Feeding

1307. Outdoor feeding is safe only in the hands of an expert. There is danger of starting robbing. It is better not to expose bee food near the apiary. If done at all it should be within, say, 1 hour of the time that flights cease, or while there is a good honey flow, unless quite weak syrup is used, say 1 lb. sugar to 4 pints of water. If combs are exposed with honey in them the bees will tear them up in their eagerness to secure the stores. I t is difficult to find any good reason for outdoor feeding, and it involves a risk of serious robbing developing, a risk which it is impossible to assess in advance.

Making Candy

1308.a Candy can be obtained from beekeeping suppliers. This candy is readily accepted by colonies. An alternative and cheaper supply is bakers fondant available from bakers shops. This has glycerine mixed with the sugar solution to maintain its mobility. It appears that bees are not adversely affected by the small quantity of glycerine in the mixture. ID

1308.   Candy is frequently sold at fancy prices. It may be made at little more than the cost of the sugar and with certainty if a thermometer is used to control the density of the hot syrup, the density being determined by the boiling-point, and the quality of the candy being mainly dependent upon the density.

1309.a   Always use stainless steel utensils when making candy.

1309.   A preserving pan is very suitable for boiling the syrup, but any convenient enamelled or aluminium pan can be used, or a well-tinned vessel.

1310.   When the right density is reached there will be rather over 6 lb. of sugar per pint (20 oz,.) of water, but if it be attempted to make the candy starting with these proportions there is considerable risk of burning the sugar. More water is therefore used and evaporated off by boiling, say 1 pint (of 20 oz.) per 4 lb. sugar.

1311.   Put the sugar into the vessel and pour in the measured quantity of boiling water, then heat slowly, well stirring. Stirring must be continued until no crystals remain. The last fine crystals may be readily seen by taking a sample of the syrup in the stirring spoon and examining in a good light. Crystals, even small ones, lying on the bottom, are liable to become scorched, and if any portion is thus overheated, producing a brownish tint, the whole may be spoiled. When the sugar is largely dissolved stirring is continued without intermission.

1312.   Boiling must be continued until a temperature of 238 to 245° F. is reached (say, 114° to 118° C.), but candy heated only to 238 F. is liable to soften over a strong cluster and fall in the container. A final temperature of about 243 F. is best, measured an inch below the surface of the liquid.

1313.   When the right condition is reached, as shown by the temperature, the syrup is allowed to cool. In order to get the right consistency, it is necessary to stir during cooling, but there is no benefit in stirring before the temperature approaches that at which crystallization commences. The syrup may be allowed to cool to 100 to 110° F. (38 to 43° C.), and must then be stirred continuously until it assumes a milky appearance. The mass stiffens as the crystals form. It should not be let become too stiff before removal into the containers used for feeding. The hot mass is ladled into the containers, which are filled so that the candy settles to a level about 1 inch below the top edge. Usually the mass gets too stiff for ladling before the job is finished. In this case the vessel should be gently warmed and the remaining candy stirred to soften it and avoid burning. The ultimate consistency is not affected by this re-warming.

1314.   It is possible to make good candy without a thermometer. To do so add 1 pint (20 oz.) of boiling water per 6 lb. of sugar in a stout vessel, stirring well and heating slowly to avoid burning. The whole of the sugar must be dissolved before boiling is allowed to set in, and finally the syrup may be allowed to boil strongly for 2 or 3 minutes. When cooling, stirring should commence when the temperature is such that the bare hand' may be held against the side of the vessel.

1315.   It is unnecessary to add any acid to obtain the correct consistency. The addition of vinegar or cream of tartar or other acids causes partial inversion of the sugar, but the bees accept acid-free sugar, more readily, and it is a question whether the addition of acid is not, in fact, harmful. Some add lemon or orange syrup, made by boiling the peel as well as adding the juice in the water added to the sugar. This may be beneficial but is not proven.

Making Queen Cage Candy

1316.   The simplest form and one thoroughly satisfactory is made by mixing disease-free honey with icing sugar. Add the honey to the sugar, using one part by weight of honey of average density to 2½ parts by weight of sugar. It is most easily made in a water bath, the water at about 110° F. Put the icing sugar in the inner vessel and add the honey, then work it together with a stick until of uniform consistency.

1317. When moulded into a cylinder 1 inch diameter and 3 inches high and at a blood temperature, 98 F., the cylinder should stand on end without the bottom expanding. If too soft, knead again and use a little more sugar.

1318. For keeping, it is a good plan to heat the honey to 145° F. in the water bath before using and let it cool to 110° F. before using as above.

The candy will then keep indefinitely in an air-tight tin, one with a "push in" lid, well pressed home. If found hard, warm in the hands and knead it like putty before using.

Robbing

General

1319. Bees are accustomed to gather their stores by laborious visits to many sources, each of which renders but a minute quantity. The finding of a source in the open, from which not only one but many bees can load up on a single visit, causes great excitement among the bees. This is especially the case if the food offered is honey, not merely nectar or sugar syrup. Access to stores, and especially to honey, outside the home, excites their cupidity. If once a colony has experience of such "easy money" it may become demoralized and seek ruthlessly for further supplies of a like kind, robbing its weaker and less aggressive neighbours and causing pandemonium and utter destruction in the apiary.

1320. The tendency to rob varies with the race and strain. The bee-keeper should not tolerate strains showing a strong predisposition to robbery. The vice is associated with the yellow races much more strongly than with the black or brown. Some of the lighter-coloured bees are tractable in this respect, but in general the goldens or yellowest races are more predisposed than the leather-coloured Italians. The American "Italian" is a mixed race and in America all "Italian" bees are suspect, not to an extent, of course, that hinders their free use on account of their many excellent qualities, but to an extent that necessitates precautions and watchfulness.

1321. Robbing is liable to be set up by any exposure of stores, and especially on cessation of a source of supply of nectar and by the temptation offered by poorly guarded stores or by leaky hives.

Detection

1322. Even if all precautions are taken, robbers will try to rob, and it is well to keep a look out by observing the behaviour of the bees. When robbers are about, the bees will be on guard and the guards unusually active, challenging all who seek entrance. Occasional combats will be seen. The robber bee hovers around the entrance much as a wasp does, not seeking familiarity with the general appearance and surroundings as does a newly emerged bee, but seeking an opportunity to dart in past the guards. The robber bees are the older bees, not the newly emerged.

1323. In the height of the day the home bees leave the hive empty and return loaded, whereas the robber bee approaches the hive empty and leaves full. Now a flying bee with empty honey sac flies with legs extended behind, whereas a loaded bee bends the hind legs to bring them forward. The observation of this distinction is particularly useful in detecting slow robbing of a weak stock in cold weather.

Hindering Robbing

1324. Some robbing and attempted robbing will occur in the best managed apiary, but if precautions are taken it should not develop into raiding and destruction. If the bee-keeper has brood disease to contend with, a check on robbing is most important, as the diseased and disheartened and weakened stock is particularly liable to attack, thus leading to the disease being spread to healthy colonies.

1325. The danger of robbing is minimized by removing temptation. A cloth of carbolic cloth (**861**) should be at hand as a temporary cover for hive bodies exposed during manipulations. When robbing is feared hives should be examined late in the day. Outer cases should be bee-tight and covers also. Robber bees are quite capable of handing out the goods to their friends outside, through cracks which will not let a bee pass. Entrances should be reduced when robbing may be expected and especially the entrances of weak stocks. The entrances of small nuclei may be reduced to 1/2 inch or less.

1326. The most comfortable time to manipulate stocks is when bees are flying freely. With difficult bees the use of a screen is handy (**896**).

The working stocks should be kept strong. Normally, if disease is kept under, the only weak stocks should be the nuclei. Other weak stocks should be united. Weak stocks do not pay.

Stopping Robbing

1327. If appreciable robbing occurs, in spite of the above precautions, it is necessary to deal with it. In a serious case drastic measures must be taken.

The robbers are hindered and home bees assisted by the use of an entrance guard. Probably the best guard is a small plate of clean glass 3 or 4 inches longer than the entrance, put to stand on the flight board and to lean against the hive, giving openings at both ends. Bees entering are hindered and confused, and this assists the guards in their legitimate business.

1328. Alternatively, if the stock is weak the entrance may be guarded with loosely piled straw or green stuff, through which the bees have to scramble. The robbers may be deterred also by the use of a garden syringe, or by the use of a carbolic cloth hung down the front of the hive, or by an automatic smoker placed to blow smoke across but not into the hive entrance. A fly repellant may be sprayed over all cracks and on the front of the hive.

1329. In a bad case, robbing may be checked by an exchange, placing the robbed hive on the stand of the robber and vice versa. If the robbed stock is a weak one it may be closed and moved to a cool place and replaced by an empty hive.

If the robber stock is known, remove its queen to a nucleus. This will cause sufficient consternation to stop robbing.

1330. Another device having immediate effect is to put a ventilated escape board under the robbed hive so that the robbers can leave but cannot enter again. The home bees will also be kept out but they will accumulate beneath the board and may be let in when all is quiet towards evening.

1331. The small bee-keeper may check a case of robbing by placing an empty super on top of the robbed stock in place of the cover and a sheet of glass on top. Over this is a board, or the cover is placed on two bricks, so that the glass is exposed, but shaded from the sun. The robbers make for the glass. In a day or two the robbers and home bees get accustomed to the glass and demoralized robbers mostly stay in the hive with the robbed bees. The author has not had occasion to try this plan, but it sounds good.

Package Bees

Use of Package Bees

1332. Package bees are used for (1) making up losses, (2) strengthening weak lots, and (3) to make increase. Loss of stocks through bad wintering renders equipment idle and leaves the bee-keeper with a store of good combs which deteriorate if not put to use. This is just the equipment required for giving package bees a good start. In certain cold regions, especially where heavy and prolonged feeling is required in the early part of the year, the practice is growing of killing off all stocks in the autumn after the harvest has been secured, and making a fresh start in the spring with package bees bought from a warm clime. These may be started in time to build up for the main harvest, being fed at first with syrup, or given stores retained for the purpose.

1333. For milder regions, where an early harvest can be secured, weak stocks may be brought up to great strength by uniting with package bees, and if these are purchased with good queens, requeening is accomplished at the same time.

Bargain with Producer

1334. The value of a package depends upon three factors: the queen, the proportion of young bees, and early delivery. Good service in all three particulars costs money, and if the packages are regularly used for conditions in which one or more of these features is unimportant, a cheaper package should be got by selection or bargaining.

1335. A good queen is not essential if packages are used for use in spring with stocks regularly requeened in autumn, nor for stocks used only for pollination purposes and discarded or sold cheaply.

1336. Where, as in orchard work, the bees are wanted for immediate outdoor work, a large proportion of older bees is not a material defect. In such a case it would be an advantage.

1337. Early delivery is not important where the main harvest is late, but is of the first importance for pollination purposes.

In average districts package bees purchased by the end of April (northern hemisphere) can equal or beat wintered stocks.

1338. Three-pound packages are most used, but for most purposes, if both young bees and early delivery are guaranteed (which involves placing orders early), a 2-lb. package will give better results in proportion to its cost.

Shipment of Package Bees

1339. Package bees have to travel long distances, and to keep down the cost of freight, should be shipped in light containers. The container need not weigh more than 36 oz. with food supply and queen cage, i.e. complete except for the bees. A 28 oz. container, properly constructed, will withstand the journey. The container is generally constructed with at least a wooden top and bottom and four uprights, surrounded with wire gauze and with a few diagonal braces, or boxed round the bottom. Food is provided in a tin can with press top, punctured with a few holes, the can being inserted upside-down through a round hole in the top of the package. The can must be secured against dropping through which may be done by soldering on a square bottom. An escape hole for the bees should be provided in the top, with a cover readily removable but securely fixed.

1340. The queen should preferably be in a cage. The cage may be fixed to a wire attached to the cover of the escape hole.

The bees may be shaken in through a cone-shaped funnel entering where the tin feeder is to be inserted. Twenty per cent. excess should be allowed for loss of weight in transit, and instructions attached to the package.

1341. Package bees are sometimes shipped with a supply of food in pieces of comb, or even on a single frame, but the risk of transmission of disease is increased thereby, and in many districts such packages are not admitted. The tin feeder is better in any case and may contain 1 lb. of syrup per pound of bees.

1342. It is convenient to have a package of a size which admits of its being placed within a hive body. A suitable size is 12 inches long, 9 inches high and 6 inches wide, outside dimensions.

Treatment on Receipt

1343. Well-fed bees are amenable to manipulation, settle down quickly and tend to form clusters for wax formation. They can control temperature readily. Starving bees have the opposite characteristics in all these particulars.

On receipt the package should be taken to a cool place therefore, preferably where the light is not too bright, and given a dose of syrup (**1291**), which may be sprinkled on, or better still; painted on, the gauze with a flat paint-brush.

The same evening the bees should be hived in the position they are to occupy, and the package should be opened only at that spot.

1344. A certain amount of trouble is experienced with package bees due to their superseding their queens, which causes a serious setback. Now the queen is off laying during transit and time must be allowed for her to reach full lay. If, however, the bees are hived on foundation and fed slowly from the commencement, the queen, if undamaged, will not disappoint the workers, as she will occupy the combs as prepared for her. A week later combs can be given as desired and rapid feeding resorted to. Alternatively the bees may be given combs well filled with honey and pollen in the first instance so that the queen is restricted to a small area over one comb in the first instance. In this case syrup feeding is unnecessary and the bees will proceed to turn the stores into bees.

If there is any appreciable flow at the time, provide a super placed over an excluder.

1345. The package is best inserted beside the combs, but may be put in an empty body beneath the one with the combs in. Open the escape hole, remove the queen cage, see that the queen is alive, remove cardboard or other covering over candy, test candy to see that it is soft, and if hard, work a pointed match-stick through it. Place cage on frames and cover all.

1346. If the queen is shipped loose with the bees, she will find her way on to the combs with the bees, but it is not advisable to examine the frames for eggs, or to disturb them, for 6 or 7 days, as there is considerable risk of the queen being balled (**56-8**) at this time.

1347. If frames of brood are available one may be inserted in each hive. This will give the stock an early supply of new young nurses and greatly assist in rapid development of a full brood nest. Moreover, in case of doubt about the queen, the comb supplied should have eggs and should be the first examined when examination is made. If the queen is not functioning, queen cells will be found on this comb. If there are none, the queen is laying somewhere and probably additional brood and eggs will be found on this same comb.

1348. The package bees should be fed on hiving, and daily until they refuse to take any, subject, however, to what is said in **1344**. The feeder should be placed on the combs, not over the package, so as to attract the bees to the combs, especially if the queen is loose.

If used for strengthening weak stocks, the package should be first well established on combs of its own and then united (**1415**).

Driven Bees

1349. Bees from skeps, driven in autumn, are sometimes purchased in packages. They should be purchased preferably at least 6 weeks before the first killing frost is due, to allow time for building up. If used for assisting weak stocks they may be purchased without queens or with old queens. If purchased without queens, they should be hived next the stock requiring them, and may be united after the next day by the newspaper method (**1409-12** and **1415**).

1350. In making up stocks late in the season, with driven bees, it is peculiarly important to secure that there is ample pollen present or there will be a serious setback in the spring.

SECTION XV

MANIPULATIONS

Manipulations Described Elsewhere

1351. Manipulations relating exclusively to queen introduction, laying workers and queen raising are described in Sections I and II. Those relating to wax, foundation and honey, will be found in Sections III, V and VIII. The use of sundry appliances is described in Sections VII and IX. Manipulation relating to swarms, other than swarm control, will be found in Section XII. The several plans involving manipulation for honey production and swarm control are in the preceding Section XIV.

All other manipulations relate to the general management and handling of bees and will be found in the present section. Further guidance is given in the index.

1352. In almost every occupation involving manual work the hands are moved quickly from place to place, but, in manipulating bees, it is essential to cultivate slow, deliberate and steady motions like those seen in a slow-motion picture. Especially should all unnecessary and all quick motions of the hands over the open hive be avoided. The operator should also avoid standing at any time in the line of flight.

1353. In any manipulations involving removal or transfer of frames, it is desirable as far as possible to place the frames again in the same relative positions as they occupied before disturbance. In a normal hive the combs containing brood will be found together, with combs of stores on either side and above, and, in particular, with combs containing pollen next to the brood. Where alteration is necessary, the arrangement left should be one conforming to the habits of the bees.

Examining a Frame of Bees

1354. If a well-loaded comb in a frame be held in a horizontal position in warm weather, or on removal from a warm hive, there is danger of its collapsing, and even if well wired it may at least become warped; furthermore, thin nectar may fall from the cells, and there is even risk also of the queen falling. In handling a frame under such circumstances, therefore, for examination of the comb or the bees on it, it is good to keep the comb always vertical or nearly so. Wired combs, not containing thin nectar can be handled in any position, but should be held over the hive if a queen is on them.

1355. To avoid the horizontal position proceed as in Plate VIII. The frame is lifted from the hive by the lugs at the top corners, held one in each hand. One side, say side 1, is towards the bee-keeper. To examine the other side, the frame should be up-ended so that the top bar is vertical by raising one hand and lowering the other. The frame may now be rotated, without tilting the comb, through half a circle round the top bar as axis. If the top bar be then brought level again, the frame will stand above the top bar with side 2 exposed to view, but upside down. The action is reversed before replacing the frame.

Examining the Combs in a Brood Chamber

1356. It is customary when the combs do not fill the brood chamber to employ a division board to close up the space at the side, and to leave just cosy room for this board when the chamber is full.

The first step then is the removal of the board, after which the combs maybe freed for examination, one at a time.

1357. As, however, the outside combs generally contain mainly honey and pollen or, alternatively, may be mainly empty, it is not essential to use a division board. An outside comb may then be removed before any have to be examined with much brood or carrying many bees.

1358. If a dummy is substituted for a division board (**686-7**), and especially one arranged to give a bee space next the hive wall as well as at the edges, it is much more easily removed than a division board, as the bees cannot fasten it at the sides or edges.

1359. A hive-tool should be used as a lever to loosen without shock the dummy or division board before removal, as well as any frames that may be securely fastened.

1360. A comb removed may be stood on the ground, on end, leaning against the hive, or placed in a spare chamber or in an inverted deep cover. It is useful to have a frame support (**879**) on which the first frame or two removed may be rested, giving more room for handling the remainder; moreover, any particular comb may be placed on it as, for example, one carrying the queen.

Lift and replace all combs carefully, so as to avoid crushing bees against the sides of the hive.

Replace all combs in their original position unless there is good reason for re-arrangement (**1353**).

Shaking Bees off a Comb

1361. Hold the comb over the hive or over the alighting board, if large enough, and jerk it suddenly and sharply downwards. If the lugs are held loosely the frame will be struck by the thumbs on the motion starting, and stopped suddenly by the fingers as the motion ends. This double jerk is very effective in dislodging the bees but scarcely possible except with frames having the long lugs employed in the British Isles. Most bees are, however, readily dislodged by a sudden downward jerk of a frame tightly gripped. If the bees are stubborn or a few remain secure, hold the frame by one end and strike the other smartly with the side of the hand, then strike the other end in a similar manner.

Bees in an upper chamber may be shaken off the combs into a lower chamber in a simple manner. Remove an end comb and shake the bees off it, then take the combs one at a time, lift about two inches, then drop sharply on to the runners, moving along each comb as emptied to make room for shaking the next.

1362. Do not shake young bees or the queen where they may fall to the ground. Shaking a queen in full lay is somewhat rough beekeeping, and unnecessary.

Brushing Bees off a Comb

1363. A large feather, such as a goose feather, or a bee brush should be used, the bees being brushed downwards off the comb, so that they are scooped off, being approached by their heads. Brushing from behind irritates them. Bees sometimes become entangled in the hairs of a bee brush; for this reason a large feather is to be preferred; its use causes less commotion than shaking and its general use is recommended. Never shake a comb which has or may have a queen cell you may want to keep.

Destroying all Queen Cells

1364. This is a manipulation frequently required, the purpose for which is utterly frustrated if a single cell is missed. The worst cell to leave is that undersized ill-placed cell which the bees so readily cover without seeming to do so. The only way to ensure success is to shake or brush all the bees off every comb in succession. All cells should be cut away as exposed, it being sufficient to make a considerable gash in each with a knife, so as to damage the base of the cell and, generally, also the larva or pupa within. If the base is damaged no repair is possible without the removal of the contents.

1365. Use a spare body if possible. Remove to one side the body with combs to be treated, and place the spare body on the stand. Treat the combs one at a time over this spare body, placing them in it as treated and in their original positions. If no spare body is available remove three combs at one side temporarily. Shake or brush the next comb over the space so left and place it next the outer wall. Continue with the others in turn. Finally, move the lot in the box bodily to their original position and shake or brush the remaining three over them, replacing them before replacement. If this is done over a body placed on a floor board the queen should not be lost or injured, but it is as well to look out for her.

It is not often necessary to perform this operation at a time when bees are likely to be bad-tempered, but it is best carried out when the bees are flying strongly and they should be well subdued first.

Destroying all Queen Cells but One

1366. This may apply to a hive or to a particular comb. The bees should be brushed (**1363**) off the comb carrying the cell to be saved, to ensure that no other and inferior cells are left. They may be shaken or brushed (**1361-2**) off the remaining combs.

Changing the Strain Throughout the Apiary

1367. One or, better, two queens of different parentage of the desired strain should be purchased. They should be "breeding" queens or "selected tested" (**307**) and used mainly for breeding purposes. It is desirable that they should not be heavily worked at brood production, as it will be seen below that their services will be required for two seasons as queen raisers at their best.

New queens must be raised (see Section II) the same year from these selected queens for every hive in the apiary, so that in the following spring there are no queens of any other strain present. Most of the virgins so raised will, however, have been mated with local drones and will· be giving progeny of mixed parentage.

1368. In the next season, however, the original strains of drones will have disappeared and all new drones will be of the selected strain. In this second season, therefore, a fresh lot of queens should be raised from the original selected breeding queens, and they will become mated with drones of the same strain. Do not raise queens from any of the locally mated queens. If any queen remains, not a daughter of the desired strain, it will be necessary to suppress her drones absolutely, during the mating period, by the use of a drone trap (**894**).

The above procedure may be used with any specially selected queens locally bred, noting the underlying principle that, for the production of bees truly of a given strain, it is essential that both the queens and the drones which have fertilized them should be of that strain.

1369. If the apiary is not large the change can be made in one season, one selected queen purchased early being used to produce large numbers of drones by the use of drone-comb, and all other drones being suppressed (**894**) until after mating-time.

1370. It should be noted that if a single queen is used for changing the strain, or even two queens, daughter of the same parent, in the end every single bee, queen worker and drone, will be the descendants of the same parents. Such close inbreeding is not to be preferred, so it is better to use two queens of different parentage from the same strain. The word strain is here used of bees having the same character and common parentage probably only a few generations back.

The drone brood should be raised first and supplied to nuclei upon which the queens are mated (see **177** and **178-82**).

1371. It should also be noted that if the first lot of virgins are mated with local drones, the resultant bees may be uncertain in temper and behaviour generally. This is especially likely if the original strain was black and the new one yellow. Nevertheless, these defects will entirely disappear in the progeny of the second lot of queens, as the drones of the first lot contain none of the genetical constitution of the local strain.

Treatment of Combs Containing Dead Bees

1372. The bee-keeper may be called upon to deal with combs containing dead bees, the bees having died during winter through lack of stores, or from cold, through reduction of strength, by loss of queen, or by disease. The first thing to do is to ascertain the cause of death, as a comb of bees in a stock destroyed by American foul brood or Nosema will be in a highly dangerous condition and should be destroyed by fire (see Section XVI). Winter losses through other brood diseases are not to be expected, so, if there are no signs of American foul brood, the combs may be utilized safely. A stock lost through American foul brood will show ample evidence besides dead bees, including many dead adult larvae, shrunk and stuck to the cells and perforated and shrunken cell cappings (see Section XVI).

1372.a Any brood frame in which bees have died is best destroyed by burning. ID

1373. *A stock dying in the hive generally smells foul, owing to some decomposition taking place, but the glue-like smell of American foul brood differs from the putrid smell of decomposed animal matter, and is generally readily distinguished (1507).*

Bees dying from cold and starvation will be found in numbers lying in the open cells with their tails outwards.

1374. If free from brood disease the combs may be laid out to air, avoiding exposure of any stores. If the combs are dry it is frequently possible to shake out most of the bees by holding the frame by the sides and striking the ends alternately and sharply on any projecting ledge.

1375. If there are not many bees in the combs and they are worth saving, they may be given, one at a time, to strong stocks to clear up. If; however, there are many dead bees, they should be rubbed off with a stiff brush or wire comb. This involves destruction of cell walls, but if the mid-rib and wall bases are retained the bees will soon build up fresh walls.

Re-arrangement of Brood Combs before the Main Honey Flow

1376. During a good honey flow the activity of nurse bees is diverted to storing and gathering, and the brood nest is gradually reduced. If, however, the reduction of breeding is forced by a rapid intake of honey in the brood chamber, one has a condition of prosperity, lack of breeding space and a sudden decrease·of nursing duties, a combination tending to swarming during the honey flow and consequent loss of harvest. The provision of super space alone may not prove sufficient to prevent swarming. Danger exists when the queen happens to be working in the centre of the brood nest and the outside combs cold-way (**647**), or back combs warm-way, are occupied with advanced brood at the time a good flow commences. This condition is likely to occur in an aggravated form in case a good flow follows an expansion of the brood nest occurring 18 to 21 days before the flow. The bees in the outermost combs will emerge, followed by those in the combs next them, so that the harvesters have excellent opportunity to occupy the outer combs and to crowd in on the queen.

1377. If this condition is found in a single-brood chamber the order of the combs, cold-way, may be reversed by lifting out each half of the brood combs and returning them reversed, so that the outermost combs with advanced brood are in the centre and those with eggs are on the outside, leaving, however, the pollen-filled store combs outside next the brood nest. With warm-way reverse all the combs *en bloc*.

1378. If some intermediate condition is found, the combs may all be re-arranged, at least to an extent to secure that the outermost combs have eggs and the youngest brood in them. This secures that the harvesters cannot use the outermost combs for nearly 3 weeks, and by that time the queen will be working again close up to them.

1379. If a double-brood chamber is in use, the combs not being large enough to give brood space in a single chamber, the same idea may be applied, save that it is desirable to keep the queen working in the upper chamber.

1380. If combs of honey have .been accumulated in the brood nest prior to an important honey flow, they should be removed and put in a super between empty combs or frames of foundation, the latter preferably between sealed combs, so that they may be drawn out uniformly and flat. The vacant spaces in the brood nest are made good with empty combs or frames of foundation, ensuring, however, that there are at least two occupied brood combs on either side of any frame or comb inserted amongst the brood.

Crowding of Bees on Eve of Honey Flow

1381. The aim is to force the bees to work in the supers. Really strong stocks usually require no such forcing. In working for section honey in localities where nights are cold some crowding is generally necessary. In such a locality it is important to secure the supers against loss of heat, especially at top **(714)**, and to use bait combs, etc. **(950)**.

1382. If, just on the eve of a honey flow, the bees are crowded into one brood chamber, re-arrange as in **1376-8**. A useful check to swarming may be obtained by putting a shallow super of frames, fitted with foundation only, beneath the brood chamber. If all goes well, the bees will work above, but if the flow is delayed and the brood nest becomes crowded, some relief is afforded by the extra space below. In these circumstances the shallow combs will be drawn out and utilized later for brood, or as a temporary resting-place for incoming nectar.

1383. In working with a single-brood chamber, the reduction of brood space is not usually practiced, the bee-keeper relying upon bringing his bees to full strength before the harvest, but reduction can be made by the use of dummies or division boards and by leaving combs of honey in the brood chamber.

Getting Comb Honey from Shallow Frames

1384. The bee-keeper with a few hives obtaining honey mainly for private use, may get comb honey more economically in shallow frames than can be done with sections. He is then saved both the labour of extraction and that attached to the production of good sections, and effects an economy in apparatus and material.

1385. He should use shallow frames without wires, fitted either with full sheets of thin foundation, as used for sections, or with starters only. The bees will take to them more readily than to sections. The shallow frames with sealed comb may be stored in a shallow super, indoors, closed with grease-proof paper or otherwise, against robbers and moths, and the combs may be cut up as required.

The bee-keeper, working for extracted honey on any scale, may of course use the above method to secure a few combs for his own use. In that case, it is best to commence harvesting with built-out combs, and to insert the super of foundation or starters for comb honey under one in which the bees are already working.

1386. With the introduction of a new method of packing comb **(538)** this method shows promise of considerable commercial value, as there can be no question that comb honey can be produced more economically in shallow frames than in sections.

Cut-comb production here. Ed.

Preparation of Section Racks

1387. When a section rack has been filled with sections It is a good plan to brush it over, top and bottom, with paraffin wax. This protects the exposed faces from staining and greatly facilitates cleaning afterwards. A soft thin brush 2 or 3 inches wide is best for the job, and the wax should be heated over a water bath to a temperature of 180^0 to 190 F. (82^0 to $88°$ C.).

Supering with a Single Brood Chamber for Extracted Honey (S)

1388. Arising out of certain difficulties in working for section honey it has been customary to add a super below one already in use. Experiment supported by experience shows that there is an actual gain in honey production by following the practice of adding all supers of empty combs above those previously furnished.

1389. The same practice should be followed also in adding supers containing foundation only in this case save for the additional labour there is some advantage in placing such a super in the first instance beneath the brood chamber, where it acts as a check on swarming. It should be moved up after the bees are well started at work on it.

1390. For use of excluders see **702-7**. Any shallow combs found to contain pollen should be retained for use as winter stores above the brood nest.

Supering with a Single Brood Chamber for Section Honey (S)

1391. The brood chamber may be prepared as in **1376-83**. If the first flow is not rapid a start may be made with a super (*b*) of shallow frames. This is lifted later above the first section rack as in (*c*), as soon as the flow has developed to a rapid flow.

To avoid travel stains and to assist in securing their occupation an additional rack should be inserted below those in use. Racks are removed as completed.

1392. If no shallow frames are used, help is obtained by starting the bees with bait sections in the first super. Bait sections are sections having partly drawn comb left over from a previous year. Sometimes a start is obtained by the use of a shallow frame designed to fit beside the sections, the bees much preferring the large surface it offers. If the bees have to be crowded into the first rack it is desirable to see that they do not lack room when once started.

1393.   Queen excluders are often used in working for section honey. They constitute one more hindrance to ready work in the sections. If the stocks are strong, and especially if deep frames are used, the excluder may be dispensed with. Two inches minimum of honey over the brood nest is a good barrier if there is no drone comb above. See also **702-7**.

*Adding an Additional Brood Chamber (**D** and **S** +)*

1394. When there is risk of overcrowding (**935**), another brood chamber should be added. The way this is done depends upon the weather and time of year and upon whether drawn combs are available or only frames with foundation.

1395. The queen always tends to work upwards as space becomes available, but in the spring, say when plum is in flower, or at any time when nights are cold, it is well to add the chamber below that in use.

When it is well occupied, say three weeks later, it may be brought above the other, thus not only giving ample room, but giving the bees a chance to complete combs to the bottom bars of the frame. This is very desirable especially when foundation is being drawn out, as it secures combs full of worker cells, but should not be done until at least one comb is occupied down to the bottom or the colony will be divided (see also **351**).

1396. If warmer nights have already arrived the additional chamber may be placed on top, but it is undesirable to place a chamber of foundation on top unless there is a steady flow on at the time, or feeding is resorted to. If there is a good flow on, add a super of combs over the foundation. If both finished brood combs and frames of foundation are available they may be alternated to advantage. The bees occupy them readily and the device is found to have merit in discouraging preparations for swarming.

1397. If only a shallow chamber is to be used to expand the brood nest it may be put on top, whether of combs or a mixture of dark combs and foundation. If no combs are available it is an advantage to have the foundation partly drawn out below as advised in **1389**.

1398. If, however, a chamber of foundation is given below the brood nest at a time when the bees are not ready to. expand into it (**935** and **1394-8**), or if it is left unfinished too long, the bees will rob the foundation for wax, cutting away the corners and edges, making useless pop-holes and places where unnecessary drone cells and hidden queen cells are liable to be built later.

Feed as in **1396** if necessary.

1399. An alternative plan of adding a second chamber is to transfer half the combs, taken as a batch from one end, into the second chamber, keeping them in the same position that they had originally, then turn this chamber end for end before putting it on top so that the two lots are one over the other. Fill up both chambers with combs or foundation. If foundation has to be used, put the lower chamber on top as soon as the bees are seen working on the outermost frame above.

Feed if necessary if you do not want the foundation robbed, as in **1398**.

*Supering with a Double Brood Chamber for Extracted Honey (**D**)*

1400. Following the addition of brood chambers for expansion of the brood nest without hindrance, as in **1394-8**, additional deep or shallow chambers are added for the harvest in the same manner as with a single-brood chamber, i.e. placed one by one on top without disturbing what is below.

1401. Combs of honey found in the brood chamber during the harvesting season, except those holding the pollen supply required for breeding, may be removed as found and put with others in the supers.

If wide-spaced combs are used generally for supering, see that there are enough combs of standard spacing to furnish winter stores and collect these into one body as occasion arises.

*Supering with a Double Brood Chamber for Section Honey (**D**)*

1402. With two or more brood chambers in use, the bees will need crowding somewhat as in **1381-3**, unless the flying bees of other stocks are added as in **1418-23**. Supering is then conducted as in **1391-3**. For use of excluders see **702-7**.

1403. If a swarm control method is employed, involving the use of an upper chamber, this chamber generally becomes ultimately part of the super space for honey. When working for sections proceed as in **1243**.

Uniting

General

1404. Any bees may be united in one colony if the conditions are made favourable by the bee-keeper. In general, bees in a similar condition, such as two swarms or two established stocks, may be united without difficulty, but if the conditions are dissimilar more care is needed to secure success, e.g. in uniting driven bees to an 'established stock.

1405. A colony objects to the entry of strange bees. The recognition of a stranger may be in part governed by scent, but it is largely determined by the attitude of the stranger, who is generally a robber. A bee fully loaded, reaching the wrong hive, or a young bee, is generally freely admitted. Such bees seek entrance as friends.

1406. *Unless nectar is coming in, the bees should he fed before uniting.* It is generally desirable to secure that the bees to be united shall have the same scent.

It is also generally desirable to secure that the added bees will note their new location.

It should be noted that late in the season, after drones are destroyed, bees are particularly liable to ball their queen (**56-8**). Uniting should thus be conducted without disturbing the queen. The newspaper method (**1409-13**) is probably the best method to use.

Uniting Scented Bees

1407. For many years after movable frame hives were introduced, it was the practice to reduce two lots to be united, to the same scent, by separating the combs and shaking flour over the bees, or by spraying them with scent. It is probable that the effectiveness of the method was largely connected with the separation of the combs and exposure of the bees to the light. Flour and scent are now much less used. When flour or scent is used, the stocks must first be brought together (**627**), if .not already side by side. After uniting the one hive is best placed midway between the positions previously occupied by the two.

Uniting after Exposure to Light

1408. Bees in a not too dissimilar condition may be readily united if thoroughly subdued and all combs separated so that all the bees are freely exposed to the light. The combs of the two lots should then be alternated, arranging to keep their general arrangement otherwise undisturbed (**1353**).

This method necessitates bringing the two lots near together, as in the last plan above. Trouble occurs if combs are not straight and fairly even in thickness.

Newspaper Method

1409. This method is the most widely applicable. The two lots to be united are placed, one on top of the other, with a newspaper between, the paper being pierced at several places with a pointed instrument so as to start holes which the bees must enlarge before they can pass through. Uniting takes some hours to effect and the bees meanwhile acquire the same scent. At a later date the two lots of combs may be re-arranged as may be found desirable. If, war-time newspapers are used, use two thicknesses.

1410. The operation is most readily carried out after the bees have settled down at the end of the day. It is desirable to take steps to secure that the bees in the upper box shall note their new location. They will generally do so after such a disturbance, but to avoid any loss of bees and undesirable drifting, it is good to leave a plate of glass over the entrance (**621**). Uniting is completed in about 12 hours.

1411. If the moved lot is put on top the temporary imprisonment and disturbance is generally sufficient to secure this. If, however, the moved lot is to be placed below, it is a good plan to confine them first for 24 hours, well ventilated, in a cool place, and then move them straight to the new site before releasing them. An entrance guard (**621**) also helps.

1412. The following notes will assist in selecting which hive should go on top, but the lower one should remain on its own stand, even if this involves two movements instead of one:
(1)   When bringing a weaker to a stronger stock, put the weaker stock on top.
(2)   When bringing a queenless lot to a normal stock, put the normal stock on top and use entrance guard.
(3)   When uniting two lots much alike, put the moved stock on top.

Use of Super Clearer for Uniting

1413. If a ventilated super clearer (**697**), such as the Shepherd, is available, in which there is a double screen arranged so that the bees cannot reach each other, and provision, accessible from without, for closing all exits, it may be used instead of newspaper, but the slide must remain closed for 2 or 3 days. The loss, however, due to this temporary imprisonment is generally greater than the loss of bee labour involved in the destruction and removal of the newspaper.

Uniting Swarms

1414. Swarms may be readily united when hiving by shaking them together, or part of a strong swarm may thus be united to a weaker one. If it be desired to choose which queen should be kept, it is well to hive the two lots side by side and search for the queen to be destroyed when they have settled down, afterwards uniting; but a sharp-eyed bee-keeper will catch the undesirable queen as she goes in, shaking the swarm containing her rather further from the entrance than the other lot, first waiting until the other lot have started running in.

Uniting Driven Bees to an Established Stock

1415. It is unwise to shake even well-fed driven bees into, or before, an established stock; they will probably fly up and be lost. They should be hived on combs first. The stock to receive them can probably spare a few combs for the purpose. Take the combs one at a time from the established lot, shaking back the bees. Choose some outside ones with some stores and at least one with eggs. Transfer the driven bees to these combs (**1215-19**), and feed if necessary. Stand this lot next the established stock and unite 3 or more days later. Before uniting, examine the combs which contained eggs. If queen cells are being raised, destroy them. These show that there is no queen. If no cells are raised there is almost certainly a queen present, which should be destroyed before uniting, unless required. If she is black and not easy to find, see **53**.

Adding Young Bees to a Stock or Nucleus

1416. Unless there is nectar coming in, the stock to be supplied and the stocks furnishing the supply, should be given feeders a day in advance. The manipulation can be carried out at any time, but the middle of the day is best. The bees in both lots are subdued in the usual manner (**853-4**). Combs carrying young bees are then lightly shaken over the hive from which they are taken, so as to dislodge the old bees. Youngsters will hold on. Then shake the latter over the combs of the hive to receive them, or better still, brush them in. If old bees are shaken in with the youngsters, most of them will fly back, some at once and others on making a flight from the entrance. Bees under about a fortnight old will remain where put. If the alighting board is wide enough, the bees may be shaken on to it.

1417. This manipulation is particularly useful in adding bees to a nucleus which may be unable to cover a comb of emerging bees. When nights are not cold, however, combs of bees ready or nearly ready to emerge, may be brushed free of adhering bees and inserted in stocks requiring them. Such combs may be shaken, but it is well not to thump them.

Adding Flying Bees to a Stock

1418. This is an important manipulation in securing a harvest and in checking swarming. From time to time special hives have been designed to facilitate this operation, but they limit the user in his choice of stocks for the purpose, and as the manipulation may be done with standard hives there is no strong case for special hives. A stock lacking in flying bees is not likely to swarm. A stock having a large proportion of flying bees at the beginning of a honey flow diverts its activities to honey-gathering, bees of a nursing age being utilized as receivers and distributors of incoming stores, and breeding is checked. It is stocks intermediate in character that are most difficult to divert to honey gathering and that are most likely to swarm. Further, so long as there is a normal condition in the brood chamber and a fertile queen there, the more harvesters the better, and it is frequently the case, especially where flows are of short duration, that more surplus may be got by using only a few of the hives, each supplied with flying bees from several others, than if a majority of the hives are used for harvesting. Ample supers must, of course, be given to those used for harvesting.

1419. The manipulation depends upon the fact that if a hive be moved so that its entrance is several feet from its original position, the flying bees will have great difficulty in finding it, and if returning with honey will be readily accepted by, and readily enter, any hive having an entrance near the original position.

1420. The manipulation is best carried out about the time a flow is expected. Swarming is further checked by removing combs of emerging bees from the hive to receive flying bees and giving them to the hive or hives which are to lose bees. These combs may be exchanged if desired for combs of eggs or young brood.

1421. The hive to receive fliers need not be the strongest. It is convenient that it should have a single body of brood arranged at the start with the youngest brood at the outside (**1376-80**).

See that the entrance of the harvesting colony is ample for the large number of fliers and that ventilation is adequate generally, and above all, see that supers are added in advance of requirements.

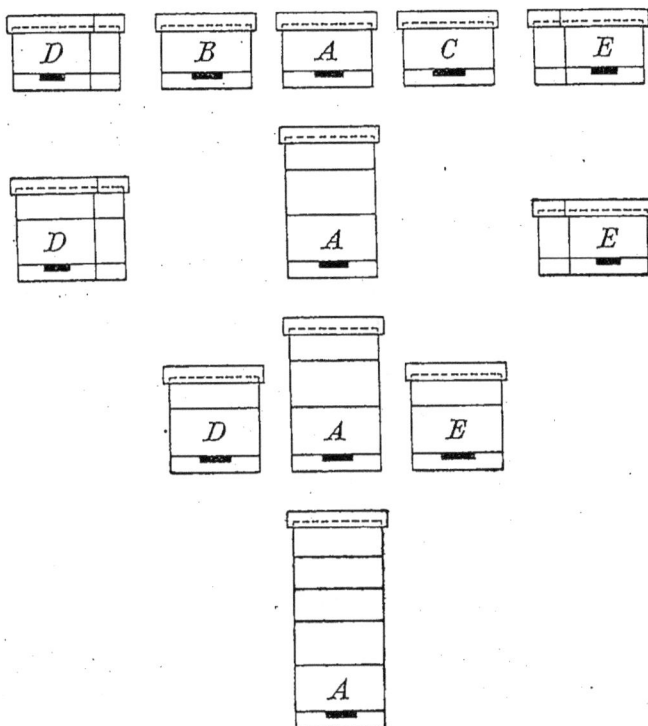

FIG. 33·-ADDING FLYING BEES TO A STOCK-Multiple HIVES.

1422. (*a*) Let there be a hive *A* (Fig. 33) ready to receive flying bees, and a hive or hives, *B* and *C*, already brought near to it, to give bees, and furthermore remote hives *D* and *E* to be used also. *B* and *C* are removed to new locations and 2 days later *D* and *E* are moved a little nearer. *D* and *E* are moved progressively (**627**) until close up and the next day removed to new locations. In this way, if desired, *A* may receive the flying bees of five hives.

1423. (*b*) A pair of hives are worked thus. Hive *B* (Fig. 34) is first moved backwards 2 or 3 feet. A day or two later it is moved across to position *B`* on the other side, and still behind, *A*. This ensures the transfer of all flying bees to *A*. Ten days or a fortnight later, the hive at *B`* is returned to *B* on the back line. It may be moved this way repeatedly while there is any flow, keeping *A* charged with a large majority of flying bees, but see that *B* does not become short of food by giving an occasional comb, or a super, to it from *A*, as may be convenient. The bees in *B* will also require water, or thin syrup, if brood production is not to be checked.

Manipulations giving Increase

When required

1424. A moderate increase is required:
(*a*)   To make up for occasional winter losses;
(*b*)   To make up for losses through disease;
(*c*)   If with growing experience, the owner feels able to handle a few more hives.
         Rapid increase is required:
(*d*)   When, starting with a few stocks, the bee-keeper is ready to launch out on a much larger

scale;

(e)   When, the apiary being as large as the district will support, it is desired to start an out apiary.

(f)   When bees are bred for sale.

Natural v. Artificial Increase

1425. Increase by uncontrolled swarming frequently prevents the taking of large crops even if all swarms are caught and hived (**1188-99**). The modern bee-keeper has swarming and increase under control to serve his ends. Increase by natural swarming hinders the development of a (so-called} non-swarming strain (see also **192-200**).

Giving Queens v. Queen Cells

1426. Whichever course is pursued, reference should be made to the section on Queen Raising, particularly to "Parentage" (**192-200**), "Requeening" (**78-83**) and "Queen Introduction" (**84-105**).

1427. Increase always involves establishment of a new colony by the provision of both bees and a queen. If a fertile queen can be given, the bees are engaged at once in tending her and raising brood. If a ripe queen cell is given, there must be an interval of probably 10 days or more before eggs are laid, whereas if the bees are left to raise a queen from the egg or young larvae, the delay will amount to, say, 20 days or more.

1428. Such a delay or check on brood raising is a serious matter if it extends to within 5 or 6 weeks of a good flow, as it then represents a serious loss of harvesters during the harvest.

1429. If, however, the check operates to hinder the production of harvesters during a lull period (**1009-10**), it can prove a source of positive economy. The check may also be utilized to hinder swarming if suitably timed.

1430. Any manipulation that involves the raising of queens in any but strong and prosperous colonies is not recommended.

Prosperity combined with favourable temperatures, such as a strong stock can secure, is what is required. Good individual queens have indeed been raised in well-found nuclei of 3 and 4 combs, having plenty of nurse bees, in the height of summer.

Feeding for Increase

1431. In securing the rapid development of a small colony, indeed, in some instances, to secure any development at all, it is necessary to feed liberally and especially so if there is not an abundance of harvesters present and a honey flow on.

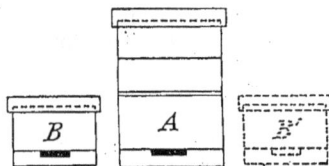

Positions B & B' are somewhat behind A

FIG. 34.—ADDING FLYING BEES TO A STOCK—PAIR OF HIVES.

1432. Now a colony well supplied with incoming stores, and having a fertile queen, is one well fitted for the economical production of comb. Do not give foundation, however, to a queenless colony.

1433. Sugar syrup alone is not a good diet for raising vigorous bees. It should be supplemented with honey if there is no honey flow, and with pollen as well if there are but few harvesters available. Sugar is good for comb-building, however (see also **1296**).

1434. If drawn combs are available, they may be used, but foundation is generally preferable for the unoccupied frames. A rapidly expanding small colony is not likely to build drone comb. A small colony must be kept warm. A frame feeder is useful to this end, all food being warmed before use (**884-7**).

Avoiding Loss of Harvest

1435. Either slow or rapid increase may be required. With either, there must be some loss of harvest. This is practically eliminated with slow increase by choosing a time (**1009-10**) when brood and a queen or queen cell can be spared. Indeed, increase can be worked in to advantage with methods of hindering swarming (**1216-18** and **1473**).

1436. For rapid increase, by choosing the right time, a few stocks may be made to give the entire increase (**1456-8** and **1466-7**).

Methods other than the most economical from the standpoint of loss of harvest are described, as they may be convenient on occasion.

Seasonal Differences

1437. For success in obtaining increase in the spring, it is necessary to bring stocks to full strength at an early date. Drones must be raised in good time, drone eggs being laid at least 37 days before the new queens are to be fertilized. Warm days are required for successful fertilization. Any stock robbed of bees for increase must not be so handled that it cannot reach full strength before the main harvest.

1438. In localities where there is a long gap between a spring flow and the main flow, increase may be made early, even in the spring.

1439. In cold localities where it is impracticable to divide stocks in the spring, spring increase may be made by the purchase of package bees from a warmer region (**1332-48**). If bought early (i.e. if ordered early) such package bees may secure surplus from a late spring flow, and may develop without difficulty for an early summer flow, or even for early summer division for a late flow, thus giving further increase.

1440. Early swarms may be purchased and will give a good account of themselves, but will need requeening later in the year.

1441. The summer is the most favourable time for making increase. Note particularly the observations in **1009-10**, *re* correct times for hindering or stimulating brood rearing. With the above in mind, increase may be made in summer with least loss and trouble, especially in the matter of queen raising, and indeed, with the beginner the difficulty is to avoid making increase.

1442. If, however, spare queens are raised in the late summer, they will then be available for autumn increase, which may then be made after the business of summer-time of securing the maximum harvest is completed.

Autumn increase, however, requires skill, good judgment and closer attention.

1443. Numerous manipulations are described below. Those for autumn use are given last and those for spring use are placed first, but strict order cannot be followed.

All the methods described, except that in **1448**, are applicable to single or double brood chamber management, although, for simplification, a single brood chamber is generally shown as the starting-point. If the bees already occupy more than one, the treatment is the same.

Increase by Robbing Stocks

1444. When a queen or ripe queen cell is available, moderate increase may be made by robbing a comb of advanced brood covered with bees from each of several hives and giving them the ripe queen cell or the queen (**84-101**). The bees must be well subdued before removal and the combs spaced apart in the new hive (**1408**) until, say, four or five combs are in, then add if necessary a comb of stores, close up, add spare combs or frames of foundation at one side, guard entrance so that bees note their

location (**621**), pack up warmly and feed as required.. This procedure secures that there shall be enough bees to cover all brood in all the hives, and there is no appreciable extension of the space to be kept warm at night.

If there is a strong stock available, well advanced in brood rearing, the new lot may be installed in its place to receive its flying bees, and the strong stock be removed to a new location, with its entrance reduced and ample packing at top. This plan has the advantage of tending to equalize stocks.

1445. When robbing stocks of a frame of bees each, insert a frame of foundation on the outside of the brood nest, thus making good the gap.

Spring Division

1446. This manipulation is used where the main harvest is late in the year, brood production being forced on the early flow. The manipulation is commenced, say 8 weeks before the main flow is due, and it is usual to unite stocks again after the flow, as there is not much time left then for them to build up again for wintering. The queen should be prolific. Procedures are given in **1447** and **1448**.

1447. As soon as one body *A* (Fig. 35) is fully filled with brood, it is stood to one side, say on its cover, and a spare body *B*, filled with spare combs or frames with foundation, is placed on the floor board. One comb or frame is exchanged for the comb in *A* containing the queen. An excluder is placed on *B* and *A* is placed on top after searching for queen cells.

Five days later *A* is again examined for queen cells which are destroyed, or alternatively, one may be retained.

Five days later again the bodies are separated *A* being removed in the evening to a new stand, so that *B* gets the flying bees. A queen or queen cell is given to *A* unless one had been retained in it.

Swarming is checked as *A* is weak in brood and *B* in flying bees.

Both stocks are then allowed to build up for the harvest, and fed if necessary to maintain breeding.

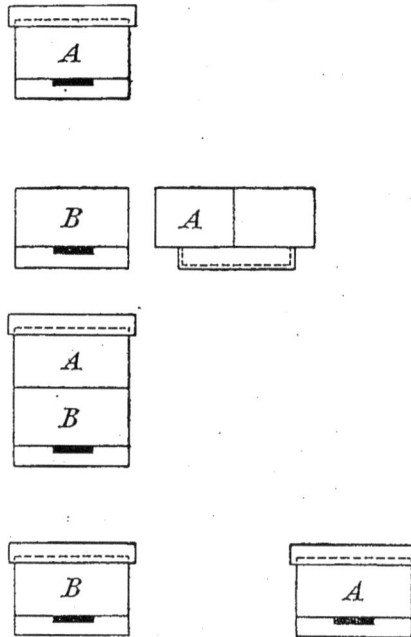

FIG. 35.—SPRING DIVISION.

FIG. 35.-SPRING DIVISION.

Spring Division (Rauchfuss Method) (**D**)

1448. Spring division gives increase without much loss of harvest, if the colony is fully developed well in advance of the main flow. The following simple method combines swarm control with increase. It is not necessary to find the queen, but a new queen is required.

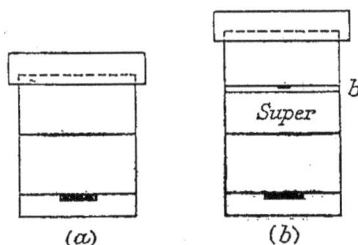

FIG. 36.-RAUCHFUSS METHOD SWARM CONTROL WITH INCREASE.

Note: - It is better to have the entrance in *b* at the side

1449. When the colony is fully expanded, occupying two chambers, as in Fig. 36 (*a*), by procedure detailed in **1394-9**, and ready for the first super, insert the super between the two chambers as in Fig. 36 (*b*), inserting above the super a board giving access to the chamber through excluder. For example, a crown board may be used with a piece of excluder zinc over the feed hole. If desired, an excluder board may be added beneath the super in the usual place, as well.

1450. Eight or nine days later, examine to see in which chamber the queen is laying and remove that chamber, as found, to a new stand to build up a new colony. Destroy all queen cells in the other chamber and requeen·. All flying bees will remain with, or return to, the lot left behind. This portion is, in fact, in good condition to receive a virgin, and one may be used if no laying queen is available.

1451. A new queen is given to ensure a quick build-up, but if the colony has built up early, and there is a considerable gap between early and late flow, one can raise a queen in the lot left behind.

1452. The chamber removed will contain the result of eight or nine days' laying at least and will spare a comb or two of brood which can be exchanged for empty combs in the other chamber when looking for queen cells. To do this put the upper chamber beside the lower or an inverted cover or other base until the exchange has been effected, and then move it to its new site.

Increase with Artificial Swarming

1453. Prepare a spare hive *B* (Fig. 37) with frames of foundation, or some empty combs and some frames (for wax building) and place next to a strong stock *A*. Insert in the middle of *B* the comb from *A*, carrying the queen, exchanging it for an empty comb or frame of foundation. Remove *A* to a new site *A'* and place *B* where *A* stood. If done in good flying weather practically all the flying bees in *A* will join the old queen in *B*. If *A* has no queen cells it must be given a queen cell or queen (**84-99**). Stock *A* will benefit by a supply of weak syrup, as there will be a lack of water-gatherers for a day or two.

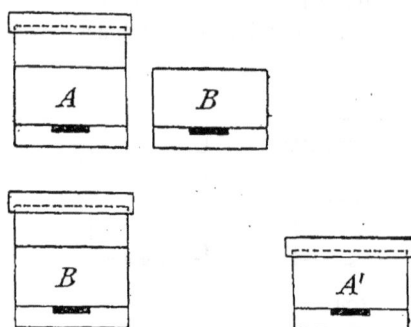

FIG. 37.-INCREASE BY ARTIFICIAL SWARMING.

Making Three Stocks from Two

1454. *A* and *C* (Fig. 38) are two strong stocks. *B* is a spare hive of empty combs or frames of foundation. Remove, say, six frames from *B* and lay aside. Remove then five combs of brood from *A*, shaking back the bees into *A*, and place the combs in *B*. Insert in the empty space in *B* one of the frames laid aside, and the remaining five to one side of the combs left in *A*. The empty combs or frames should be placed on one or both sides of the brood in both *A* or *B* but not in the middle of it (**145**). Now remove *C* to a new location *C'*, and place *B* where *C* stood. *C* then gives flying bees to *B* and *A* gives brood. *A* or *B* will require a ripe queen cell or queen.

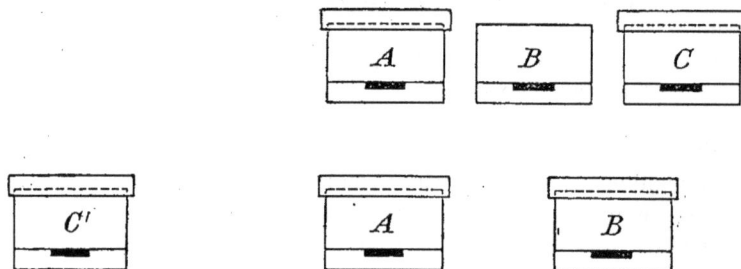

FIG. 38.-MAKING THREE STOCKS FROM TWO.

Multiple Increase

1455. When queens, or ripe queen cells, are available, a stock may be broken up into 2-comb nuclei each in a separate hive, each supplied with combs or frames of foundation and well fed until able to support itself. Take the most mature brood first and 10 days later take the remainder, except two combs. The excess of bees left in the parent stock will make up for the delay in maturing of the youngest brood which is left behind.

Continued Increase from One Stock

1456. If the nuclei above described are assisted by the heat of a strong lot, they develop even more rapidly. This leads to a modification, now to be described, which embodies a form of artificial swarming, and which may be put into operation in late spring as soon as a stock has brood and bees well filling at least one body.

1457. Prepare a spare body with frames of foundation. Exchange two frames of foundation for two frames from the stock containing emerging bees. Place these combs in the middle of the spare body between the frames of foundation and place it upon the stock without any excluder. Cover up warmly, feed if necessary and leave for 3 weeks. If then the queen is not already in the upper body, place her there and remove the lower body to a new stand so that the upper body receives all the flying bees. The lower body removed will require a queen or queen cell and if the latter given there will, of course, be considerable delay in building up again. Unless there is a flow on at the time, give weak syrup to the removed lot.

As soon as the old upper body is full of bees and brood the operation may be repeated, and so on.

1458. Alternatively, if the weather has turned warm and queen cells are available, the contents of the lower body may be divided into nuclei and built up rapidly as in **1483-4**.

Nuclei from Stock Superseding

1459. A strong stock superseding its queen may be conveniently divided into several nuclei, each having one queen cell and built up as in **1483-4**.

Continuous Production of Nuclei

1460. A strong stock may be made to give a 2-comb nucleus every 5 days, if fed between flows, the combs removed being replaced by foundation. One comb should have emerging bees and the other brood, less advanced, so that the stock is not too depleted of prospective nurses.

1461. The nucleus must receive a queen or queen cell and must be fed, preferably by a frame feeder, so that the top may be packed warmly. Add frames of foundation a little in advance of requirements.

1462. It is better in the late summer to use three combs instead

1463. It is desirable to close the entrance with perforated zinc for 24 hours after forming the nucleus, so that it shall not lose too many flying bees, shading the nucleus while closed.

Combining Increase with Queen Raising

1464. Some increase is generally required to make good any losses. A few nuclei may be used for queen raising, and when all queens required have been raised these nuclei may be rapidly built up into stocks.

1465. If these nuclei have been formed in the first instance as a check on swarming by removal of brood combs (**1213-14**), they will be ready in time for dealing with a number of cells from a queen-raising stock and will still be ready for rapid building up into new stocks, an economical arrangement. Any insufficiently built up may be united with each other or with other stocks weak from any cause.

Rapid Autumn Increase

1466. A stock may be built up for winter from even one comb in the 6 or 7 weeks preceding the first killing frost if a young queen commencing to lay is available, but it is more usual to commence with two combs of mature brood well covered with bees. These may be hived with two combs of food and a set of empty combs and well fed. There must, of course, be pollen available, and any late autumn flow is helpful. Stores for winter must be added later.

1467. To force increase from a single comb conservation of heat is essential, and as most of the bees will be required in the hive, a continuous supply of food must be maintained without fail. Increase can be made, however, with a single comb commencing with an empty comb on either side of the brood comb, a dummy feeder, a division board, warm packing at top and a very small entrance. The entrance should be closed with perforated zinc for 24 hours after making up the nucleus so that flying bees will note their new location. In 3 weeks insert between the combs a frame filled only half-way down with foundation so that the cluster is not broken. Insert another similar frame every 5 or 6 days and finally two empty combs, one on either side.

The frame feeder must be kept going all the time preferably with honey and sugar, and stores will be required for the winter and early spring.

Use of Pollen Combs and Driven Bees when Making Increase

1468. When extracting, be sure to save any combs containing pollen. They are most valuable for wintering recently made colonies, as it is easy to provide syrup, but difficult to provide a new colony with enough pollen for early brood development.

Such combs may be used to great advantage, when hiving driven bees. Put them on the outside of the chamber or above, when making up with combs to receive the driven bees, and give ample stores for wintering.

Again queen-right nuclei can be built into stocks late in the season if plenty of pollen is available, and can be helped out with surplus brood combs from any stray stock, for example from upper chambers removed before all brood has hatched out.

Using Old Bees for Autumn Increase

1469. A week before the end of the main honey flow in late summer, proceed to make an artificial swarm as in **1453**, except that a new queen in a cage must be given to the flying bees, the previous queen being removed with the brood, and combs given, not foundation.

The old bees put in several days' work at the main harvest, and if there is an autumn flow, will build up stores without assistance.

1470. A plan having the same effect is to prepare a 4 or 5comb nucleus of mature brood, place it above an excluder for the rest of the day so that it is well covered with nurse bees and remove it to a new site in the evening, giving a young queen in cage. This lot will require to be well fed.

For this plan the bees left with the original stock will be mostly old bees, and they will need stimulation to raise brood.

Nuclei in Cool Weather

1471. Nuclei must be kept warm to make rapid progress. There may be considerable loss of hot air under the lugs of division boards. There must be no place which the bees cannot readily seal. Felt can be used under the lugs, but if British metal frame ends are used and the division board is properly made, the frame ends should make a close fit with the top of the division board and the top surfaces should be flush all over to receive the quilt.

1472. The nucleus may be hived over the parent or other strong stock, being placed over a ventilated super clearer (**700**) with its through ways closed, or over some other board· having a double screen and a special entrance (**671-2**). This allows warm air to rise from the stock below, but the top entrance must be very small or there will be much loss of heat. Moreover, even with a double screen there may be antagonism between the queens. A floor of thin three-ply wood, framed, with entrance, would let through a lot of heat to the nucleus without letting out warm air at night.

When the nucleus has built up, precaution must be taken on moving it, to avoid loss of flying bees, unless it be decided to let the old stocks have them. They will soon join the old stock if not hindered (see **621**).

Increase when Using Upper Chamber for Swarm Control

1473. When requeening after demareeing, as in **181-111**, after the queen is fertilized, the bees and combs in the top super may be removed to make increase. In this case, unless precaution is taken (**621**), the flying bees will join the stock below. There is no objection to such addition, however, unless the honey flow is over.

The same principle may be used whenever queen cells are raised in an upper chamber in any of the manipulations involving the use of such.

Transferring

General

1474. Moving stocks in hives is dealt with in Section VI. We are here concerned only with transferring bees from one hive or colony to another.

1475. With modern hives with interchangeable frames, stocks of bees are easily moved from hive to hive, but where bees are established on fixed combs, as in skeps and box hives, or in old trees and the roofs of buildings, special steps have to be taken to transfer them to modern hives. Have all preparations made and plans thought out before commencing work. A sharp look-out must be kept for disease.

Bees Transferring Themselves from Skep or Box Hive

1476. The simplest method is to let the bees build downwards into the hive body, and is best applied in the spring when the brood nest is expanding.

1477. Place a hive body, fitted with empty combs, or frames of foundation, where the skep or box hive stood, the entrance occupying as near as may be the position previously occupied by the skep or box. Place a piece of American cloth, or failing this, a piece of thin board, over the top, with a 3-inch hole in the middle. Place the skep or box hive on this, closing up with rag any space at the edges, where the fit may not be good. Cover over with an empty body and a cover, or with some weather-proof material, and leave alone until the brood nest is extended so that the queen is working in the combs below. When she is found so occupied, place a piece of excluder over the hole. Three weeks later all the brood in the old skep or box will have emerged and an escape board or super clearer may be substituted for the excluder so as to get all the bees down below. The skep or box may then be removed and the combs utilized as described in the next section.

Transferring from Skep or Box by Driving

1478. This is the most satisfactory and expeditious method. It is best carried .out in the spring, on a warm day, when there is not much honey coming in. The shade temperature should be about 60° F. or more.

The bees are to be driven from their combs, into an empty skep or box, where they will cluster like a swarm. They are transferred later to their new home like a swarm (**1115-19**).

1479. First subdue the bees with smoke, and then invert the box or skep on a secure foundation. A bucket affords a convenient support for an inverted skep. An empty skep or box is now placed over this, secured to it on the side remote from the operator and tilted and fixed so as to give an opening of not more than 8 inches at the front. See Plate IX, facing p. 331, in which the bucket is replaced by an empty skep with flat top.

1480. Driving irons and staples are used, the irons being about 9 inches long in the straight part with ends bent at right angles for 1½ X 2 inches and sharpened to a point. Skeps are readily secured with these. Boxes take more contriving, as they must be made secure to withstand what follows. It is useful to tie a cloth round so as to close up and cover the joint at the back so that bees do not climb outside.

1481. The combs in the inverted hive should run from front to back so that the bees may readily leave and climb where the upper receptacle touches at the back.

1482. Now drum steadily on the lower hive with both hands, or with two sticks, one each side, so as to jar the bees, but not too hard. The bees become alarmed for the security of the combs and soon begin to climb upwards, forming a cluster in the upper receptacle. Watch for the queen rising. The operation generally takes 10 to 20 minutes.

1483. When all the bees have risen, detach the upper receptacle and place it where the hive is to stand permanently. To ensure that the queen is present a piece of black paper or American cloth may be placed under the hive and examined 10 minutes later for eggs, which the queen will drop. To see the queen go up, watch the dividing line between the hives. She may be caught if desired. If missed, the bees may be driven yet again into a third box, or she may be caught when or after placing in permanent hive.

1484. Now proceed to cut out all the combs and fit them in empty frames, provided they are free from disease.

For removal of combs, knives are used. The most useful is one having a short blade at right angles to the handle. The handle passes between the combs and the short blade cuts them off close to the wall of the skep or box. The blades may be made of strip iron 6 inches X ¾ inch X 1/16 inch, fixed to a long handle, 1½ to 2 inches of the end being bent at right angles and the edge sharpened.

1485. To fit the combs in the frames, lay the empty frame on a flat surface on three tapes. Cut the comb with sufficient straight length at the top to fit against the cross-bar, then tie the tapes so that they pass around the frame and comb from top to bottom. If the comb does not reach the bottom the middle tape may be passed above the top bar and round the bottom of comb, omitting the bottom bar. Cotton tape is frequently used, but the gummed strip paper used for securing parcels is more convenient. The bees will soon secure the combs in the frames and cut away the tapes or paper. Drone brood and drone comb should be cut out.

1486. The combs should be arranged in a hive body in their normal position, brood at centre, pollen next and honey outside. The bees may be hived on them like a swarm (**1115-19**).

1487. The bees may be given access within the hive to odd pieces of comb containing honey, the cappings being bruised. These may be placed in an empty super above a super clearer with the free-way open, and removed for melting when cleared of honey.

If there is a honey flow, postpone driving, as the bees may be drowned in the nectar shaken out. Under these circumstances transfer the flying bees (**1418**) to a hive run for honey and drive the remainder when the heavy flow has ceased.

Combination Method

1488. The above methods may be combined, the bees being driven and hived on foundation, preferably with one frame with a little brood added from another hive. The perforated board or cloth is put on with a queen excluder above, and the skep or box hive, from which the bees have been driven, placed on top and well packed to keep it warm. Work is commenced at once in the hive and the emerging bees will leave the hive above and finally bring down the honey so that the beekeeper has only empty combs to deal with. To secure that the honey is finally removed, there should be a space between the combs above and the board below, and only a small hole, say 1 inch or less, in the board or cloth. If, however, a flow intervenes, it may be desirable to insert a super of combs or foundation below the skep. Do not attempt to keep the skep or box especially warm. Drones above the excluder should be released, say, every seven days.

Last of all, the empty combs may be cut out and the best of them secured in frames, but not given at a time drone comb is likely to be built (**312-15**).

1489. If at all possible, first drive out the bees and then get access to and remove the combs, dealing with them as above. To drive, it will be necessary to secure an exit at the top as driven bees always go upwards. They may be smoked up from below and vibration employed as well.

1490. Where driving is not possible it will be necessary to subdue the bees well, then open up the nest and cut out as in a box hive. The difficulties may be considerable and the bees bad-tempered. This is hardly a job for a novice. Receptacles for honey and comb are required and a damp towel for sticky hands.

1491. A less expeditious but safer method may be used, and must be used in case it is not practicable to obtain access to the combs.

Contrive a hive fixed over the entrance so that the bees must pass through the bottom of it, but not so arranged that access cannot be got fairly readily to the original entrance. To get the queen, the bees must be allowed to swarm into this hive, after which the original entrance is covered with excluder so that bees can travel through but queens cannot. Queens hatched within will never leave. The brood will hatch and the bees join the swarm and finally fetch out all honey. Provide ample storage room.

1492. A modification consists in covering the entrance with a cone escape so that the bees can get out but not in, then place the contrived hive over the entrance and insert a frame of brood and eggs, and preferably a young queen as well. The colony within will die out, all flying bees joining the new colony. After all signs of life have ceased within, substitute excluder for the cone escape and let the bees rob the old combs. Provide ample storage room.

SECTION XVI
DISEASES AND PESTS

Diseases

General

Readers should be aware that this whole section should be considered with care. In UK they should consult the National Bee unit www.nationalbeeunit.com. Knowledge of bee disease has advanced significantly since the original publication for example foul brood is viral and not bacterial and sac brood is no longer rare in Europe.

1493. It is most important that the bee-keeper should have some knowledge of the nature and symptoms of the more serious and widespread diseases to which bees are prone. Ignorance may not only lead to disappointment, but may result in serious financial loss to the beekeeper himself and to other bee-keepers in the district. Were it not for disease, profitable bee-keeping would be a simple matter, but in certain districts it is necessary to wage constant warfare against prevalent diseases. Everywhere constant watchfulness is essential, for certain of the diseases may wipe out whole apiaries in a few months if not checked in time, and may reappear and repeat the lesson if inadequately dealt with.

1494. For this reason in most countries the interests of bee-keepers are guarded by legislation, and every bee-keeper should be familiar with any local legal requirements. Where legislation is most effective all bee-keepers have to be registered and certain diseases are notifiable. In several countries provisions are made against the removal of bees from infected areas and against the importation of bees, with or without certificates. These apply even to queen bees. Restrictions are also put upon the importation of honey and of honey combs.

1495.a In Great Britain notifiable diseases (namely American Foul Brood [AFB], European Foul Brood [EFB], Small Hive Beetle and Tropilaelaps) must be reported to government appointed bee inspectors and treated under their supervision. Beekeepers in local associations are able to join an insurance scheme (BDI) that will compensated beekeepers for the necessary destruction of colonies. ID

1495. *In Great Britain valuable work on the control and treatment of diseases is being done under the auspices of the reconstituted British Bee-Keepers' Association, provision being made for inspection, treatment and compensation.*

1496.a It is advisable to never feed honey to colonies of bees. ID

1496. *The spores of American foul brood can live for long periods in honey. No honey should be used for feeding bees unless it is known to be disease free.*

Examination for Disease

1497. Examination for signs of brood disease should be made a definite part of the system of management. The experienced beekeeper is always on the look out for disease, but experience enables him to carry on with a minimum of disturbance of his bees and he may thus fail to see the first signs of brood diseases in stocks not systematically examined throughout for the purpose. Complete examination of all stocks should be made every spring and again in the autumn. It is best made at a time when there is some honey flow so that the bees are not disposed to rob, as robbing is the commonest way by which disease is spread. Thus it is necessary to make the autumn examination before the last flow ceases.

1498. This work should not involve much extra disturbance, as the examination for disease should be made part of the general examination of stocks in the spring, and again in the autumn, when making arrangements for wintering. In regions free from brood diseases it is not important to examine every comb of brood, especially in autumn, save in the case of stocks recently introduced into the apiary or still having stores obtained from elsewhere and which may carry disease. All old combs of doubtful origin should be marked on receipt, for replacement as soon as possible.

1499. This periodic examination is for brood diseases, but without opening the hive dysentery may be observed in the spring and diseases attacking adult bees may be observed at any time, though generally most noticeable in the spring. Apart from examination of the brood and cappings the principal signs of disease are to be found in loss of activity and loss of numbers, loss of activity amounting even to partial paralysis of wings and even legs.

1500. The symptoms and progress of the more important diseases are well marked and are set out in sufficient detail below with particulars of appropriate treatment. In cases of doubt a final scientific diagnosis can only be made by the use of the microscope and of expert knowledge. Expert reports can be obtained for a small fee, or even free of charge, and no attempt is made here to deal with microscopical and bacteriological examinations, these being beyond the capacity of the average practical bee-keeper.

Many diseases of minor importance are as yet but imperfectly understood. These are dealt with in outline below with practical notes.

1501. As certain brood diseases are highly infectious, it is impolitic to examine a healthy stock after a diseased stock has been handled, without first well washing the hands and disinfecting any tools used.

Examination of Dead Bodies

1502. There may be a loss by death of thousands of bees per day per hive in the active season, by normal wear and tear. These bees in the main die in the surrounding country, being unable to reach home after their last effort. Dead bees observed on the alighting board or in the neighbourhood of the hive should be particularly examined.

A bee dead by stinging, be it queen or worker, is found in a characteristic attitude, lying on its side with abdomen curved and legs drawn up underneath.

Starving bees may come out in numbers, too weak to fly, and die on or before the alighting board.

Bees dying of disease in the neighbourhood of the hive show some signs described under the headings of the several diseases. Where a number of young bees are found dead, the cause may be cold winds; others are obviously drowned by rain. The bee-keeper soon becomes familiar with the common signs.

Brood Diseases in General

1503. In brood diseases the larvae are attacked and death occurs before or shortly after the pupal stage is reached. These diseases have all been described as foul brood, but several different diseases are now clearly distinguished, of which the important ones are American foul brood, European foul brood and Sac brood (sometimes called pickled brood). Larvae may die from other causes, as, for example, chill. Chilled larvae generally die in large evenly distributed batches, whereas diseased larvae generally die irregularly, giving a patchy appearance. Diseased larvae never die before the fourth day, hence the presence of young dead brood (brood not old enough to substantially cover the bottoms of the cells) in even batches, changing to grey and eventually to black, indicates that there has been chilling.

Distinguishing Symptoms of Brood Diseases

1504. The brood diseases of major importance are American and European foul brood. Sac brood is rare, at any rate in Europe. These three are dealt with together for diagnosis and others under "Minor Brood Diseases" following. The causes of these diseases are carried in food and enter the larvae by their mouths. The several diseases then take different periods to develop, or to reach the stage at which any marked symptoms present themselves in the larvae. This incubation period averages just under 3, 6 and 7 days respectively for European foul brood, Sac brood and American foul brood. Death follows quickly and we then find the following distinguishing characteristics:

(1) With European foul brood most of the larvae die before sealing and before attaining full size;

(2) With Sac brood most of the larvae die about the time of sealing;

(3) With American foul brood all larvae die after sealing is due, although some may not be sealed, and many dead pupae are found, and cells with sunken and perforated cappings.

1505. It will be seen that the most characteristic feature of (3) is the large number of dead pupae. These will be found lying on their backs, frequently with their tongues in the air. An occasional pupa may be found dead, however, with Sac brood. The characteristic feature of (1), viz. the death of numbers of larvae about 5 days old, still coiled in their cells, may not be evident, as good house cleaners will clear away the dead larvae short are present, further signs must be sought by examination of the dead larvae. In Sac brood, larvae recently dead will be found like sacs of fluid, with skins sufficiently tough to admit of the larvae being withdrawn whole. The sac when broken emits a granular fluid. The larvae, lying lengthwise in the cells, frequently have their near ends turned up.

Sac brood is relatively infrequent and unimportant.

1506. American foul brood has a characteristic disagreeable smell rather like glue. There is generally no smell with the other diseases, but sometimes there is a putrid smell of decay with European foul brood, in the presence of putrefactive organisms, such as *Bacillus alvei*, and a "sour" smell in other cases.

1507. Brood dead of American foul brood disease becomes viscous and on inserting the point of a match and withdrawing it the rotten substance draws out like rubber even to 2 inches or more. European foul brood sometimes shows some ropiness also, as it is called, but not to the same degree. The difference may be more readily distinguished by putting a little of the decayed mass into a drop of water on a piece of glass, when it will be found that the mass either remains ropey and draws out slightly like chewing gum or it softens up like gruel. If, however, the substance has a putrid rather than a glue-pot odour, it may be European foul. brood even if appearing ropey.

Incidence and Effects

1508. European foul brood and Sac brood are more likely to show in the first half of the season. American foul brood is more likely to show in the second half of the season.

1509. Sac brood, by causing a high death-rate amongst larvae, seriously hinders the development of the colony. European foul brood is more deadly, and, if spread by robbing, may seriously deplete, or destroy, the whole apiary. American foul brood develops more slowly but necessitates thorough treatment and prolonged care as the spores of the disease are most difficult to kill. It is, therefore, the most insidious of the three.

1510. All brood diseases are spread from hive to hive if robbing occurs, or by drifting bees. This affords strong reason for preventing both robbing and drifting (**601-2, 776, and 1327-31**).

1511. In European foul brood, the cappings become contaminated and become a source for spread of the infection when handled by the bees.

Destruction of Germs and Virus

1512.a   In Great Britain any colony with confirmed American Foul Brood will be destroyed together with all the frames and any honey in the hive. Any treatment to destroy AFB spores in the honey is a waste of time and it is better never to feed honey to colonies of bees. ID

1512. *The spores of American foul brood can be destroyed in time by boiling. If suspended in honey the honey should be boiled for at least half an hour. The spores will actually withstand boiling for shorter periods. Honey taken freshly from a diseased stock should be diluted and boiled for three-quarters of an hour. The effects of various agents are shown in the following table. Scientific tests show that there is but little help to be expected from the use for brood diseases of drugs and disinfectants whether given to the bees or placed in the hives. The vapour of naphthalene, carbolic, and eucalyptus may have some effect upon moulds, and probably stimulate the bees to ventilate more energetically and to house cleaning generally, but it is questionable if the return justifies the attention and whether the attention might not be better directed to direct attack on the disease, and to the use of vigorous stock, especially vigorous young queens, and the selection of good house cleaners.*

Further information is given in Table XXI below:

TABLE XXI
DESTRUCTION OF DISEASE GERMS AND VIRUS

Disease.	In Honey.			In Water.			In Air, Dry.		
	Heated.	In Shade.	In Direct Sunlight.	Heated.	2% Carbolic.	In Direct Sunlight.	In Dark.	In Direct Sunlight.	100° C.
American Foul Brood	100° C. at least ½ hour	Years	6 weeks	100° C. 12 min.	Some months	?	Years	45 hours	10 min.
European Foul Brood	80° C. 10 min.	8 months	4 hours	65°C. 10 min.	6 hours	6 hours	1 year or more	35 hours	?
Sac Brood	70°C. 10 min.	30 days	6 hours	60°C. 10 min.	3 weeks	6 hours	3 weeks	7 hours	?

Treatment of Brood Diseases in General

1513.a   Both AFB and EFB are subject to the Bees Act 1980 and subsequent Bee Disease and Pests Control Orders and any suspected case has to be notified to the National Bee Unit. in England (and similar organisations in Scotland/Wales. GR

1513. *The bee-keeper must make himself familiar with any laws or regulations applying*

to disease and its treatment, in his district. He should also make it his business to co-operate with those who are endeavouring to control disease, whether by law or by voluntary cooperation.

Certain information on treatment is given in the following paragraphs, but it is not intended in any way as a substitute for obtaining expert advice or for co-operation with neighbours; both so necessary in the common effort to combat disease.

Treatment of Sac Brood

1514. This disease generally disappears without treatment, but in a bad case it is desirable to requeen with a young queen. Italian bees are better house cleaners than the black races. If the stock is weak give one or two combs of Italian brood ready to emerge, or requeen with Italian queen.

Treatment of European Foul Brood

1515 to 1525 describe methods to treat AFB and EFB which are no longer legal in England. Both AFB and EFB are subject to the Bees Act 1980 and subsequent Bee Disease and Pests Control Orders and any suspected case has to be notified to the National Bee Unit for diagnosis and treatment. Several of the chemical treatments described are also potentially dangerous and use chemicals which are now difficult if not impossible for most beekeepers to obtain. GR

1515.a European Foul Brood should be treated according to the legal practices that apply in the country. In Great Britain the disease is notifiable and is treated under the supervision of and Appointed Bee Inspector. EFB will remain on the comb for some considerable time and primarily affects the brood. An effective treatment if the infection is mild is the Shook Swarm technique for removing all brood frames and brood in the colony. More serious infections are better treated by destruction. ID

1515. *The bees fight the disease by carrying out the dead larvae. The work is helped by caging the queen, or, in a bad case, by removing her for 2 or 3 weeks and giving combs of emerging brood from a resistant strain. Replace the queen by a vigorous young queen, preferably Italian, unless recently requeened from a good strain. Unite all weak colonies. To prevent a fresh outbreak, encourage neighbouring bee-keepers to act; avoid robbing; keep all stocks strong by uniting weak ones and winter them strong. If, however, there is any serious doubt as to whether the disease is European or American foul brood, treat as a case of the American disease.*

Treatment of American Foul Brood

1516.a In many countries American Foul Brood (AFB) is notifiable and enforced destruction is the only option. Where this is not policy antibiotics have been used but these can only delay the onset of symptom of AFB as the spores of AFB are extremely robust and long lived. Once the use of antibiotics ceases the disease will re-emerge. ID

1516. *Remember that this disease is spread by robbing and reappears either by robbing from neighbouring apiaries or from spores remaining in honey in diseased stocks or from hives that once contained diseased stocks.*

1517. A strong case can be made out for total destruction of affected stocks. If the bees themselves are to be saved, which may be justified if treatment is prompt and all possible precautions taken against robbing, there are means of separating them from the brood and stores. Again, combs and hives have been successfully treated, working on a large scale, but there is as yet no known method of treating diseased brood not involving destruction.

Bacilli larvae are intolerant of the presence of other bacilli, and the author therefore suggests that in future, research may lead to the discovery of some harmless bacilli the introduction of which would prevent the development or further development of American foul brood.

Saving the Bees

1518.a   The above method is commonly known a Shook Swarm except that there is no need to stupefy the bees with saltpetre. Instead each frame with bees is shaken into the new brood box to dislodge all the bees. A queen excluder is placed under the brood box to prevent the colony from absconding. ID

1518. *Probably the safest method is that of stupefying the bees and transferring them to clean combs and stores. The bees should be treated at dawn, before flight commences, or at eve, when flight has ceased.*

Insert, in advance, an empty super below the brood chamber, to provide a space into which the stupefied bees will fall. Insert in the smoker a piece of sacking, 8 inches square, which has been saturated with a saturated solution of saltpetre and thoroughly dried. Set alight this fuel and immediately smoke the stock heavily from below and then under the quilt or crown board, using two lots of fuel if the stock is a large one. In a few minutes all the bees will be stupefied and fall to the bottom.

Place over them the clean hive with frames fitted with foundation and provide a rapid feeder, well filled. The bees, on recovering, will occupy the clean frames and the old hive base can be dealt with, together with the bodies previously removed with brood and stores. The woodwork should be thoroughly scorched with a blow-lamp and the combs totally destroyed by fire and buried in the pit dug for their destruction.

1519. Another method much practised but less reliable is that of shaking. The bees are transferred to a clean skep, at a time when robbing is unlikely, by shaking them, after which they are allowed to consume any stores carried with them. The hive is moved to one side and a skep placed on the old site, prepared to receive the bees as for receiving a swarm (**1115-16**). The stock is then smoked and the bees shaken before the entrance of the skep. The bees are left to consume their stores and care for themselves and on the third day transferred to a new clean hive like a swarm. They will then require feeding.

Formalin Treatment

1520.a The use of formalin as a destroyer of foul brood has found favour in the U.S.A., its use as gas being the latest development and likely to prove the best as well as by far the cheapest method. Formalin is a gas soluble in water and is sold as a solution containing 40 percent. formalin, hereafter described as formalin solution. The use of formalin avoids the necessity of destroying combs. Great care must be exercised to avoid robbing from the combs before they are treated. Honey absorbs and holds formalin and is then poisonous to the bees. Honey should be washed out with a syringe, after scratching the cappings, before treating with formalin, and no reliance should be placed on treatment of sealed cells. Every cell should be open. See that no bees can get access to the honey washed out. KF

1520.b The treatment of Foul Brood and other diseases with formalin is highly dangerous and may be illegal in certain countries. You should access for extensive details http://jem.rupress.org/content/6/4-6/487.abstract for further details. JC

1520. *The use of formalin as a destroyer of foul brood has found favour in the U.S.A., its use as gas being the latest development and likely to prove the best as well as by far the cheapest method. Formalin is a gas soluble in water and is sold as a solution containing 40 percent. formalin, hereafter described as formalin solution. The use of formalin avoids the necessity of destroying combs. Great care must be exercised to avoid robbing from the combs before they are treated. Honey absorbs and holds formalin and is then poisonous to the bees. Honey should be washed out with a syringe, after scratching the cappings, before treating with formalin, and no reliance should be placed on treatment of sealed cells. Every cell should be open. See that no bees can get access to the honey washed out.*

1521. A formalin-alcohol solution, known as Huntzelman's solution, is employed, consisting of 20 per cent. formalin solution and 80 per cent. alcohol, which has great penetrating power. The combs are soaked in the solution under cover for 48 hours, drained and put aside in a warm place until required. If stood in the cold a chemical change may occur which makes it difficult to evaporate the disinfectant, and indeed to avoid this it is better to wash the combs after draining, a garden syringe being suitable for the purpose.

1522. A formalin-water solution, with salt, soap or glycerin sometimes added, is much cheaper than formalin alcohol, and may be employed successfully provided steps are taken to ensure penetration. A good mixture consists of 1 part formalin solution in 4 parts water with ½ lb. salt added per gallon. Subject each comb to a small powerful jet of water so that all seals of cells are broken. Leave for 12 hours, then wash again. Put through an extractor and finally immerse in the formalin-water solution for at least 24 hours. Put through the extractor and dry in moving air.

1523. Empty combs may be treated by soaking in

Glycerin	1 part
Formalin solution	5 parts
Water.	94 parts

and then washing and drying.

Formalin and Chlorine Gas

1524. Formalin gas is the active agent and may be applied directly. Honey combs with honey extracted and brood combs with all seals broken, may be subjected to the gas by putting them in a closed chamber· and inserting say 2 oz. of formalin solution per cubic foot of space. The hive bodies themselves have been used as a chamber for the purpose and thus disinfected at the same time, all joints and openings being sealed gas-tight so as to maintain the concentration of gas. Joints may be sealed with paraffin wax just melted. Large cracks may first be covered with gummed paper, as sold in strip rolls for fastening parcels. Care must be used to make a tight joint at the top. The formalin should be inserted in a shallow tray or poured on to some rag and the whole left exposed in a warm place for 2 weeks, or longer unless the temperature is high.

1525. The process is more conveniently applied, and with more certain results, to combs collected in metal or paraffined wooden tanks, of a size to take ten or more combs, fitted with covers having a gasket or joint of rubber, or waxed joint or fitted with an oil seal. If all cell seals are broken and the combs washed with a syringe and dried in the extractor, they may be treated in 7 days. It is important to wash them on removal. If given to the bees while wet with absorbed formalin many bees may be poisoned, and if not washed the formalin may be subject to a chemical change, making its removal difficult.

The changed formalin insoluble in water can be removed with ammonia solution. It is good to add a little household ammonia to the water used for the final washing.

The rate of sterilization varies but little between room temperature and blood heat, but falls off seriously at much lower temperatures. It is a mistake to attempt treatment of combs with seals unbroken or containing honey as the time of treatment must be very much longer and is most uncertain. With a ten-comb tank used as above 1 oz. of formalin solution per cubic foot is sufficient, but if a large tank or chamber is used the amount should be increased to allow for lack of uniformity, and 2 oz. per cubic foot is not too much for 100 combs or more. Avoid keeping the room at an even temperature as condensation after evaporation requires changes of temperature.

The evolution of the gas is assisted by adding permanganate of potash to the solution, and if this is done 500 c.c. of formalin and 240 grams of permanganate are said to be sufficient for 1,000 cubic feet of space.

Minor Brood Diseases

1526. Chilled brood has been dealt with above (**1502**). It is not a disease although sometimes mistaken for such.

Pickled brood is another name for sac brood.

Chalk brood and Stone brood are due to fungi, the former to *Pericystis apis* and the latter to *Aspergillus flavus*. Full-grown larvae, filling the cells, turn mummified, whitish or yellowish and fluffy, and then shrink to dark corpses. There is no ropey or sticky stage. These diseases generally affect small patches of brood and the bees clear them up in well-ventilated hives. Stone brood is very rare in Great Britain, if indeed it occurs at all.

Requeening with a young and vigorous queen is commonly effective in dealing with minor brood diseases.

Diseases of Adult Bees in General

1527.a Dysentery is often associated with some other disease such as nosma where the gut is not performing as it should. Acarine is not such a problem in Europe any more at present but could return again in the future. However Nosema is a major problem especially as there are now 2 strains that have been identified (*Nosema apis* and *Nosema cerana*) the second has jumped species from the Asian honey bee (*Apis cerana*) and spread throughout the world.

1527. Many minor diseases of adult bees, and diseases in a mild form, cause serious inroads on the population of hives, shortening the effective life (**18**) of the occupants and having a serious effect therefore upon returns. The use of hardy and vigorous strains, good house cleaners, and the observance of hygienic measures by the bee-keeper, go far to provide a remedy or preventive, but it is necessary to observe and guard against certain particular diseases which will require individual preventive measures and treatment. Dysentery is perhaps the commonest. Acarine disease is still widespread in Great Britain and parts of the continent of Europe. Nosema, at one time troublesome in Great Britain, is still troublesome in Germany, Switzerland and elsewhere, being probably widely spread with other diseases where not yet recognized.

Dysentery

1528. Under normal conditions the contents of the bowel are discharged during flight and at a distance from the hive. If the bowel becomes overcharged during weather too cold to admit of flight it may be discharged within the hive. It may be discharged on and around the hive when the bees are so incommoded by an overloaded bowel that free flight is impossible. The normal discharge has the colour of yellow ochre. The malignant form is dark brown, but is not serious unless occurring in summer or autumn.

The bowel content is derived from nitrogenous and other ingredients in the honey and from pollen. Honey dew (**451-3**) and low grades of honey are unsuitable for winter food because they cause overloading of the bowel (see Winter Feeding, **1303-6**). As soon as the temperature permits, the bees should be given a clean hive, and even before this any unoccupied outside combs may be removed temporarily. The hive should be packed down well and the bees given warm syrup and diluted honey if short of stores. Spring dysentery is prevented by proper attention and management in autumn.

Dysentery is also associated with other diseases, as Nosema and fungoid poisoning.

Acarine Infestation

1529. Acarine disease is caused by a mite, *Acarapis woodi*, which, entering the spiracles, breeds in the trachea, or breathing tubes, in the body of the bee, especially in the forepart of the thorax, living on the juices of the bee's body which it obtains by puncturing and suction. The damage to the tissues and the poison of the excrement and dead mites and the choking of the spiracles by the dead and living mites bring about the death of the bee.

1530. Long before death ensues and, in fact, shortly after infestation, other symptoms appear. The injury and poisoning cause loss of normal control of the flight muscles, so that the wings assume an unnatural position, standing out from the body, separated, and at odd angles, and the bees crawl.

1531. Eggs laid by the mites produce larvae which develop into the perfect insect, male and female. The adult female may leave the bee in about 15 days, but usually a second generation of mites is necessary to cause overcrowding before the females emerge to seek fresh victims; thus the most dangerous period is about 30 days after the first infection. Large numbers of bees may be infected before marked external symptoms appear.

1532. Infection generally occurs in the first four days after emergence and requires actual contact with an infested bee. After this, by changes in the hairs at the entrance of the spiracles, the bee quickly obtains immunity, and infection after 6 to 9 days is quite negligible. This feature of early development of immunity is important to note. Any system of management tending to separate emerging bees from the older workers, such as the Demaree system, helps greatly to reduce the spread of the disease, but unfortunately such procedure is not possible in the early stages of development of the brood nest. One can, however, stimulate brood production, thus at the same time making good the losses and wearing out as many as possible of the old bees before they reach the dangerous stage.

1533. Drones are frequently infected and may become carriers of the pest. It is established that queens also may become infected in the early days of their virginity.

1534 With the shorter-lived and more prolific strains, considerable acarine infestation is often met with, producing apparently no serious consequences. On the other hand, with the longer-lived strains and where no steps are taken to control and check the disease, the disease may reach such a stage by mid-summer that the whole apiary may be wiped out.

The mite does not willingly leave the bee, save to crawl on to another, and if separated from its host, probably dies in less than 30 hours.

Diagnosis of Acarine Infestation

1535. The external signs of the disease are not easy to interpret, as there may be many mites and no crawling, or but few mites and almost immediate signs of wing trouble.

The only sure test is by microscopic examination of individual bees, and even then it is difficult to prove complete freedom as an examination of even 200 bees all found absolutely free of mites gives only a low degree of probability against a 1 per cent. infection. While a negative result is thus difficult to prove, a positive result is readily obtained where other symptoms are present. The examination of a few crawlers or dead bees will show acarine at once if it is present. Where a sample is to be taken from apparently healthy bees to reduce doubt, 20 to 30 bees should be secured, avoiding taking young bees. Mature bees can be caught at the entrance by partially closing one end with wire gauze or fine perforated zinc, and taking the bees which accumulate there.

1536. Dissection to expose the prothoracic spiracles is the usual procedure, and examination under a power of not less, than, say, thirty. The bee should be laid on its back on a piece of cork or soft wood and secured by two pins, passed through the thorax from behind, near the abdomen. The head is removed, then, using a small-bladed knife, the forepart of the thorax is stripped off. There is a collar-shaped piece, near the front, which will come away whole, freely exposing the principal trachae.

In Great Britain and elsewhere, one can secure the expert examination of samples of bees on payment of a nominal fee, or even free of charge.

Treatment of Acarine Infestation

1537.a Acarine only lives on adult bees so the brood and hive stores can be given to another colony once the adult bees have been destroyed. Shaking all the bees out of the hive and putting another empty hive on the original site will attract all the bees and separate the brood. Once the adult bees have reformed on a dummy hive they can be destroyed with a small amount of petrol. ID

1537. In a really bad case, with many crawling bees and dysentery, the stock had better be destroyed, but the brood and food can be saved, being placed over a strong stock after the bees have been separated. For this purpose they should first be stupefied with saltpetre, chloroform, or ethyl chloride.

1538.a Acarine infestations seem to be susceptible to treatments used to control varroa (Varroa *destructor*). In particular Apiguard© with the active ingredient of thymol and the pyrethroid treatments have been shown to be effective. ID

1538. The treatment for acarine infestation consists in the employment of vapours which hinder the mites from finding their way to the trachea of the young bees and also injure them. Further, certain manipulations which have the effect of separating old and young bees (**1166-70**) are decidedly helpful; indeed, it should be possible to effect a cure, by any system of management by which emerging bees were definitely prevented from contact with infected bees, until say six days old. If some emerging brood were raised to an upper chamber above a screen, and, either incubated by a strong stock below, or helped out by the addition of some combs of healthy bees from another hive, further batches of brood could be added later, thus building up an entirely healthy upper lot, which, however, would require a small exit of their own after, say, the first week, and a new queen or queen cell about the same time. On establishment of the new queen the old queen would be destroyed and the lower lot allowed to die out, any supers, first cleared of bees, being moved to the top.

Frow Treatment and the Like

1539 - 1543: These treatments are no longer appropriate for use in beehives. Frow mixture contains substances which are carcenogenic. Other chemicals suggested may contaminate honey and leave residues in the hives. GR

1539.a Frow mixture is highly carcinogenic and it is advised that this mixture is never used in beekeeping. ID

1539. Vapour is produced by the evaporation of certain drugs or a mixture of substances, inserted on a felt pad at the entrance, placed over the feed hole and covered, or inserted in small containers furnished with wicks. An empirical mixture introduced by Frow, and known by his name, has been widely and successfully employed, made to the following formula:

Nitrobenzene	4 parts by weight
Safrol	1 part by weight
Petrol	2 parts by weight

During war-time an alternative has been successfully employed:

Nitrobenzene	6 parts by weight
Ligroin	3 parts by weight
Methyl salicylate	2 parts by weight

In Russia methyl salicylate has been substituted for the safrol in the original formula, thus:

Nitrobenzene	2 parts by weight
Petrol	2 parts by weight
Methyl salicylate	1 part by weight.

1540. *No doubt, other, and better substances will be found by research. Already some success has been obtained with turpineol, at Rothamsted, in doses of 10 to 50 ml. (say 1 to 2 ounces), the mites being killed without injury of the bees and brood. It can be applied on pads from over the feed hole, covered to prevent wasteful evaporation, and repeated in a fortnight, but the technique requires further development and during the war turpineol is no longer readily obtainable.*

1541. *During treatment with Frow mixture stocks are liable to be robbed, in part due to the effect of the drugs on the bees, and in part due to the disturbance of the sense of smell. The trouble is much less with methyl salicylate and turpineol. It has been found in Austria that robbing can be prevented, with Frow mixture, by its use in November or end of October, all light being excluded so that the bees will not fly. This requires some ingenuity, as ventilation must be maintained. A closely fitted labyrinth entrance, painted black throughout within, and so fitted or covered as positively to exclude light at the joints, will exclude light and permit ventilation.*

Robbing is always minimized by treatment in cool weather, November and February being chosen for this reason and because breeding is reduced (see also 1543).

1542. *Some bee-keepers advocate the annual treatment of all stocks. Others experience some infestation but do not treat at all, apparently without material loss. Much depends upon the district, the bees and the system of management employed. Generally it has been advised to spread the treatment over a period, applying consecutive small doses, six to ten, say on alternate days. On the other hand some have used a single application successfully, and this should certainly be sufficient where only a preventive is required.*

For an average stock, a single dose of 60 minims (4 ml.) (one teaspoonful) may be used in November or February, the pad being removed after 10 days. The pad must be kept away from the cluster. Again, 6 doses of 20 minims, on alternate days, have been used in February, the last remaining on. If applied later, the individual dose should be halved and applied over twice the period.

Methyl Salicylate Treatment

1543.a   see **1538.a**   ID

1543. *Treatment by methyl salicylate alone is not universally successful, but bee-keepers in a large way, using sufficient, have obtained success. Camphor has been used in mild cases.*

Methyl salicylate may be introduced in flat tin lozenge boxes, filled with cotton wool and having, say, half a dozen holes punched in the lid with a large bradawl. Two to four tins are used according to size of hive, pushed to the back of the floor board, using about half an ounce of methyl salicylate in each. Put them in in November and leave until spring, then renew, but remove before harvest-time. There is very little risk of robbing.

1544. After drug treatment it is desirable to requeen in case the queen may have suffered. Requeening is also desirable in manipulative treatment provided the new queen is free from suspicion. A queen emerging in a recently formed nucleus in which there are no old bees is practically free from suspicion except in a badly affected apiary where there are many drones about.

Nosema

1545. Nosema is a disease caused by the presence of a protozoon which multiplies in the digestive tract, being found in the chyle stomach in such numbers as to give it a white colour. In bad cases the bees attacked may be seen crawling from the hive, being unable to fly, and there may then be much dysentery. The bees attacked may die in 2 or 3 days and are found with their legs gathered or drawn up to the body. The disease is widespread but not as a rule serious in its consequences. It manifests itself more especially in early spring and especially in bad weather. If, however, it occurs late in the year the stock will probably die out before spring. Various drugs have been tried but no worthwhile treatment is known yet.

1546. The disease is spread by spores and possibly through fouling of the source of drinking water. Any quilts or frames fouled by dysentery associated with Nosema should be burned or disinfected as for foul brood, and the soil around the hive dug over and sprinkled with paraffin. Honey from an affected hive, heated for 1 hour to 140° F., is considered safe for use as bee food.

Amoeba Disease

1547. Amoeba have been found multiplying in the malpighian tubes in association with Nosema, and thus aggravating the diseased condition. They generally disappear in April. For treatment, see Nosema. This disease has been noted in Great Britain and the United States of America and is common on the continent of Europe.

Paralysis

1548. The symptoms of paralysis include some loss of locomotive power. The affected bees tremble and spread their legs and wings. Frequently there is loss of hair, notably on the abdomen. Sometimes the bees may be seen hustling diseased bees out of the hive and over the flight board.

There appears to be an infectious type, for which, dusting the bees with flour of sulphur is helpful. There is an hereditary type, necessitating requeening, and a nitrogen deficiency type in which loss of hairs is due to their becoming brittle, but loss of hair has also been attributed to *Bacillus gaytoni.* With the infectious type it is good to remove the floor board, supporting the hive so that the bees have to enter and leave from below the cluster. Diseased bees fall to the ground and cannot crawl or fly back.

In general, cure is assisted by seeing to the pollen supply, by requeening and by good ventilation.

Poisoning by Fruit Spraying and Otherwise

Information in the following paragraphs is now obsolete **1549-1551.** MG

1549. *Bees may be poisoned by poisonous pollen and honey, or intoxicated, for example, by honeydew from certain lime trees, but a more serious risk is that from arsenic and other poisonous sprays used to protect fruit trees from caterpillars.*

Before alleging loss of bees by fruit poisoning it is well to have an examination made

for the presence of arsenic, but the principal evidence of spray poisoning is to be found in a sudden marked loss of a large proportion of the flying bees at a time that spraying is in use in the neighbourhood.

1550. A bee dying of arsenic poisoning is found with hind legs stretched out, tail raised and tongue extended. Before death, the bee, trying to fly, falls to the ground on her back. Most of them die before reaching the hive. It is said that a stock is helped by feeding with syrup in which has been incorporated one per cent. of a half per cent. solution of ferric hydrate. The risk of poisoning is reduced by providing a safe water supply, and particularly by supplying sweetened water by use of a frame feeder.

1551. No serious risk need be feared from fruit spraying, save where arsenic (lead arsenate) is used. The risk is reduced if the spray is applied only before the pink-bud stage or after the flowering stage, but is increased if the spray falls on flowers beneath the trees, for example, dandelions.

Undoubtedly the best protection for the bee-keeper is to be got by the addition to the spray of substances repellent to the bees, of which at the time of writing the best are, one per cent. lime sulphur, or one twentieth of one per cent. of nicotine sulphate. These will give adequate protection provided the bees have access to other sources of water and pollen.

Pests

General

1552. Bee-keepers have to contend with numerous pests: the bear, sloth and skunk which may upset the hive; the birds and other creatures which eat the bees; the moths whose larvae attack the wax; the humble mouse which finds its abode in the warm hive, and ants, small flies and other creatures which cause destruction and loss in various ways. The larger pests need no description and do not call for treatment in this work.

Braula Caeca

1553. The Braula Caeca or bee louse is a creature about the size of the head of a pin, reddish-brown in colour, which fastens itself on the thorax of the bee very securely by means of its claws. It is sometimes thought to suck the juices of the body, but in any case has actually been observed taking food from the bee's tongue.

1554. It lays its eggs on the wax cappings of honey comb to which they are firmly secured. The egg is about 1/32 inch (3/4 mm.) long and white in colour. On hatching, the larva forms in the capping a tunnel nearly 1/32 inch diameter, more plainly visible from the inner than the outer surface of the capping. The larva is about 1/12 inch (2 mm.) long and white in colour. The pupa is about three-quarters this length.

1555. It is not clear just when the louse attaches itself normally to a bee. They can crawl and if dislodged would no doubt find a new host in the hive. One has been seen attached to a bee on emergence from its cell, and this may be the normal course.

1556. As a rule the pest is most active in July and August, but dies out during a cold winter. Tobacco has been used to suffocate and dislodge them, but it is difficult to avoid suffocating the bees and queen and most difficult to affect a real clearance by its use. The principal danger from this pest lies in injury to the queens, which may be found covered with them. If the pest is found present the queen should be examined and assisted if necessary.

1557.a Do not use formalin! Freezing comb honey will destroy braula larvae (and wax moth eggs/larvae) GR

1557.b   see **1538.a**   ID

1557. The larva may be destroyed in honey comb by exposure for an hour to formalin gas.

Treatment for acarine has also been found effective. A little creosote on the smoker will cause them to drop.

1558. This louse has been commonly thought to be blind, but, in fact, it has minute eyes at the outer corners of its triangular head.

Moths

1559. The study and treatment of wax moths is so closely associated with the care of combs and wax that it has been dealt with in Section III.

The Sphinx or Death's Head moth is frequently listed as a pest, but is seldom found in numbers. By some it is regarded as seeking entry to the hive for warmth. In South Africa these moths have been found gorged with honey.

Birds

1560. There are numerous bee-eating birds, some catching the bees on the wing and others visiting the alighting board which they tap, thus luring the guard bees to their destruction. If the latter activity proves troublesome a coarse net should be spread over the entrance and alighting board. This will not seriously impede the bees.

1561. The Bee-eater found in Australia is a handsome bird with a bad name. Its depredations are outdone by those of the Masked Wood Swallows, even as they are outdone by the Sordid Wood Swallows. Common swallows in Europe have been accused and are chased away by bees, but it is doubtful if they account for many, as they prefer smaller insects.

1562. The shrike or butcher bird is a great insect eater, with a predilection for bees, but is probably outdone by the tits. Common sparrows sometimes develop the habit and are then very destructive, being quite unashamed. Tits may be diverted by offering them fat. Ducks, as a rule, are harmless, but better kept out of the apiary. Fowls may be disregarded.

Toads, Dragon Flies, Wasps, Spiders and Lizards

1563. These all take toll of living bees, according to their kind, the first mentioned being persistent and secretive. Among the bee-eating dragon flies may be noted *Aeschna cyanea*. The Mutillae are bad wasps, but like the dragon flies mentioned are not widespread. The Mutillae work at night, leaving many dead bees in the hive.

Ants

1564. Ants of various kinds and sizes from the white ant downwards are very troublesome in some districts, entering the hives in endless numbers for honey. They are kept out by using hive-stands with metal legs standing in vessels of heavy oil. These require frequent inspection, as a single twig or leaf acting as a bridge will soon be utilized. Sometimes bands of sticky substance are put round the legs as used to protect apple-trees from codlin moths, but in case of a persistent attack the ants will bridge the substance with their dead bodies, forming a path for others to follow. Coal tar and axle grease may be tried and are frequently effective (see **667**).

1565. In Australia supports of "Jarrah" or "Red Gum" are used, the ants, including white ants, having a great objection to these materials.

1566. Ants' nests maybe destroyed by pouring in carbon disulphide or by a solution of permanganate of potash followed by a little formalin solution. The permanganate releases the formalin gas in quantity.

1567.a There are many proprietary ant killers that should be used rather than formulating one with highly poisonous chemicals. ID

1567. *Ants may be poisoned, the bees being protected, by covering the poison with a cover, perforated to admit ants only. Make a solution of 1 lb. of sugar in 1 quart of water, adding 125*

grains of sodium arsenate and on cooling add a teaspoonful of honey. This is highly poisonous and must be treated with appropriate respect.

Rats and Mice

1568. Rats are seldom troublesome save in the honey house, where they should be dealt with in the usual way. Mice are troublesome if they enter hives, where they may be attracted by the warmth, as they cause much destruction in the winter, and if they die in the hive still further nuisance is caused. They are kept out by using entrances not more than 3/8 inch (1 cm.) high. Where they abound it may be desirable to use metal entrances.

Drosophila Ampelophilos

1569. This is a small black fly which enters the hive and lays its eggs on the honey cappings. The larvae feed on the honey and regurgitate their food, producing a damp-looking patch on the cappings. This foul mess causes fermentation of the honey and may cause dysentery. A strong stock in good fettle will clear out the mess, but if noticed the bad portions may be uncapped, the cappings being burned. The resulting excitement assists in accelerating house cleaning, but entrances should be reduced somewhat to guard against robbing.

SECTION XVII
INVENTIONS AND DISCOVERIES

Foreword

1570. An inventor is one who has created something which is neither expressed in nor clearly inferable from the prior art.

Many things seem obvious after they have been done, and in the light of the accomplished result, it is so often a matter of wonder how they so long eluded search and set at defiance the speculations of inventive genius.

Knowledge after the event is always easy, and problems once solved present no difficulties; indeed, *may be represented as never having had any.*

Introduction

1571. It is hoped that practical bee-keepers will welcome this brief sketch of important discoveries and inventions. It is admittedly imperfect and the writer would welcome evidence supporting the inclusion of other names in addition to, or in place of, those given and to fill in the more obvious blanks, and especially the names of the first to propose other important appliances and the more important manipulations. It is introduced mainly because discoveries and inventions are so frequently wrongly attributed in current literature.

If asked to name the two greatest discoveries the author would suggest the discovery that queens could be raised from worker eggs, by Swarnmerdam, and of parthenogenesis, by Johann Dzierzon, although some would mention the prehistoric discovery that honey bees could be "subdued" by smoke. For the two greatest inventions he would name the Langstroth frame with bee space all round, and the centrifugal extractor by Hruschka.

1572. In this section the endeavour has been made in each instance to name the first discoverer or inventor, but in connection more especially with the earlier work, pioneers are mentioned who were working independently at the same period, as they did not have the facilities we enjoy for world-wide exchange of ideas, and were, in fact, the pioneers in their own countries.

Natural History of the Honey Bee

1573. Bee-keeping having been practised from immemorial times, some knowledge of bees is prehistoric. Aristotle noted the division of labour among the bees and allocation of particular duties to individual bees, even to the working of one kind of flower. He noted the carrying of bee bread on the bees' legs, and the deposition of honey in the cells by regurgitation and the ripening of it. He even toyed with the suggestion, one of several alternatives already proposed, that the queens and drones might be females and males, mated by copulation. Smoke has been used from the earliest times for subduing bees.

1574. In 1586 *Luis Mendez de Torres* (Spain) described the queen as mother of all the colony. At this time the "big bee" was generally regarded as the king. In 1609 *Charles Butler* (Great Britain) observed the "king" bee laying eggs. In 1630 *Cleaters* described the queen as mother of all the colony. In the seventeenth century *Jan Swammerdanz* (Dutch), born 1637, settled the sex of bees by dissection under the microscope, an instrument not previously available, and described queen raising from worker eggs. In 1720-40 *Rene A. F. de Reaumur* (France) discovered the function of the spermatheca, and in 1792 *John Hunter* Great Britain) made the discovery independently. In 1789 *Anton Janscha* (Serbia) and later *Francois Huber* (Switzerland) observed and recorded the mating flight and return.

In 1845 *Johann Dzierzon* discovered parthenogenesis, of which an important experimental proof was given in 1919 by *Gilbert Barrett* (Great Britain) when he raised normal queens from drone eggs after fertilizing them by hand, thus making also a contribution to controlled parentage.

This procedure should be verified by others.

Adam G. Schirach (Germany) showed between 1760 and 1770 that queens could be raised from worker larvae and in 1855 *Leuckart* (Germany) showed that the production of a queen or worker from the same egg was controlled by the food given. In 1902 *F. W. L. Sladen* (Great Britain) published an account of the scent organ in bees.

Queen Raising

1575. In 1870 *Carl Weygandt* (?) introduced grafted worker larva for queen raising. In 1874 *J. L. Davis* (U.S.A.) suggested grafting, independently practised in 1882 by *G. M. Doolittle* (U.S.A.). In 1878 *W. I. Boyd* (U.S.A.) suggested the use of natural queen cups stuck on bars. In Germany, however, *Wm.Wankler* had used artificial cell cups in 1883. *Eugene L. Pratt* (U.S.A.) introduced the wooden cup with flanged support and the spiral wire cell protector, and *J. L. Davis* (U.S.A.) used queen cages in 1899.

1576. Pursuing another line, about 1780 *Anton Janscha* (Serbia) and in 1880 *O. H. Townsend* (U.S.A.), and in 1881 *Th. W. Cowan* (Great Britain), cut down the edges of brood comb to expose larva of suitable age for queen raising and the latter enlarged the mouths of cells containing selected larva. In 1880 *J. M. Brooks* (U.S.A.) followed in 1882 by *Henry Alley* (U.S.A.) cut strips of comb containing larva of suitable age and mounted them on the cut edge of comb and on wood strips in a frame. In 1899 *Pridgen* (U.S.A.) cut out individual cells containing larva, mounting them separately. In this century *Barbeau* (Canada) used folded metal supports for securing such cells, for which later *Perret Maisonneuve* substituted wooden cell cups accessible from the back, and suitable for reception also of queen cells, making ingenious tools for use therewith.

1577. Instrumental insemination was first practised and described by *N. W. McLain* (U.S.A.) in 1886 and carried recently to a high degree of perfection by *L. R. Watson* (U.S.A.).

1578 Laying workers appear to have been noted first by *Johann Riem* (Germany) about 1770-80,

Wax, Comb and Honey

1579. In 1609 *Charles Butler* (Great Britain) observed that comb was built from scales of wax. In 1691 *Martin John* (Prussia) observed wax scales carried in pockets on the bees' abdomens. In 1720 *Herman Hornbostel* (Germany) repeated the observation and published a suggestion as to their origin in 1744. In Great Britain _John Thorley_ observed wax scales carried in pockets in 1744, and in 1771 *Francois X. Duchet* (Switzerland) suggested that they were of glandular origin. *John Hunter* demonstrated that they originated in an oily glandular secretion in 1792. In 1793 *Francois Huber* demonstrated the glandular production of wax and its production in quantities on a sugar diet.

1580. In 17 12 *John Maraldi* (Italian) demonstrated the geometry of comb building. *Gottlieb Kretchmer* (Germany) first made foundation in 1842. He used waxed cloth moulded between rollers. *Jean Mehring* (Dutch) 1857 used pure wax cast between metal moulds, and *A. I. Root* (U.S.A.) in 1876 used a metal roller press. *Otto Schenk* in 1872 exhibited foundation with projecting starters for the side walls and *John Long* (U.S.A.) in 1874 produced similar material. *D. S. Given* (U.S.A.), 1879 or 1881, produced wired foundation made in a press, but it was not until 1892 that *E. B. Weed* (U.S.A.) produced sheet wax in long lengths for use between rollers.

1581. Little is recorded about honey, but *Nussbaumer* (Switzerland) made an important contribution in 1910, tracing the fermentation of honey by certain sugar tolerant yeasts operative even in honey of high density.

Hives, Frames and Appliances

1582. Hives have been made in numerous forms from the earliest times. The desire to control the individual combs, a feature of the first importance, has controlled hive design in the last 100 years, but was in evidence long before that. As early as 1675 is recorded the use in Greece of basket-shaped bodies, open top and bottom, having wooden bars across the top for the support and removal of individual combs, the whole being covered with clay. In 1768 *Thos. Wildman* (Great Britain) described the use of similar bars with a box hive, and in 1780 9 *John Keys* (Great Britain) described the use of slides between the bars, a construction employed in the *Stewarton* hives, and in 1790 *Della Rocca* described self-spacing bars with space between for the bees to pass to an upper chamber, the ends of the bars resting in rabbets cut in the side walls.

1583. In 1683 an anonymous writer, initials *J. A.* (Great Britain), described the use of four-sided frames with sloping sides, the top bars resting in channels cut in the side walls and the sloping sides resting against the hive walls to which they would be securely fastened by the bees. In 1773 *Daniel Wildman* (Great Britain) employed complete rectangular frames with slides between the top bars in place of his father's plain top bars with slides, but the sides of the frames touched the walls. Meanwhile, in 1819, *Robert Kerr* (Great Britain) used a frame consisting of one top and two side bars but fitting the hive sides. Hanging frames of various forms without spaces at the sides were described by *Major Munn* (Great Britain) in 1834, *Prokopoertel* (Dutch) in 1841, *Johann Dzierzon* (Silesia) in 1845 (used in 1837 by him), by *Clement* in France, and by *Debeauvoys* (France), the last named being square. Then, finally, we have *L. L. Langstroth's* (U.S.A.) great discovery in 1851 that contrary to their natural habit in building, the bees would tolerate a "bee space" for the whole length of separated side bars, and his invention of the hanging frame with space all round, which received the treatment common to so many really great inventions, being described as not good, then not new, then not invention, a thing anybody might have done, and then brought into universal use.

1584. Langstroth first discovered that the bees would keep reasonably clear a space of 3/8 of an inch above the frames and under the crown-board. It was about a year later, and on October 31, 1851, that he conceived the greater invention, which he patented in 1852. Enquiry has shown, however, that *Debeauvoys*, above cited, had published in 1851, in the third edition of his Guide de L'Apiculteur, the use of rectangular frames having six millimetres clearance to the side walls so as to ease their removal and reduce wax-moth; thus his conception may have anticipated Langstroth, to whom, however, falls the whole credit for successfully introducing this great invention into world-wide use against severe and most unfair opposition.

1585. *Moses Quinby* (U.S.A.) was responsible in 1866 for the introduction of the large-size frame later given world-wide popularity by the Dadants.

Quinby also improved the bellows smoker in 1875, placing the bellows beside the fuel container where it has since remained, though *Bingham's* (U.S.A.) 1877 addition of a gap between the nozzle and the fuel chamber enabling the fuel to smoulder was perhaps equally important.

1586. The movable division board appears to have been introduced by *John Hunter* (Great Britain) in 1792. *Francois Huber* (Switzerland) invented his well-known observation hive described in 1780. *Jas. Lee & Son* (Great Britain) introduced the plinthless case (**834**) in 1898.*P. J. Prokopovitsh* (Russia), in 1841, described self-spacing sections, with edges cut away to form bee-ways, also a slotted honey board. In 1870 *G. T. Wheeler* (U.S.A.) introduced tin separators. In 1876 *A. J. Cook* (U.S.A.) made sections of basswood, and in 1879 *James Forncrook* introduced the grooved one-piece section.

Perforated queen excluders were introduced by the *Abbe Collin* (France) in 1849.

1587. To *Francesca de Hruschka* (Austria) is attributed the invention of the centrifugal extractor in 1865, but it is said that a radial form was in use in France about 186o, and was mentioned by *Hamte* in I867. In 1874 *Thos. W. Cowan* exhibited a radial form and in 1875 introduced the self-reversing type.

1588. The useful cone escape appears to have been developed, from a crude affair first made from mosquito netting, to a neat, metal device, by *Charles Dibbern* (U.S.A.) in 1890. Mechanical trap escapes were developed in the 1870's, but it was not until 1891 that the widely used *Porter* escape was conceived. Probably this will be largely superseded by a type without moving parts (**701**).

Bee Diseases

1589. Disease in various forms has been observed from the earliest times; indeed, it could hardly escape observation. Moreover, methods of management have been devised to cope with it. The old fashion of destroying the weakest stocks annually had hygienic advantages. But it is only within the last 1oo years that bee diseases, along with most other diseases, have been scientifically studied.

1590.a   The bacteria causing American Foul Brood is now commonly known as Paenebacillus *larvae*.

European Foul Brood is considered to be caused by the bacrterium Mellissococcus *pluton.*

There is now doubt that the Isle of Wight disease was caused by Acarine (Acarapis *woodi*) and that the major symptoms noted were due to the virus Chronic Bee Paralysis Virus(CBPV). In Great Britain the decline of acarine and CBPV occurred simultaneously and there may be a link between the two. However Dr Rennie's work showed that some colonies with Acarine did not suffer the symptoms of the Isle of Wight Disease whereas some colonies showing the symptoms did not have acarine. The work on the Isle of Wight Disease was done before viruses could be easily identified.   ID

1590. In 1874 Cohn indicated that foul brood was due to a bacillus. In 1885 Frank Cheshire (Great Britain) attributed foul brood (European) to Bacillus alvei. In 1903 G. F. White (U.S.A.) attributed American foul brood to Bacillus larvae and practically simultaneous T. W. Cowan (Great Britain) associated it with a bacillus he called burri, and Albert Maassen (Germany) to a bacillus he called brandenburgiensis.

In 1909 Dr. Enoch Zander (Switzerland) discovered the protozoon responsible for "Nosema" disease.

In 1920 Miss E. J. Harvey observed a mite in the tracheae of bees suffering from "Isle of Wight" disease, which was shown to be pathogenic by P. Bruce White, both working under the direction of Dr. J. Rennie. The disease caused by the mite "acarapis woodi" is now known as Acarine Disease.

THE NEW CLP REGULATIONS

(Classification, labelling and packaging of substances and mixtures.)

Beekeepers are not immune to the tidal wave of new regulations that seem to be a constant part of modern life. One of my jobs in our Association is to oversee the purchasing, storage and safe distribution of chemicals to students taking the BBKA microscopy course and various wax and candle making classes. Others in the Association are similarly responsible for the wide range of chemicals commonly used in modern beekeeping by our members. Being responsible for the storage and distribution of chemicals means that it is essential to be up to date with the new regulations that are due to be fully in place by 2017. We are in an interim period from this month (December 2012) until June 2015 when all substances must begin to be classified and labelled according to CLP regulations. *(CLP is changing the classification system for identifying and describing chemical hazards in Europe. It means that labels on bottles and packets of chemicals, safety data sheets and other documents are changing.)*

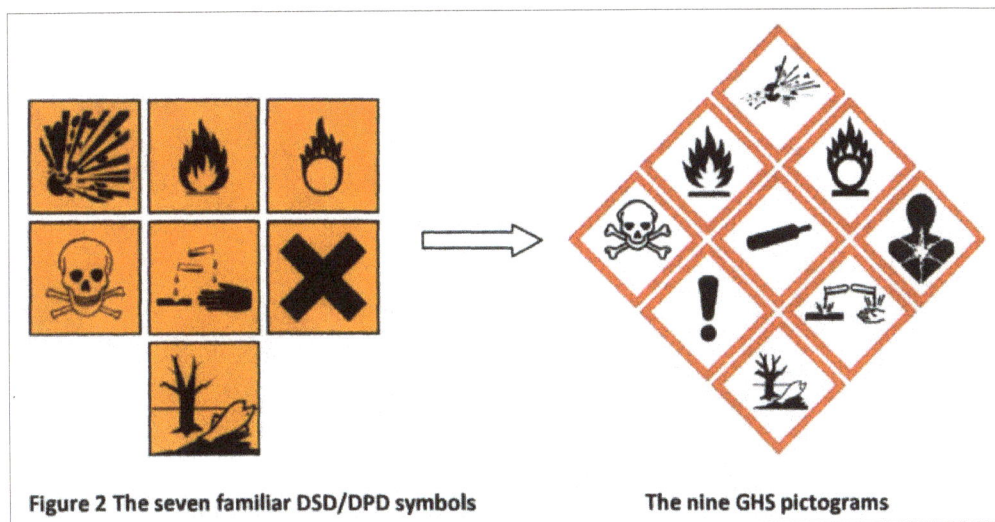

Figure 2 The seven familiar DSD/DPD symbols   **The nine GHS pictograms**

The old system (CHIP – Chemical Hazard Information and Packaging) with its orange hazard symbols supported the DSD and DPD (the Dangerous Substance *(ie. Chemicals)* Directive and the Dangerous Preparations *(ie. Mixtures of chemicals)* Directive). Anyone who has handled chemicals over the last few decades will be familiar with the **symbols** and the **risk** and **safety phrases** on the sides of bottles and packets. They appear on dishwasher tablet boxes, metal polish tins, drain cleaner and a host of other household products, as well as laboratory chemicals.

In this brave new world you will begin to see GHS/CLP Pictograms which form part of the **G**lobally **H**armonized **S**ystem of Classification and Labelling of Chemicals (GHS). Two sets of pictograms are used – one for the labelling of containers and for workplace hazard warnings, and a second (not discussed here) for use during the transport of dangerous goods.

What do these new pictograms mean?

The table below shows the main categories of new CLP pictograms and the current Hazard Symbols which they are replacing.

Current CHIP Symbol	New CLP Symbol (Pictogram)	CLP Hazard Class	Number
Explosive		Explosives Self-reactive substances and mixtures, types A, B Organic peroxides, types A,B	GHS-01
Highly/Extremely flammable		Flammable gases, aerosols, liquids or solids Self reactive substances and mixtures Pyrophoric liquids and solids Self-heating substances and mixtures Substances and mixtures, wich in contact with water emit flammable gases Organic peroxides	GHS-02
Oxidising		Oxidising gases, liquids and solids	GHS-03
No current symbol		Compressed gases, liquids and solids Liquefied gases Refrigerated liquefied gases Dissolved gases	GHS-04 NEW
		Corrosive to metals Skin corrosion Severe eye damage	GHS-05
Toxic/Very Harmful Toxic		Acute toxicity (Cat 1 - 3)	GHS-06
Harmful/Irritant		Acute toxicity (Cat 4) Skin and eye irritation Skin sesitisation specific target organ toxicity Respiratory tract irritation Narcotic effects	GHS-07 NEW
No current specific symbol Use either		Respiratory sensitisation Germ cell mutagenicity Carcinogenicity Reproductive toxicity specific target organ toxicity Aspiration hazard	GHS-08 NEW
Dangerous for the environment		Hazardous to the aquatic environment	GHS-09

The New Look Labels.

These new pictograms will be accompanied on a chemical label with various other items of information which amplifies their meaning with **Hazard Statements**, and handling instructions now known as **Precautionary Statements**.

Other information should include the name and address of the supplier, the name and product identifier (usually an EC No.), the quantity of the product in Grams or Kilograms, or if a liquid, in Millilitres or Litres. There must also be a 'Signal Word' of either 'DANGER' or 'WARNING' for less hazardous materials.

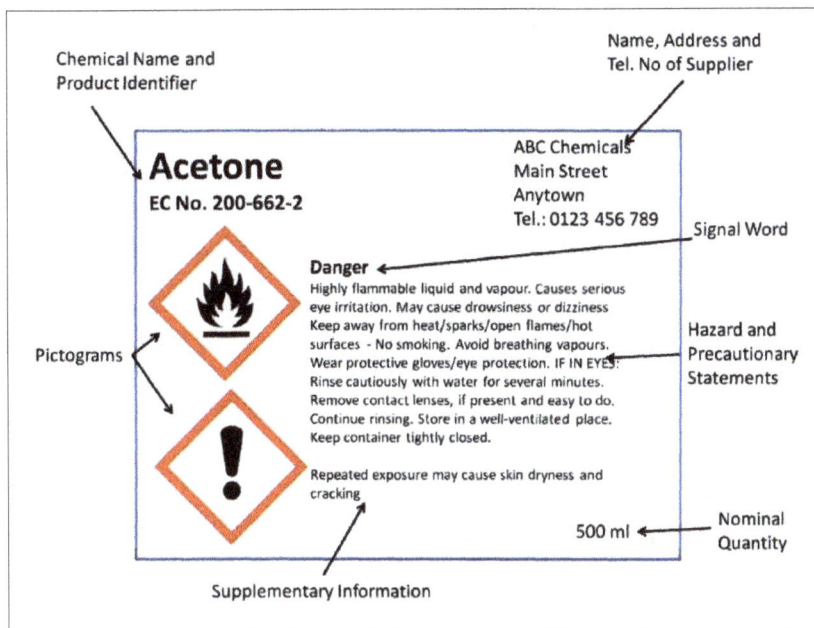

The diagram shows a chemical label for Acetone with the following labelled parts:

- Chemical Name and Product Identifier → **Acetone** EC No. 200-662-2
- Name, Address and Tel. No of Supplier → ABC Chemicals, Main Street, Anytown, Tel.: 0123 456 789
- Signal Word → **Danger**
- Pictograms
- Hazard and Precautionary Statements → Highly flammable liquid and vapour. Causes serious eye irritation. May cause drowsiness or dizziness. Keep away from heat/sparks/open flames/hot surfaces - No smoking. Avoid breathing vapours. Wear protective gloves/eye protection. IF IN EYES: Rinse cautiously with water for several minutes. Remove contact lenses, if present and easy to do. Continue rinsing. Store in a well-ventilated place. Keep container tightly closed.
- Supplementary Information → Repeated exposure may cause skin dryness and cracking
- Nominal Quantity → 500 ml

Hazard Statements These are assigned a unique numerical code which can be used as a handy reference when translating labels written in other languages. They have the form 'Hnxx' where H stands for 'Hazard Statement', a number 'n' and 'xx' is a sequential numbering scheme.

Where 'n' = 2 this is for physical hazards,
        3 is for health hazards and
        4 for environmental hazards.

So for example :
> **H241** means 'heating may cause a fire or explosion',
> **H318** 'causes serious eye damage' and
> **H402** 'harmful to aquatic life'.

Worryingly there are no environmental hazard statements that are concerned with land animals or flying insects! (The old CHIP system had a risk statement – **R 57** – **'toxic to bees'.**)

Precautionary Statements

Again these are assigned a unique numerical code. They have the form '**Pnxx**' where P stands for 'Precautionary Statement'. Xx is a sequential numbering scheme and the value 'n' refers to:

1 General Statements
2 Prevention Statements
3 Response Statements
4 Storage Statements
5 Disposal Statements
1 Some examples:

> **P102** – Keep out of reach of children
> **P235** – Keep cool.
> **P270** – Do not eat, drink or smoke when using this product.
> **P313** - Get medical advice or attention
> **P410** – protect from sunlight.

Safety Data Sheets

The CLP Hazard and Precautionary Statement Codes (and during the interim period the old DSD codes) should be present on **Safety Data Sheets (SDS)**.

These sheets set out the Hazards and Precautions in clear language and are often sent with any chemicals supplied by a company. If beekeepers are supplying chemicals onwards to other beekeepers then it is good practice to include a copy of the relevant SDS along with the bottle or container. If no safety sheet is sent then it is a good idea to request one from the supplier company.

If you look up the CLP regulations for Acetone then you will find the following table of information:

Summary Of Classification and Labelling

Harmonised classification - Annex VI of Regulation (EC) No 1272/2008 (CLP Regulation)

General Information

EC Number	CAS Number	Index Number	International Chemical Identification
200-662-2	67-64-1	606-001-00-8	acetone propan-2-one propanone

ATP Inserted / Updated: CLP00

CLP Classification (Table 3.1)

Classification			Labelling			Specific Concentration limits, M-Factors	Notes
Hazard Class and Category Code(s)	Hazard Statement Code(s)	Hazard Statement Code(s)	Supplementary Hazard Statement Code (s)	Pictograms, Signal Word Code(s)			
Flam. Liq. 2	H225	H225	EUH066	GHS07 GHS02 Dgr			
Eye Irrit. 2	H319	H319					
STOT SE 3	H336	H336					

Signal Words	Pictograms
Danger	! Exclamation mark Flame

Translated, this document gives you all the information needed to write a label on a bottle of acetone.

You can see the alternative names and EC number across the top. The Hazard Statement codes translate as follows:

H225 – Highly flammable liquid and vapour.

H319 – Causes serious eye irritation.

H336 – May cause drowsiness or dizziness. (STOT SE 3 means Specific Target Organ Toxicity – category 3 - the lowest level.)

The Signal Word is DANGER and the pictograms that apply in this case are the **'Exclamation Mark'** and the **'Flame'**.

(**EUH066** is a Supplementary Hazard Statement meaning 'Repeated exposure may cause skin dryness or cracking'.)

This article on CLP regulations was first published in BBKA News in May 2013 pp 29-31.
This information is to be found at the following web site:

Here is an EU based search site that advises you of the labelling and special phrases that should be used for any specific chemical:
http://echa.europa.eu/information-on-chemicals/cl-inventory-database

Under 'Search Criteria' you can enter the common name of the substance. Tick the legal disclaimer box and click 'Search'. Scroll down and under 'Search results' find the chemical you are looking for and click 'View'. A new window will open. Hold the cursor over the 'Hazard Statement Code' to see the relevant warnings. Appropriate pictograms are also displayed.

These new regulations are being introduced gradually over several years. There are about 70 new Hazard Statement Codes and over 130 new Precautionary Statement Codes. The vast majority of these will have little immediate relevance to beekeepers as we only tend to handle the more common and less dangerous chemicals. However substances such as oxalic acid and acetic acid, for instance, can be extremely hazardous if used incorrectly. We do need to be aware of hazards associated with chemicals, protect ourselves (and our bees) accordingly and know what to do if things do go wrong.

Here are some of the other web sites I visited in researching this article

1. This document explains the changes fairly clearly:
    http://www.hse.gov.uk/pubns/indg350.pdf

2. The required codes and required graphics (now called pictograms) are given in the two attached spreadsheets. These are available here:
    http://esis.jrc.ec.europa.eu/index.php?PGM=cla
    under the tab CLP/GHS Tables 3.1 and 3.2.

3. The risk phrases are defined here:
    http://www.ilpi.com/msds/ref/riskphrases.html

4. and the Safety phrases are here:
    http://www.chemie.fu-berlin.de/chemistry/safety/s-saetze_en.html

5. The hazard statement codes are here:
    http://www.ilpi.com/msds/ref/hstatements.html

6. and the numbered pictograms are included in this document:
    http://www.penarth.co.uk/about/Guides/CLP-Hazard-symbols-for-Resource.pdf

Besides the information gained from the above web sites I am grateful to the following publication for allowing the reproduction of some diagrams and tables:

'Occupational Safety and Health and the Chemical Classification Labelling and Packaging Regulation' by the R.P.A.Consortium. Dec 2011.

Wikipedia –GHS hazard pictograms

Finally I would like to thank Mike Rowbottom (Chairman Harrogate and Ripon Beekeepers) who searched out so many useful sources of information.

John Chandler,
Harrogate and Ripon Beekeepers Association

SECTION XVII
INDEX

NOTE.-Numbers refer 10 paragraphs.

Acarine disease, cause of, **1529-3**
  diagnosis, 1535-6
  in relation to management, 1021
  manipulations to control, 1537-8
  treatment of, 1537-44
Achroia grisella. *See Wax* moth Africa, South,
sources of honey, 471 African bees. See Bees
Albino bees. See Bees
Alley, Henry, 207, 1576
America, honey /low, 417-19
  judging honey, 570
  sources of honey and pollen, 463-7
American foul brood, 1503-13, 1516-25
  cause of, 1590
  destruction of germs, 1508, 1512
  incidence and elfects, 1508-10
  relation to management, 1022 treatment, 1516-25
American hives. See Hil'es Amceba
disease, 1547 Anti-freezing mixtures, 580
Ants, 667, 1564-7
Apiary, barrow, 878
  equipment, 614 fencing,
  586 fires, 585 /loods, 585
  hives in, 599-604, 916-19
  honey house, 605-10
  in garden, 597
  in town, 597
  location of, 521-2, 592-8
  neighbours, 588
  pollen and water supplies, 583
  size of, 613
  stands, preparation of, 598
  starting an, 612-14, 616-18
  vibration, 585
  wind,584-6
Applianees, eare of, 897-901
Australia, bee farm site, 590
  bee range, 590
  frames, 791, 794-5 lourees of
  honey, 469
  use of petrol tins for honey, 541

Babies, honey far, 447
Balling of queen, 56-8, 921
Barbeau, 1576
Barrett, Gilbert, 221, 1574
Barrow, apiary, 878
Baume. See Hydrometer
Bee bob, 1094-5
Bee brush, 875, 1363
Bee houses, 660
  Bee space, 336, 677-8, 683, 778, 802, 1583-4
  Bee-keeping, partnership in, 619
  starting, 612, 616-18
  statistics, 428-30
  Bees, African, 150
  British black, 116, 124, 143 buying, 612, 616-18
  Carniolan, 116, 138-40
  Caucasian, 16, 141-2
  Central European, 16
  Cyprian, 16, 135-7

dead, 924, 1372-5, 1502, 1549-51
division of labour, 7, 23-5, 1150, 1573
drifting, 30, 277, 601-2, 776,1570
driven, 1349-50
Duteh, 116, 146
Eastern, 153
Egyptian, 149
examining, 1354-60
feeding. See Feeding
/light. See Flight of bees
French, 116, 144
German, 148
German Swiss, 145
hardy, 117,981
homing, 602-3, 621
Indian, 152
Italian, 16, 129-134, 231
longevity, 18-21, 88, 1011-13, 1146-7
management of. See Management
manipulating. See Manipulation.
moving. See Moving bees
natural his tory, 1573-4
of mixed sex, 6
Bees, races and strains, 116-128, 1025, 1367-71
  renting, 619
  Spanish, 147
  subduing, 853-63
  Syrian, 151
  See also Queens, Drones, Worker bees
Beeswax, bleaching, 389
  carrying in, 1356
  colour, 314, 370, 391-2
  exhibiting, 402
  extractors, 394-9
  judging, 403
  moulding, 401
  production, 318-23, 370-4, 396
  properties of, 375:"8, 381
  purity, 353-4
  rendering, 390-400
  sorting, 393
  strength, 301
  testing, 387-8
  See also Comb building, Wax moth Bingham, 1585
Boyd, W. 1., 1575
Brace comb, 316
Breeding queen, 192
British black bees. See Bees
British standard frames, 338, 788-92, 1006-7
British standard hives, 736, 745-55
Brood, care of, 484
  diseases, 1503-26
  spreading, 931-2
  stimulating production of, 931-2, 949, 1009-14,
  1294-30, 1435-6
Brood chamber, divided, 338
  single 'V. double, 789, 906-10
Brood nest, growth and arrangement of, 53. 337. 340,
  783-5, 949, 1015, 1283, 1353
Brooks, J. M., 1576
Burr comb, 317

Butler, Chas., 1574, 1579
  Canada, honey flows, 417-19
  honey standards, 490
  registration of apiaries, 589
  sources of honey, 468

Candy, 920, 1308-15
  queen cage, 1316-18
Cappings, 127, 326-8, 400
  drainings, 400
  uncapping. See Extracting
Carbolic cloth, 861
Carnolian bees. See Bees
Caucasian bees. See Bees
Cellar wintering, cellar temperatures, 994
  conditions, 982, 989', 999-1001
  construction of cellar, 990-3
  moving bees into, 995
  moving bees out, 1002
Cells, dimensions of drone and worker, 329-34
  disadvantage of large, 333
  importance of size of, 333,343
  thickness of cell walls, 333, 343
Chalk brood, 1526
Cheshire, Frank, 1590
Chilled brood, 1503, 1526
Chlorine gas for foul brood, 1524-5
Chloroform, use of, 862
Chunk honey, economics, 948-54
  production of, 503-6, 944-6, 1144, 1176
  storing, 536
Cleaning metal parts, 397-9
Cleaters, 1574
Clement, 1583
Clothing for bee-keeper, 846-8
Cluster, form and size of, 337, 783-5, 978-80
Coating wood and metal with wax, 900-1 Cohn, 1590
Collin, Abbe, 1586
Colony, definition of, 617, 1150-5
  evolution of, 1059-60
Comb, cutting, 352
  defective, 927, 957
  holders, 879
  indicators, 877
  pollen clogged, 345-7
  preservation of, 357-69, 400, 930, 1372-5
  repairing, 350-2, 344
  size of, 335-41
  spacing, 796-802
  use of old, 342-4, 425-7, 966
  utilizing natural, 1484-6
  value of, 425-7
weight of honeycomb, 248-9, 335, 422
  weight of wax comb, 348
Comb building, brace, 316-17
  burr, 316-17
  by swarm, 312
  colour,311
Comb building, dimensions of cells, 324-5, 329-31,342
  drone comb, 312-15, 340, 1434
  economical, 343, 414
  foreign substances in, 353-6
  from honey and sugar, 318
  in nature, 782
  inclination of walls, 324
  irregular, 315
large combs preferred, 321, 336-8, 778-81
method of, 308-11, 321, 325-6, 335-6, 343, 351, 414,778-82, 1579

repairing, 344, 350-2
supplying beeswax for, 355, 780
temperature,29
to bottom bars, 315, 351
worker, 312, 315, 340, 351, 1434 See also Cappings
Comb honey, economics, 425-7, 944-7
  packing, 538
  taking, 473
  See also Sections
Cookery and Confectionery, drinks, 577
  honey for, 487, 571
  ice cream, 578
  icing, 576
  neutralizing acid of honey, 574
  preserves, 577
  sauces, 577
  sugar equivalent of honey, 573
  value of honey, 571
Cowan, Thos. W.,207, 1576, 1587, 1590
Cyprian bees. See Bees

Dadant, 1585. See also Modified Dadant
Davis, J. L., 1575
Debeauvoys, 1583, 1584
Demareeing, 234, 244, 261,1226-48
Dextrose, 432, 435-40, 443-4
Dibbern, Charles, 1588
Discoveries, 1570-90
Disease, examination for, 928, 14931502
  See also Acarine, Amee ba disease, Disappearing disease, Discoveries, Dysentery, Foul 'Brood, Fungi, Hairless bees, Minor brood diseases, Nosema, Paralysis, Pickled brood, Sac brood, Septicemia, Spring sickness, Stone brood
Division boards, 677, 687, 1357-9, 1586
Docility, 120
Doolittle, G. M., 1575
Driven bees, 1349, 1415
Driving bees, 1349-50
Driving irons, 1480
Drone, colour of, 7, 291
  drifting, 30
  duties of, 31, 1575
  escape above excluder, 245, 707
  flight, 29, 32
  importance of, 192
  in hive, 31, 36
  potency, 4, 29, 32, 176-7
  stages in development of, 1
  starvation of, 34
  traps, 894-5
  See also Theory of swarming
Drone breeding, in general, 1075
  feeding, 32
  parentage, 7, 30, 109, 175-7, 199
  reduction of, 35, 293-4, 799, 957
  selected, 293-4
  See also Theory of swarming
Drone combo See Comb, also Cells
Duchet, Franois X., 1579
Dummies, 704, 1357-9
Dutch bees. See Bees
Dysentery, 1528
Dzierzon, Johann, 1571, 1583

Egg, weight of, 9, 48
  laying. See Queen laying size,48
Egyptian bees. See Bees

Escape boards, 477-84, 697-701
Ethyl chloride, 863
European foul brood, brood disease, 1504-7
  destruction of germs, 1512-13 incidents and
  effects of, 1508-11 management with, 1020,
  1515 symptoms of, 1504-7
  treatment of, 1515
Excluders. See Queen excluders Exhibiting honey and
wax. See Grading, also Beeswax
Extracted honey. See Honey
Extracting honey, 491-503
  centrifugal force, 495-6
  grading, 486-90
  heather honey, 501-13
  temperature, 491
Extracting honey, uncapping, 492-4
Extracting wax, 394-99
Extractors, care of, 500
  design of, 496
  invention of~ 1571, 1587
  use of, 491-503

Feeders, 884-7
Feeding, candy, 920, 1287-9
  feeders, 884-7
  for increase, 1431-4
  honey supply, 934-6, 940, 1144, 1282-5
  in autumn, 959-63, 967-77,1303-7
  in general, 1265
  in spring, 449, 920, 968-70
  in summer, 1301-2
  in transit, 1274
  maintaining stores, 934-6
  nuclei, 264, 271, 282
  outdoor, 1307
  pollen substitutes, 1277-81
  pollen supply, 347,1144,1275-81
  stimulative, 1281, 1294-1300
  sugar syrup, 960, 1286-1315
  water supply, 985-6, 1266-74
Fermentation, cause of, 516, 1581
  prevention of, 436-7, 517-22, 1206
  risk of, 434, 436, 475, 507
  use of fermented honey, 523
Flight of bees, 17, 409-12, 421, 735, 915
  mating flight. See Queen, fertilization of
  See also Swarming
Flowers. See Honey, sources of
Foul brood. See European, American Foundation,
  care of, 832-5
  choiee of, 832-5
  cutting, 815
  drone, 340
  drawing out, 320, 322, 349
  fixing, 207-14
  in sections, 827-31
  invention of, 1580
  manufacture of, 834, 836-43
  method of use, 309, 807-21
  strength of, 835
  thin, 832-3
  value of, 425-7
  weight of, 348-9
  wired, 820-1
Frames, assembling, 805-6
  British standard, 338, 788-92, 906-7
Frames, comb area in, 795
  deep brood, 643, 654, 789, 792
  dimensions of, 789, 791-6

divisible, 267-71
ends, 800-1
invention of, 1682-5
Langstroth, 338, 643, 790--1, 793, 796,800,907,
  1571, 1583-4
long top bars, 793
materials, 763
modified Dadant, 338, 643, 785, 787-9,791,793,
  796,907
parallel v perpendicular, 320
proportions of, 789
shallow brood, 643
shallow supering, 790-2
size of, 335-41, 777-81, 786-88, 847-51
spacing, 796-803
special top bars, 779, 792, 808-10
studs, 802
supports for, 878-9, 1360
weight of wax in, 348
wiring, 816-20
See also Foundation French bees.
See Bees

Galeria mellonella. See Wax moth Gatinais, 144
German bees. See Bees German Swiss bees.
See Bees Gilbert Barrett, 221, 1574
Given, D. S., 1580
Glossometer, 15
Gloves, 852
Grafting. See Queen raising Granulated honey,
as food, 1284
  packing, 542
  showing, 560, 570
  treating, 512-14, 525, 527
Granulation, fine crystals, 509
  large crystals, 508
  prevention of, 512-14
  rapidity of, 510
  stages, 436, 507
  tendency to, 511
Gynandromorphs, 6

Hairless bees, 1548
Harvey, E. J., 1590
Heather, moving to, 634
  See also Flowers, Moving beee Hive-tool, 874, 359
Hives, alighting boards, 668
Hives, arrangement of frames, 647-9
  bee house, 660
  British National, 746-9
  British standard, 736, 745-55
  brood chambers, 677-8
  capacity in relation to fecundity of queen, 50-1
  colour of, 277, 776
  creosoting, 766
  cubical form, 786
  Dadant, 736, 743-4
  division boards, 677, 687
  dummies, 686
  entrances, 277, 673-6,720,729-32, 760-1
  escape boards, 697-701
  examining, 853-61, 916-29, 956-8, 1356-60
  excluders. See Queen excluders fillets, 661-3
  fioors, 280, 644, 664-5, 669-72, 726,733
  frame runners, 681-2, 684
  glass walled, 757
  inner covers, 688-91, 7J6, 919
  installing, 598-604
  invention of, 1582-8

keeping out bees, 696
keeping out damp, 765-8, 983-6
keeping out water, 694-5
Langstroth, 736, 741-2, 1583-4 legs, 666-7
"Long idea," 650-4, 736
materials, 762-3
middle entrance, 757-61
nucleus. See Nucleus hives
observatory, 756-7 ornamental, 653
outer cases, 658-60
packing. See Packing hives painting, 764-76
propolization, 681-4
quilts, 688-90, 715
re-painting, 773-5
roofs, 698, 976
seal es, 880-3
seetion racks, 685
shrinkage, 679
single v. double walls, 655-7
size of, 637-41
skeps, 146, 636
special boards, 102,280,644,669-72
standardization of, 737
stands, 598, 666-70
Hives, sundry, 745-5
    super clearers, 697-702 supers, 680
    temperatures, 26-30
    temporary covering, 864
    top entrance, 757-61
    types of, 642--6, 736, 738-40
    ventilation. See Ventilation of hives
    weight increase, 412, 417-21
    winter passage, 975
Honey, acidity, 450
    analysis, 439, 442-4, 532
    anti-freezing mixtures, 580
    blending, 458,487, 551-7
    bottling, 499
    chunk. See Chunk honey clarifying, 527-9
    colour, 455, 486-90
    density, 435-41
    diastase in, 531-2
    economics, 948
    fermentation. See Fermentation fiavour, 458
    fiow. See Nectar fiow food value, 445-9, 532
    fuel value, 448
    grading, 486-90
    granulation. See Granulation heating, 524-6
    house, 605-10
    in cooking. See Cooking in toilet preparations, 579
    judging, 569-70
    labelling, 543-6, 550
    medicinal value, 445-9
    objectionable, 459
    packing, 537-42
    pollen in, 457
    press, 501
    productian statistics, 417-21, 423-30
    refractive index, 438
    ripening, 498, 1513
    selling, 527-31, 547-58
    showing, 560-8
    SOurces, 454-72
    sterilizing, 524-6, 1285
    storing, 533-6
    sugar equivalent, 573-5
    surplus, 321, 1144-8
    taking, 473-85, 939-42
    uncapping, 493-4

use in cooking, 571-80
whipped,515
yield per acre, 423-4
Honey. Su also Comb honey, Extracting, Sections
Honeydew, characteristics, 452
    collection, 453
    detection, 452
    sources, 45 J
    use of, 453, 961
    Hopkins, 207
Hornbostel, Herman, 1579
Hruschka, Francesca de, 1571, 1587
Huber, François, 1574, 1579, 1586
Hunter, John, 1574, 1579, 1586
Hydrometer, 441
    Increase, avoiding loss of harvest during, 1435-6
    by artificial swarming, 1453
    by robbing stocks, 1444-5
    continued, from one stock, 1456-8
    feeding for, 1431-4, 1468
    giving queens v. queen cells, 1426-30
    in general, 943
    in relation to district. See Management
    multiple, 1455
    natural 'v. artificial, 1425
    nuelei, continuous production of, 1460-3
    nuelei from stock superseding, 1459
    nuelei in cool weather, 1471
    queen raising combined with, 1459, 1464-5
    rapid autumn, 1466-7
    seasonal differences, 1437-43
    spring division, 1446-52
    swarming, 1197.
    See Swarming three from two, 1454
    use of pollen combs, 1468
    using old bees for, 1469-70
    when required, 1424
    with demareeing, 1473
    with swarm control. See Manipulations

Incubators, 274
Indian bees, 152
Indicators, comb, 877
Insurance, 588
Invalids, honey for, 447
Inventions, 1570-2, 1575-88
    judging honey, 570
Isle of Wight disease, 1590
Italian bees. See Bees J. A., 1583
Janscha, Anton, 1574, 1576
John, Martin, 1579
Kerr, Robert, 1583
Keys, John, 1582
Kretchmer, Gottlieb, 1580
Langstroth, hive, 736, 741-2, 1583-4
    frames, 338, 643, 789-91,793, 796, 800, 1571,
        1583-4
Larvre, age for queen raising, 8, 209, 216
    feeding, 10, 23, 45-47
    size of, 8
    stage, 1-3
    weight of, 9-10
Laws and Regulations, 436, 443, 488, 558-9,588,
        1099-1100
Laying workers, appearance, 109
    discovery of, 114, 119, 1578
    drones, 109
    fertility, 107
    partially fertilized, 150

production of, 107-8, 142, 149-50
suppression of, 84, 111-15
See also Theory of Swarming
Lee, Jas. & Son, 1586
Leuckart, 1574
Levulose, 432, 435-6, 440, 443-4
Life of bee, length of, 18, 99, 1146-7
"Lang idea" hive, 652-4, 736

Maassen, Albert, 1590
MacLain, N. W., 1577
Maisonneuve, Perret, 220, 225, 1576
Management, autumn, 956-77
    cellar wintering, 989-1002
    close of nectar flow, 955
    comb honey from shallow frames, 1384-6
    control, of swarming, 1149
    dead bees, 924, 1372-5, 1502
    disease. See disease by name
    driven bees, 1349-50, 1415
    feeding. See Feeding
    in general, 1068-9, 1139-43, 1156-8
    in relation to disease, 1018-22
    in relation to districts, 1023-57
    in relation to flow periods, 1006-10, 1026-53
    in relation to longevity of bees, 1011-13
    large apiaries, 1016
    package bees. See Package bees
    production ,of bees. 1014-15, 1143
Management, production of chunk honey, 944-6, 1144,
    1176
production of comb honey, 944-6, 1144, 1176, 1384-7
    production of extracted honey, 944-6, 1144, 1176
    queen raising. See Queen raising robbing.
    See Robbing
    seasonal management, 911-13
    single 'V. double brood ehambers, 789, 906-10
    spring time, 914-36
    subduing bees, 853-61
    summer, 937-55
    swarms. See Swarms
    two queens in one hive, 1177-80, 1264
    uniting. See Uniting bees winter, 978-1002
Manipulations, bee brush, 1363
    before honey flow, 1376-83
    changing the strain, 1367-71
    checking swarming by removal of brood, 1213-15
    checking swarming by removal of brood and queen,
    1216-18
    checking swarming by removal of queen, 1211-12
    dead bees, 1372-5, 1549-51
    delaying fiight of swatms, 1186
    Demareeing, 234, 244, 261, 1226-48
    destroying queen cells, 1187,1207-10, 1364--6
    diseovering parent hive, 1181-2
    double hives, 1163-5
    doubling without inerease, 1200-3,
    1219-25
    driving. See Transferring examining combs in a
    hive, 1356-60
    examining frame of bees, 1354-5
    frames, 921, 1354-5
    getting comb honey from shallow frames, 1384-6
    getting pollen sealed over, 1281
    giving room in brood nest, 1394-9
    in general, 910, 1352
    increase. See Increase
    reswarming, preventing, 1185
    section raeks, preparation of, 1387

shaking bees, 240, 1341-2
subduing bees, 853-62
transferring. See Transferring uniting.
See Uniting
See also Swarms
Maraldi, John, 1580
Marking bees, 69-73
Mating . See Queen fertilization
Mehring, Jean, 1580
Metal parts, care of, 897-901
Middle entrance hives, 757-61
Miller, Dr., 207
Minor brood diseases, 1526
Modified Dadaut, hives, 736, 743-4
    frames, 338, 643, 785, 787-9, 796, 1585
Moving bees, a swarm, 635, 1112-4
    aids to location, 621
    critical distance, 627
    heavy hives, 632, 916
    in general, 620, 988
    in skeps, 892-3
    in winter, 988
    long distances, 625, 633, 1148
    package bees, 1339-42
    preparations for, 628-31
    to heather, 634, 1143
    travelling-boxes, 890-1
    weather, infiuence of, 622-4
Munn, Major, 1586

Nectar, composition of, 431-2
    density of, 413-16, 431-4
    flow, close of, 955
    flow, good and bad years, 400, 615
    gathering, 409-16
Nest. See Brood nest, Cluster
Nosema disease, 1545-6, 1590
Notes, keeping, 829
Nueleoli in queen raising, 270-3,289
Nueleus, definition of, 617-18
    making, for queen raising, 190
    management in queen raising, 264-6
    self-supporting, 262
    use of in making increase, 1434, 1459-61, 1470-2
    use of in week-end bee-keeping, 1057
    See also Nueleoli
Nueleus hives, construction of, 263, 267-9
    use of in queen introduction, 92
    use of in queen raising, 190,262-9
    Nursery cages, 275
Nussbaumer, 1581

Observation hives. See Hives
Orchards, bees in, 411, 592-6, 1336
Out apiaries, Iocation, 591-2
    queen raising in, 187-9

Package bees, buying, 1334-8
    in orchards, 595
    management of, 1340-8
    shipment of, 1339-40
    use of, 1001, 1332-3
Packing hives, amount of packing, 710-13
    at bottom, 725
    at sides, 717-18, 984
    at top, 714-16
    in general, 708-9
    tarred paper, use of, 723-4
    winter cases, 719-24

Paralysis, 1548
Parthenogenesis, 45, 1571, 1574
Partnership in bee-keeping, 619
Pechaczek, 207
Pests, ants, 1564-7
    birds, 1560-2
    braula caeca, 1553-8
    dragonflies, 1563
    flies, 1569
    in general, 1552
    lizards, 1563
    mice, 156S,
    moths, 1559
    rats, 1568
Pests, spiders, 1563
    toads, 1563
    wasps, 1563
Pickled brood, 1526
Poisoning bees, 596, 1549-51
Pollen, clogged combs, 345-7
    food value, 445, 454
    gathering, 411, 583, 917-18,1276
    importance of, 415, 1275-6
    sources of, 455-9, 460-72
    substitutes, 1277-81
Porter, E. C., 1588
Pratt, Eugene L., 281, 1575
Pridgen, 1576
Prokopoertel, 1583
Prokopovitsh, 1586
Propolis, colour, 405
    inclusion in wax, 406-7
    on lugs of frames, 681-4
    removing, 408
    sources of, 356, 405
    use of by bees, 121, 137,356,414
    use of metal to avoid parts sticking together, 800
    See also Vaseline
Protection of the person, 846-52
Punie bees, 150
Pupae, 1-3

Queen, activities of, 36
    balling of, 56-8, 921
    breeding queen, 192
    cages, 94-5, 156, 162
    call of, 41
    catching and holding, 45
    cells. See Queen cells clipping, 66-8, 1121
    colour, 43
    development, 45-6
    emergence, 1. See also Queen raising
    excluders, 680, 688, 702-8, 789,
    958, 1393, 1586
    failing, 54-5
    feeding, 47
    fertilization, 175-7, 285-6, 306, 1574, 1577
    finding, 59-64, 1232-4
    importance of, 78
    introduction, 85-105
    laying, 37, 45-6, 48-9, 202-4,288,
    921,931-3,955,964
    longevity, 42, 198
    1055 of, 17, 74-6
    marking, 55, 69-74
    mating, 4, 33, 36, 38, 45, 175-7, 192-201,
        274-89,290-306,1574, 1577
    points of a good, 42-3, 192, 195-6,
    203,307

sale of, 307
    stages in development, 1
    sting of, 40
    supersedure, 54-5, 74, 81, 157-70,
    165, 167-70, 1344
    tested,307
    traps, 894-5
    two in colony, 106, 965, 1151,
    1177-80, 1264
    untested, 307
    wing clipping, 66-8, 1121
    See also Virgin queen, Requeening, Queenlessness
Queen cells, artificial, 218-21, 1575
    destroying, 1187, 1207-10, 1364-6
    evolution, 1063-5
    giving, 1426-30
    good, 221, 252
    protection, 256, 347-8
    starting and raising, 157-70, 205-45
    supersedure, 161, 166-70

Snelgrove, L. E., 98, 1171-2
Spacing of frames. See Frames
Spanish bees. See Bees
Spraying in orchards, 596, 1549-51
Spring cleaning, 926-30
Stamina, 117, 119,216.
See Longevity Standardization, 737-8
Starvation, 415, 1373
Stewarton, 1582
Stimulating. See Brood and Feeding
Sting, when potent, 23
Stings, 846-52, 865-72
Stock of bees, definition of, 617-18
Stone brood, 1526
Storing honey, 523-6
Sucrose, 432
Sugar, cane, 432, 443
    fruit, 432, 443
    grape, 432, 443
    inverted, 432, 443, 1289, 1295
    syrup, 1286-93
    tables,435
Super clearer, 477-82, 697-702, 1413
Supers, attracting bees into, 322, 339
manipulation of, with swarms, 1120
provision of. See Management, Hives pollen in, 485
    removal of, 939-42
    stuck together, 483
    See also Section racks
Surplus, 321, 1144-8
    removal of, 480, 939-42
Swammerdam, Jan, 1574
Swarm, Swarms, Swarming, age of bees in, 1085
    artificial, 1129-38, 1162
    attracting, 1096-8
    box, 238-41, 1110
    cessation of, 1124
    checking, 1161
    clipped queens, 1121
    control of, 1149, 1159-75, 1181-1263
    delaying flight of, 1088-90, 1186
    destroying queen cells, 1187, 1207-10
    discovering parent hive, 1181-2
    encouraging, 1123, 1125-6
    feeding, 1120
    final flight of, 1092, 1099-1100
    hindrances to, 370, 1127-8
    hiving, 1115-20, 1183-1206

impulse to, 158-60
in inconvenient place, 1106-8
Swarm, influence of drones, 158,1070-3
  influence of weather, 1083, 1088- 90
  instinct, reducing, 177
  issue of, 1083-5
  issue of casts, 1086, 1089, 1183-4
  legal ownership, 1099-1100
  manipulation of supers, 1120
  moving, 635, 893, 1113
  natural, 1129-32
  net, use of, 11 09
  preparations for, 1075-81
  preventing casts, 1183-4
  propensity to, 123
  propensity to, reducing, 172, 196-7, 204
  queen raising under impulse to, 15- 60, 1082, 1125-6
  reswarming, 1185
  returning, 1122
  returning to parent hive, 1188-92
  settlement of, 1091-5, 1101-3
  signs of, 1075-81, 1087
  Snelgrove system, 1171-2
  taking, 1103-1110   .
  theory of. See Theory of Swarming
  time of, 158-9,958, 1083-4
  uniting, 1414
  weight of becs in, 9
Syrian bees. See Bees.

Tarred paper, use of, 723-4
Temperature, comb building, 28
  control, 28, 734-5, 978-82
  egg-Iaying, 28
  fatal to brood, 29
  flights, 409-12
  hive, 26-30
  wax secretion, 29
Theory of swarming, brood food, 1061, 1067-9
  cause of swarming, 1066-74
  checking swarming, 1069, 1211-8
  dependence upon drone raising, 1070-3
  dependent queen, 1061-5
  evolution, 1058-60
  excess of drones over queens, 1060
  fertility of queens, 1060, 1062, 1064
  fertilization of females, 1060
  flight of old queen, 1066
  laying workers, 1065, 1067-8
  non-swarming bees, 1065
  social habit, 1061
Theory of swarming, temporarily suppressed female,
        1060, 1063
  utilizing excess of brood food, 1063, 1067-9
  worker caste, 1059, 1061-5
Thorley, John, 1579
Toilet preparations, use of honey in, 579
Tongue length, 15-16
Torres, Luis Mendez de, 1574
Townsend, O. H., 1576
Transferring, combination method, 1488
  from roofs, old trees, etc., 1489-92
  from skeps or box-hives, 1476-88
  in general, 1475
Travelling-boxes, 889-91, 1339-42
Tropics, sources of honey, 472

U .S.A., judging honey, 570
  packing honey, 540

pure food laws, 436, 443
  standards for honey, 488-9
  sources of honey and pollen, 463-6
Uniting, driven bees to established stocks, 1415
  Bying bees to established stock, 1418-23
  in general, 924, 1404-6
  light, use of, 1408 newspaper method, 1409-12
  scent, use of, 1407
  super clearer, use of, 1413
  swarms, 1414   ;
  young bees to stock, 1416-17
Vaseline, 483, 682, 779, 876
Veils, 849-51
Ventilation, 714-16, 728-33, 782, 985-6, 993, 1128
Virgin queens, fertilization, 4, 33, 36, 39,45, 175-7,
        192-201,274-89, 290-306, 1574,
        1577
  introduction of, 89-91, 105
  production of. See Queen raising and Queen
        superseding
Virgin queens, selected, 307 See also Drones

Wankler, Wm., 1575
Water supplies, 583, 1266-74
Watson, L. R., 1577
Wax. See Beeswax
Wax making, 23, 29, 318
Wax moth, 361-9
Weather forecasts, 416, 938
Weed, E. B., 1580
Week-end bee-keeping, 1054-7 Weighing hives, 880-3
Weygandt, Carl, 1575 Wheeler, G. T., 1586
White, G. F., 1590
White, P. Bruce, 1590
Wild man, Daniel, 1583
Wild man, Thos., 1582
Winter, cellars, 989-1002
  feeding. See Feeding losses, 981-4
  passages, 973-5 protection, 133,710-17
  Worker bee, attitude in flight, 17
  colour, 127-53
  conservation of energy, 22
  development, 8, 11, 46
  docility, 120, 138, 141,853-63
  duties of. 7,23,1150
  evolution of. See Theory of swarming
  flight of, 23, 409-12
  honey sac content, 17
  hybrid, 125, 138
  length of life, 18-21, 1011-13
  load carried, 17
  points of good strain, 118-123
  sex, 1059, 1061-5, 1574
  size, 12-13
  size of larvre, 8
  tongue length, 15-16
  weight of, 12-14
  See also Bees
Worker combo See Comb

Zander, Euoch, 1590